TECHNOLOGY AND ENGINEERING

DE DIVERSIS ARTIBUS

COLLECTION DE TRAVAUX
DE L'ACADÉMIE INTERNATIONALE
D'HISTOIRE DES SCIENCES

COLLECTION OF STUDIES
FROM THE INTERNATIONAL ACADEMY
OF THE HISTORY OF SCIENCE

TOME 49 (N.S. 12)

BREPOLS

PROCEEDINGS OF THE XXth INTERNATIONAL CONGRESS OF HISTORY OF SCIENCE (Liège, 20-26 July 1997)

VOLUME VII

TECHNOLOGY AND ENGINEERING

Edited by

Michel LETTE and Michel ORIS

BREPOLS

The XXth International Congress of History of Science was organized by the Belgian National Committee for Logic, History and Philosophy of Science with the support of :

ICSU
Ministère de la Politique scientifique
Académie Royale de Belgique
Koninklijke Academie van België
FNRS
FWO
Communauté française de Belgique
Région Wallonne
Service des Affaires culturelles de la Ville de Liège
Service de l'Enseignement de la Ville de Liège
Université de Liège
Comité Sluse asbl
Fédération du Tourisme de la Province de Liège
Collège Saint-Louis
Institut d'Enseignement supérieur "Les Rivageois"

Academic Press
Agora-Béranger
APRIL
Banque Nationale de Belgique
Carlson Wagonlit Travel - Incentive Travel House
Chambre de Commerce et d'Industrie de la Ville de Liège
Club liégeois des Exportateurs
Cockerill Sambre Group
Crédit Communal
Derouaux Ordina sprl
Disteel Cold s.a.
Etilux s.a.
Fabrimétal Liège - Luxembourg
Generale Bank n.v. - Générale de Banque s.a.
Interbrew
L'Espérance Commerciale
Maison de la Métallurgie et de l'Industrie de Liège
Office des Produits wallons
Peeters
Peket dè Houyeu
Petrofina
Rescolié
Sabena
SNCB
Société chimique Prayon Rupel
SPE Zone Sud
TEC Liège - Verviers
Vulcain Industries

D/2000/0095/119
ISBN 2-503-51158-9
Printed in the E.U. on acid-free paper

TABLE OF CONTENTS

Presentation 9
Michel LETTE - Michel ORIS

Part one
PRACTICE AND THEORY OR TECHNOLOGY AND SCIENCE : AN EQUAL FOOTING DIALOGUE

De la Pirotechnia of Biringuccio. A Key to the Craftsman Production : Technical Questions and Aspects of the Organization of Production 13
Vasco LA SALVIA

Science and Technique in the Work of Galileo 27
Fernando L. CARNEIRO

James Watt's Application of Scientific Technology to Industry 37
Richard Leslie HILLS

Industry and Industrial Relations within the Laboratory. The Material Conditions of the Joule-Thomson Experiments 49
Christian SICHAU

Science industrielle et seconde industrialisation en France. Une proposition d'organisation rationnelle des relations entre la science et l'industrie (1880-1914) 61
Michel LETTE

Changes in the Concepts of Flight in the First Decade of the 20th Century : The Pioneering Work of Alberto Santos-Dumont 81
Henrique LINS DE BARROS - Mauro LINS DE BARROS

A Pre-study for the Historiography of the Philips Research Laboratories .. 95
Marc J. DE VRIES

Histoire, interaction et union de la technique électronique et de l'optique physiologique 107
V.M. TCHESNOV - M.G. YAROCHEVSKY - M.A. OSTROVSKY

Part two
Technology as a Bridge between Economy and Society

Imprimerie et transfert de technologie à la Renaissance ... 115
Marie-Claude Déprez-Masson

Les méthodes d'analyse des métaux précieux et les voyages de Martin Frobisher au Canada (1576-1578) ... 125
Bernard Allaire - Réginald Auger

The Transfer of Indian Textile Printing Techniques to Europe ... 133
Walter Endrei

Inventivité technique et naissance d'industrie innovante en Belgique (1860-1910) ... 139
Michel Oris

Calling into Question the Current Knowledge about Zénobe Gramme and his Inventions ... 163
Philippe Tomsin

Modern Influences on Rural Building in Northwest Germany (1880-1930). The Effect of Vocational Training ... 171
Michael Schimek

Le rôle de l'entreprise et de l'entrepreneur dans l'introduction du béton précontraint. Eugène Freyssinet et les Entreprises Campenon Bernard ou l'histoire d'une rencontre ... 185
Dominique Barjot

Le poteau et le voile : essor et apogée du béton armé dans le logement collectif en France ... 193
Dominique Theile

Une problématique pour une histoire des infrastructures urbaines à travers le cas de Montréal (1845-1960) ... 201
Robert Gagnon - Dany Fougères - Michel Trépanier

Les sciences dans le processus d'exploitation des ressources naturelles au Brésil au XIXe siècle ... 207
Heloisa Maria Bertol Domingues

A Study of Controlled Nuclear Fusion in Japan Concerning Arguments about the Policies of Research Organization ... 225
Eisui Uematsu - Sigeko Nisio

Car Industry, Technology, and Society in the USA. The Fight against Air Pollution between 1950 and 1990 ... 231
Joop Schopman

Un enfoque actual de la tecnología ... 245
Claudio Katz

Part three
SCIENTISTS, SOLDIERS AND MANUFACTURERS : THE DYNAMICS OF MILITARY TECHNICAL DEVELOPMENT

Sixteenth and Seventeenth Century Arms Production in Gipuzkoa 265
Ignacio M. CARRION ARREGUI

Mathématiques, théorie et pratique dans l'artillerie de la marine espagnole au XVIIIe siècle. Les manuels 281
Encarnación HIDALGO CÁMARA

La influencia de las escuelas de armería de Liège y Saint Etienne en la creación y posterior desarrollo de la escuela de armería de Eibar.... 291
María Cinta CABALLER VIVES

Science, Technology and the Armaments Industry in Great-Britain (1854-1914) 309
Marshall J. BASTABLE

An Unknown Naval Accident and the Development Trajectory of the *Kanpon* Type Turbine in Pre-war Japan 317
Miwao MATSUMOTO

Contributors 327

Présentation

Michel Lette - Michel Oris

La présente livraison des Actes de la section *Science, Technology and Industry* du Congrès International d'Histoire des Sciences, tenu à Liège du 20 au 26 juillet 1997, voudrait contribuer à compliquer un peu plus encore, s'il était besoin, l'illusoire dichotomie technique/société.

La sélection des textes a été faite avec à l'esprit la volonté de pratiquer la complexité de liens que les sciences sociales s'efforcent de repenser avec une vitalité salvatrice depuis au moins l'émergence des *Social Studies* au début des années 1970. Corrélativement, ces dernières ont généré des attentes auxquelles les historiens des sciences et des techniques se sont manifestement habitués. Un dialogue s'est établi. Non sans douleur. La crise n'a été visiblement qu'une étape de croissance. Au demeurant, la diversité des textes présentés et la multiplicité des approches témoigneront ici que nous vivons toujours sous le régime du renouveau des études historiques, sociologiques et anthropologiques sur les sciences et les techniques. Il faut s'en réjouir et participer à la multiplication des bilans, penser aux réorientations, tourner le regard vers les zones obscures.

Faut-il insister ? En dépit de notre intitulé *Technology and Engineering* (pour le moins globalisateur) il n'est question ni d'une chaîne linéaire, ni d'une hiérarchie, ni d'une filiation depuis la science vers la production industrielle ; et moins encore d'une juxtaposition de secteurs d'activité autonomes. L'image, s'il faut absolument en retenir une, ne doit pas être forcément la plus simpliste. Compliquons ce qui paraît trop simple. Au mieux concevons l'ensemble *Technology and Engineering* comme un réseau complexe de liens à redéfinir en permanence et plutôt mal tricotés, avec ses mailles renversées et ses noeuds. Tirez sur n'importe quel bout et vous serez certain de détenir le fil d'une explication de leurs relations. Le sommaire du volume donne, à lui seul, une illustration stimulante du propos. Il doit inciter à ne pas perdre de vue l'impératif de repenser les éternelles recompositions qui s'opèrent dans la structuration des liens entre la science, la technologie et l'industrie. Non seulement pour les périodes dites “ d’industrialisation ” pour lesquelles la question reste à l’évidence pertinente, mais également pour les temps où il semble que la notion plus générale d’économie soit plus opérante. L’attitude est d’autant plus raisonnable que l’on doit déjà penser au processus de “ dé-industrialisation ”. A titre d’exemple, la mise en inactivité forcée d’une partie croissante de la population oblige, à elle seule, à mettre au clair nombre de notions et d’outils mobilisés pour comprendre les rôles en interaction des sciences, des techniques et de la technologie dans la structuration et l’organisation des sociétés modernes. Autant dire que les schémas explicatifs deviennent plus que jamais impératifs pour appréhender notre propre modernité. Fin de siècle ou promesses vertigineuses d’une technologie qui suscite éternellement l’inquiétude ? Les historiens des sciences, de la technologie et de l’industrie se trouvent, de fait, mis en demeure de contribuer à expliquer ; rassurer aussi. De telles perspectives les invitent à poursuivre le dialogue avec les spécialistes des sciences humaines impliqués (mais ne le sommes nous pas tous ?) dans la problématisation des processus d’organisation des sociétés industrielles.

Le recueil s’organise ainsi. Nous ne tenterons pas de retracer dans le détail ce qui a été dit lors des exposés et des discussions, lors des journées consacrées aux interactions science/technologie/industrie. A partir de la lecture des textes qui ont depuis, rappelons-le, décanté et se sont enrichis de la réflexion post-congrès, il appartient au lecteur de reprendre et de prolonger les débats. La manière de faire a au moins l’avantage de restituer le rythme et le ton des travaux présentés à Liège. Il n’est question ni d’inventaire, ni d’état des lieux. Plutôt doit-on considérer l’ensemble comme un matériau, une matière à réflexion sur l’utilité d’élaborer d’autres grilles d’analyse et de nouveaux schémas.

Pour une lecture aisée, nous avons divisé l’ensemble en trois parties. La première, *Practice and Theory or Technology and Science : an Equal Footing Dialogue*, consacre l’interaction et les échanges entre le théorique et l’appliqué ; le savoir et le savoir faire ; la science et la technologie, l’un et l’autre étant considérés comme participant à la fabrication des savoirs.

Dans le prolongement de la partie précédente, la seconde, *Technology as a Bridge Between Economy and Society*, se focalise sur la notion de technologie proposée comme interface privilégiée entre l’économie et la société. La multiplicité des termes et des expressions employés par les auteurs pour qualifier une réalité, en apparence identique, rend compte, une fois de plus, que la notion ne se laisse pas saisir aisément.

La dernière partie, *Scientist, Soldiers and Manufacturers : the Dynamics of Military Technical Development*, s’impose par la nature et la qualité des textes. Il y est question de regarder le terrain militaire comme opportunité d’éclairer les choix scientifiques et techniques.

PART ONE

Practice and Theory or Technology and Science
An Equal Footing Dialogue

De la Pirotechnia of Biringuccio
A Key to the Craftsman Production : Technical Questions and Aspects of the Organization of Production

Vasco La Salvia

Introduction

Within the body of the technical treatises tied to a specific branch of craftsmanship which flourished especially from the second half of the 16th century, the *Pirotechnia* of Biringuccio is the earliest printed work to cover the whole field of mining and metallurgy[1] ; moreover, the *Pirotechnia* has not to be regarded as a mere compilation but, in contrast to this, as an original work which embodies all the fire-based crafts[2]. Nevertheless, this technical treatise, as the other contemporaneous works such as those of Agricola and Palissy, cannot be considered as the improvised creation of individual workshop-masters ; it rather has to be regarded as a source where *les traditions techniques des arts industriels qui ont subsisté sans interruption dans les souvenirs professionnels* fix in written form experiences going back to much earlier times[3]. Therefore, the work of Biringuccio has to be considered, on the one hand, as the results of the long-lasting workshop traditions and, on the other, as the first attempt to establish a new production system which can be considered as indicating the traces of the emerging capitalistic economy. Actually, Biringuccio ideas focus on the importance of carrying out operations on metals

1. C.S. Smith and M. Teach Gnudi (eds), *The Pirotechnia of Vannoccio Biringuccio*, New York, The American Institute of Mining and Metallurgical Engineers, 1942, x ; F. Brunello, " Vannoccio Biringuccio e il trattato *De la Pirotechnia* ", in *Trattati scientifici nel veneto fra il XV e XVI sec.*, ed. E. Riondato, Vicenza, Neri Pozza, 1985, 32.

2. I. Guareschi, " Vannoccio Biringuccio e la chimica Tecnica ", *Supplemento annuale alla Enciclopedia di Chimica Scientifica e Industriale*, 20 (1903-1904), 425.

3. M. Berthelot, *La chimie au Moyen Age*, Paris, Nationale, 1893 (reprint Osnabrück 1967), vol. I, 1.

for profit and for use, on the advantages of large-scale operations, and on the importance of the use of power-driven machinery in place of human labour whenever possible. Thus, according to him, the availability of adequate water power became the first thing to evaluate in establishing a smelter. Fuel and transportation were considered as the next relevant requisites[4].

However, also if there was clearly an attempt to create a new production system, there were still several technical problems directly connected to the craftsmen mode of production : concerning firearms, the most relevant one was tied to the difficulty in avoiding and eliminating the differences between the masters' working techniques notwithstanding the similar qualities of the materials, as Biringuccio stated : " In order to make his work easier, every master of any art whatever keeps always to the road that he has learned or that his skills or good judgement has shown him to be the best. Although there are various methods used in making moulds for guns..., nevertheless almost the same road is travelled in all "[5]. This quotation, apparently, by using the term almost point to a situation consistently characterized by approximation rather than precision that is, to a world where measures, calculations, and uniformity of the producing procedures are not always the same thing[6]. Again, on the one hand, it has to be stressed that Biringuccio made a continuous use of the scale (as it is stressed by the general context of his work and especially by the figures presented such as that at page 47r. of his work)[7] and that he wrote that whatever is promised by the assay should be obtained in large scale operation, having established through this method the general rule of quantitative chemistry and the law of fixed reacting proportions. Moreover, according to Biringuccio, the balance for weigthing assays and furnace charges and the pen for computing them were as important as the furnaces themselves[8]. On the other hand, within the pages of his very same work, it is always possible to recognize the great extent of freedom that the craftsmen had while operating inside their own workshops ; this situation led to a complete lack of uniformity : in fact, concerning the production of moulds for casting, Biringuccio stated that " This is sometimes called by the skilled workmen the male, sometimes the core. Some make it one way and some another ; each precedes as he has learned or as his judgement or ability dictates[9], and I told you above that there are as many ways of making this mould as there are wits or opinion of masters "[10] ; moreover, in respect to measures and calculations regarding a

4. C.S. Smith and M. Teach Gnudi, *op. cit.*, xiv.

5. C.S. Smith and M. Teach Gnudi, *op. cit.*, 234.

6. A. Koyré, *Dal mondo del pressappoco all'universo della precisione*, Turin, Einaudi, 1992, 87-111.

7. I. Guareschi, *op. cit.*, 441.

8. C.S. Smith and M. Teach Gnudi, *op. cit.*, xvi.

9. *Ibidem*, 221.

10. *Ibidem*, 244.

crucial innovation he proposed such as the bell-scale was, the same author said that " It has been discovered by skilled bell founders, more through experience than from geometrical calculation (although calculation does enter), that a certain relationship of dimensions in both large and small bells makes the tone and the weight almost certainly what is desired. Among themselves bell founders have made a rule of the measurement and have called it the bell scale. (…) This scale is of great enlightenment if one does not have a bell already made for comparison. To do this, they have taken as their guide and basis, the rim (sound bow) of the bell which they wish to make, that is, the place where the clapper strikes with the head to sound. (…) Make it attractive and ornamented by following the measurements of the scale, and with your good judgement and the art of design "[11], and that " Just as I have told you that it is impossible to give an exact rule for the bell scale, so I say the same concerning the clappers… "[12]. These quotations testify to the fact that the quality level of craftsmen production was still basically tied to skills and working experience. Actually, as stated by Pistofilo Bonaventura in his Oplomachia, " nobody could be considered as a skilled craftsman in its own art if he is not able to properly use all the necessary tools to master the very craft "[13]. The practice of the craft was still thought as a monopoly of specialized people who kept their working procedures closely secret and, in fact, there are various accounts of secret processes, miraculous springs of water, poisoned ores, and such-like things which were employed to attain fine quality metals[14]. The following citations of Biringuccio and Bossi testify to the survival of the idea of the craft as a close world : " this art and work is not well known, so that no one can practice it who is not, so to speak, born to it, or who does have much talent and good judgement "[15] ; and " those people, who are trying to succeed in this operation, are to be regarded mainly as curious gentlemen rather than real practicioners and experienced people ; and actually, many are the things that a man think he could achieve but, when he attempts to put them into practice he also shall realize that the way to be followed is full of difficulties and faults, (...) fact that I had the occasion to realize only while experiencing the craft "[16].

11. *Ibidem*, 260-261 ; the bell-scale was used to establish the right proportions between the dimensions of the very bell and its sound.

12. C.S. Smith and M. Teach Gnudi, *op. cit.*, 272.

13. Pistofilo Bonaventura, *Oplomachia*, Siena, Ercole e Agamennone Gori, 1621, 3, the translation from Italian is mine.

14. C.J. Ffoulkes, *The Armourer and his craft from the* XIth *to the* XVIth *century*, New York, Blom 1967, 13.

15. C.S. Smith and M. Teach Gnudi, *op. cit.*, 213.

16. G. Bossi, *Breve trattato d'alcune inventioni che sono state fatte per rinforzare e radoppiare li tiri de gli archibugi e moschetti*, Anversa, 1626, 78, the translation from Italian is mine.

Casting metals for making artillery

After the spreading of gun powder, to each and single state it became essential to have at its own disposal skilled craftsmen in the art of casting, which was then generally identified with the art of making cannons and other firearms[17]. As a matter of fact, in respect to the given period, together with Bossi. " We have to take for granted that war is one of the most important thing in the world ; the destinies of dominions, kingdoms and empires as well as freedom and serfdom, misery and happiness are directly tied to war "[18].

Concerning the craft of casting, according to Biringuccio, there were five golden rules to be followed in order to succeed : " But to conclude, there are five steps in this art and you cannot fail in one of them, nor should you, for on each depends the whole. The first is to make the moulds… ; the second is to bake them well ; the third is to arrange and seal them up well in the pit ; the fourth (the most important) is to melt well ; and the fifth is to put enough material into your furnace so that you fill your moulds to overflowing "[19].

The casting of cannons was divided in three different phases. The first was the construction of a clay mould which was then subdivided again in three different parts. With regard to moulds-making, the first crucial step was to carefully select the clay ; the choice was again completely based on the craftsman skill : " Since this is a very necessary thing, you must try to have the best kind (of clay) and one that resist the fire well. It must be disposed to receive the metals well, must make a neat casting, and not shrink or break with cracks on drying or backing… I believe that there is little that can help you, since the clay in itself has no colour or visible sign that I know of to show how satisfactory it is. (...) Their quality depends on what the trial shows "[20]. Concerning the shape of the moulds, one of these was the exact reproduction of the external feature of the cannon, that is the jacket ; the other was the model of the breech ; the last, was the core which represented the space to be occupied by " the empty hole in the middle of the gun where the powder is found and through which the ball is shot "[21], that is, the muzzle : " To make this it is necessary to take two things into consideration. (...) One is what you are to make it on so that it may be supported and keep true. The other is what composition of clay will stand up during casting and yet not be very difficult to extract from the body of the finished gun. For the first, it is found that only an iron spindle of a suitable thickness is able to support the weight of the clay without

17. A. Bertolotti, *Figuli, Fonditori e Scultori in relazione con la corte di Mantova*, Milan, 1890, 45.

18. G. Bossi, *op. cit.*, 60-61, the translation from Italian is mine.

19. C.S. Smith and M. Teach Gnudi, *op. cit.*, 260.

20. *Ibidem*, 218.

21. *Ibidem*, 238.

bending under the heat of the fire or wobbling as it is turned on is journals or when handled "[22]. In order to build a functional weapon, it was necessary to properly support the core so that it could suffer the casting without bending, as stated by Collado : " Fifth advice. If it is given the case that the iron bar which has to sustain the clay core is well positioned inside the very core that is, it is firm and straight and resistant enough to suffer the casting without bending, the cannon would result perfect ; on the contrary, if these conditions would not be respected the pieces would have the inside not straight, whether much or just a little it depends (...) however, this fault is evident and because of that this artillery will always shoot wrongly and will suffer cracks, granted that the metal is not well distributed all along its body "[23].

These three very different parts were then assembled together and united and reinforced by using iron bars and hoops, and then placed into a pit. Again, the core has a relevant role, within the three different moulds it has to be located exactly in the middle of the others : " For it is a thing of great importance to the profit of the owner as well as to the honour of the master, because he who does not locate the core in the middle makes the gun weak and incapable of shooting straight, and shows great lack of skill as a master. In short, it is very necessary to do this well "[24]. However, before casting the metals inside the moulds the latter have to be backed : " it would be an obvious sign of ignorance to wish to cast without baking and, indeed, without baking very well. (…) usually all the moulds are baked either from the outside or from the inside with charcoal or flames from dry wood as is most satisfactory for the craftsman "[25]. This backing was intended to dry the clay and to avoid the problems which the moisture could cause during casting : " About casting artillery, proper moulds shape, with many important advises which are necessary in order to masterfully work the artillery. Chapter XIIII. Concerning casting artillery, (…) it has to be taken for granted that both the Germans and Flemish are extremely skilled (…) the Germans always cast their artillery inside old and dry moulds completely without moisture "[26].

The second phase consisted of casting the metal into the moulds ; afterwards, the mould had to be broken in order to extract the artifact. Therefore, each and every artillery piece had to be cast with a particular and single operation. In respect to this phase, it has to be stressed that, during this very given period, there was the spreading of the reverberatory furnaces the main advantage of which has to be considered as being the larger quantity of melting at one time ; however, concerning these type of furnaces, scant archaeological

22. *Ibidem*, 240 ; moreover, see 236, 238.

23. L. Collado, *Pratica manuale di arteglieria*, Venezia, Pietro Dusinelli, 1586, 20v., the translation from Italian is mine.

24. C.S. Smith and M. Teach Gnudi, *op. cit.*, 246.

25. *Ibidem*, 250 ; on moulds backing see also 251-252, 254.

26. L. Collado, *op. cit.*, 18v., the translation from Italian is mine.

evidence had been yet disclosed[27]. The third and last phase, was the polishing of the inside of the core granted that the casting could not be precise ; this operation was often mastered through the usage of an iron implement mounted on a long bar usually actioned by hydraulic power. However, this method was scarcely precise and, thus, could also cause and enhance flaws which occurred during the casting[28].

The guns'casting method was mainly a continuation of the medieval bell-founding techniques. In fact, bell-founding techniques consisted of the very same basic procedures, that is, to prepare the clay moulds and then bake them, and finally to melt and cast the metal. In the case of bells, there are only two moulds to be manufactured : one is the cope (the outer mould) and the other the core (the inner part)[29]. Moreover, since relevant archaeological evidence has been found concerning bell-founding, this example could be useful in order to evaluate methods and structures of both artistic and military bronze casting. Actually, within the fire-based techniques, there is always a certain grade of similarity : this analogy is to be considered as much important as much of the materials and the production procedures are similar. These craftsmen had to have knowledge of the qualities of the furnace charge, of the features of the workshop tools, of the characteristics of the wax used for the models and of those of the clay used for the moulds which had to be at the very same time ready to be modelled and resistant to cracks[30]. As a matter of fact, many written documents of the given period yielded evidence of the close relationship between all the fire-based producing activities. The casting-house of the Marquis of Mantua provides a relevant example of this situation. Actually, all the craftsmen charged with the duty and responsability to direct the mentioned smelter, were at the same time expert in casting both bells and statues in addition to guns. This was namely the case of Federigo Calandra (chief of the casting-house since 1498), of Ippolito Calandra (chief after 1512), of Giorgio Albenga who directed the smelter somewhat after 1586, and of his successor Belisario Cambio known as Bombarda[31].

As far as the rough material is concerned, the late Middle Ages gives evidence of an increasing usage of non-iron metal alloys, particularly bronze[32].

27. T.F.C. Blagg, " Bell-founding in Italy : Archaeology and History ", in *Papers in Italian Archaeology I : Lancaster Seminar*, H. MCK. Blake, T.W. Potter, D.B Whitehouseeds (eds), Oxford, BAR 41 (1), 1978, 425.

28. T.K. Derry and T.I. Williams, *Storia della Tecnologia*, Turin, Einaudi, 1977, 179.

29. T.F.C. Blagg, *op. cit.*, 425.

30. T. Mannonni and E. Giannichedda, *Archeologia della Produzione*, Turin, Einaudi, 1996, 306-311.

31. A. Bertolotti, *Figuli, Fonditori e Scultori in relazione con la corte di Mantova*, Milan, 1890, 46-53.

32. P. Benoit, P. Dillmann, P. Fluzin, " Iron, cast Iron and Bronze. New Approaches of the Artillery History ", in *The Importance of Ironmaking : Technical Innovation and Social Changes. Papers presented at the Norgberg Conference on May 8-13, 1995*, Gert Magnussoned (ed.), Stockholm, Jernkontorets Bergshistorika Utskott, 1996, 241, 245-246.

These copper alloys, which had already a long lasting tradition regarding bells and statues production, during the 15th century began to play a relevant role also in guns production[33]. However, both archaeometallurgical analyses and written documents yield evidence of the fact that until the middle of the 15th century artillery making was dominated by wrought iron production, and also cast-iron had a limited use in manufacturing artillery. Thus, the blast furnace revolution and the consequent spreading of the indirect method in ironmaking did not immediately influence the western European artillery production procedures[34]. Moreover, when during the second half of the 15th century appeared a new type of artillery, which had a lighter weight and a higher grade of mobility, again cast-iron was not often used and, in contrast to this, bronze imposed itself as the most important rough material. This alloy is resistant and easier to be melted and worked. It was then possible to cast pieces in one block and, therefore, to avoid the problem of the junction between the chamber and the barrel. The main problem regarding cast-iron was the low quality of the material, granted that was indeed difficult to control the Carbon content inside the pig iron which often resulted to be too brittle to stand the explosion without suffering cracks ; in addition to this, there was the problem of how to control the rust formation. The importance of bronze is also evident considering the personal of the artillery at the end of the 15th century. The records of the king of France and of the city of Rennes give evidence of the number of the persons employed : in both cases, the non-iron smiths represent the vast majority of the workers being, moreover, well paid ; the blacksmiths were less numerous : in Rennes they were still furnish same pieces but in respect to the royal artillery, they were only serving as workers to maintain the metallic part of the carriages[35]. Nonetheless, the usage of copper alloys presented problems related to the provision of this metal, which indeed is not defused as iron and, therefore, was more expensive. In addition to this problem of copper supply, there was also the smelter custom to alter the alloy they were given by the donators using scrap material in order to save good quality copper for their successive works : “ Thertinth Advice. It is often given the case that the smelters, in order to spare good quality copper, add during casting other metals which are not suitable to attain the proper alloy that is bronze, brass, and/or lead ; these bad quality metals are collected by the smelters through refining the workshop waste materials ; however, the cannons worked with this method will be of no use being spongy, brittle, too poor in metal. Therefore, it shall be ordered that for 100 lire of pure copper 10 lire of pure tin should be added ; this would be a perfect alloy... ”[36].

33. T.K. Derry and T.I. Williams, *op. cit.*, 170.
34. P. Benoit *et alii*, *op. cit.*, 243.
35. *Ibidem*, 245-246.
36. L. Collado, *op. cit.*, 21r., the translation from Italian is mine.

Thus, it has to be stressed that the problems connected to copper supply were considered as to be less relevant and of easier solution rather than those tied to the control of rust formation and the Carbon diffusion into the iron, that was, for the craftsmen the question of steel brittleness. As the following quotations testify, these were problems consistently considered by the craftsmen of the given period : " About the invention of Gun Powder, Artillery and about the method and the period during which these terrible machines were first used. Chapter XIII. (...) Some of these artillery pieces were mastered by welding together by hammering various iron sheets and reinforced and united through thick iron hoops ; other consisted of cast iron ; the latter method was for a certain period taken as the best one but, as time went by and having achieved a better knowledge of the craft of producing artillery, of the way of using cannons, and of its effects, it was found out that the cast iron artillery was not that good and that it was also dangerous because it often suffered cracks ; therefore, after several experiences and investigations it was then discovered the proper alloy of metals (which now we know is crucial) and this alloy is namely bronze or however it might be called "[37] ; and " the empty space cannot be easily cleaned and it is also difficult to dry it after you have washed it ; therefore, as time goes by, the rust will eat it and thus the core will suffer cracks "[38].

However, there was a field within artillery production which consistently underwent a radical change thanks to the " blast furnace revolution " ; namely, the indirect method offered an effective and cheap projectile : Tilly had a daily consumption of about 12.000-18.000 cast-iron projectiles during the siege of Magdeburg in the year 1631. Indeed, the cast-iron cannonballs did not totally and immediately impose, and lead was used for small size calibre[39]. In respect to the projectiles question, Biringuccio said that at his time the modern guns : " shoot balls of iron that are smaller than those of the bombards but are discharged more frequently, and have a greater effect than those of the bombards since they are of a harder material "[40].

CRAFTSMANSHIP AND ITS ORGANIZATION OF PRODUCTION BETWEEN TRADITION AND INNOVATION

The problem of achieving uniformity while producing artifacts, was particularly relevant for firearms making. In fact, in order to avoid such a problem, Collado in his work about practical artillery stated the following : " Third Advice. (...) It shall be issued that during all casting a general rule should be carefully followed and it shall be ordered to the smelters not to cast any other

37. L. Collado, *op. cit.*, 18v., the translation from Italian is mine.
38. G. Bossi, *op. cit.*, 12, the translation from Italian is mine.
39. T.K. Derry and T.I. Williams, *op. cit.*, 171.
40. C.S. Smith and M. Teach Gnudi, *op. cit.*, 224-225.

class of pieces but those listed within the given general rule… "[41]. In order to solve the problem of the above mentioned variety in production, it was suggested by Biringuccio that bad luck " is nothing but ignorance or carelessness " and that " the founder can assure Fortune's favouring him by careful attention to details "[42]. This attitude towards this patrimony of knowledge and rules of the craft, which had to be respected and had been passed down orally, did not easily allow the formation of changes within working procedures : as a matter of fact, each and single technical achievement was regarded as to have been the result of the strict respect for the traditional production procedures, the specific working tools' shape, and the peculiarity of the specific — special — rough material. The particular structure of this knowledge did consistently influence also the surviving possibilities of the very craft since the latter would, actually, disappear as soon as the conditions able to maintain the constant transmission of the traditional information through times and generations would not be granted[43].

Thus, since this patrimony of technical procedures was totally empirical and not written, the trend was to identify calculation and exactness not with a specific method but with accuracy : " Be careful also not to let yourself be carried away by impatience into wishing to force the process more than the art allows by its nature "[44]. However, this strict rules respect, did not automatically provide good results : there is, in fact, evidence of contracts stating that the craftsmen should not be paid until their artifacts were not proven to be functional[45]. Again, Biringuccio ideas are between innovation and tradition. On the one hand, he realized the importance of exactness and calculation, and the foolishness of the misleading concept of bad fortune ; on the other, the only parameter he knew and regarded as useful to achieve good results was the careful attention to details : " These effects (that is, faults) seem to be produced by fortune because of the absence of some information necessary to the intelligence… For, in truth, lack of skill or too little diligence in the art, and I am of the firm opinion that every error proceeds from these and not from fortune ; and if you always use the necessary care to make the means perfect the end will never be in doubt… Thus, to conclude, whoever wished to exercise this art well and with certainty must do everything with exactness. if you are careful nothing will ever happen that your judgement does not first point out to you, although a sure result cannot be shown before the end, and finally, therefore, you must plan to spare no labour or expense and to be careful and very patient in every

41. L. Collado, *op. cit.*, 20r., the translation from Italian is mine.

42. C.S. Smith and M. Teach Gnudi, *op. cit.*, xv.

43. G. Stabile, " La torre di Babele : confusione dei linguaggi e impotenza tecnica ", in J.C. Maire-Vigueur e A. Paravicini Bagliani (eds), *Ars et ratio. Dalla torre di Babele al ponte di Rialto*, Palermo, Sellerio, 1990, 264, 269.

44. C.S. Smith and M. Teach Gnudi, *op. cit.*, 258-59.

45. T.F.C. Blagg, *op. cit.*, 429-430.

details in order to bring all means which you are to use to perfection "[46]. According to Biringuccio, instruments and craftsmen skills are considered as to be equivalent in respect to the producing performance. The discussion about bells-production clearly testify to this situation ; with regard to bell-design, Biringuccio stated that to refine the drawings the craftsmen should proceed " according to judgement or with compasses "[47]. Moreover, as far as the bell-scale is concerned, it is possible to understand that this new instrument of calculation was not followed as a scientific method : as a matter of fact, either each and single master applied to the scale his personal arrangements, either they did not follow it " with the necessary degree of accuracy "[48]. Summarizing, it is possible to state that in the given period the basic criterium to achieve uniformity and good results was then yet not the scientific calculation as it is intended within modern concepts but accuracy, a production system grown on a daily workshop experience and on a hit-or-miss basis, and based on the careful attention to each and single details of the traditional producing procedures : " If you have not been the workman yourself, you must at least have been an active helper in this and in every other part in order to be able to observe everything without a fault. (...) But if you are the arteficer the process is made easier "[49].

However, craftsmen repeated production could achieve a credible uniformity[50] since there are examples which testify to an early concentration of the production activities. The case of Venice arsenal is, possibly, a representative fact to clarify the relationship between craftsmen, technology, science and large scale production in the given period. Generally, these types of great economic structures and relevant production centers, such as the Venetian arsenal was during the 15th and the 16th centuries, had always been in connection with concentration of technological tools and development of technology. Therefore, by being part of the production activities, science and technology were, therefore, forced to enter the process of modernization[51]. Between the years 1524-26 and 1540, the Venetian arsenal underwent a complete restructuring. New workshops for casting and shaping metals were built and it was, moreover, issued that all fire weapons had to be produced only in these new buildings. This situation can be regarded as the attempt to organize and fix the economic trend which tended towards the organization of a system of functional planning of the production centers. This restructuring led to a terrific production improvement, almost doubling the quantity of the produced firearms as stated by Nicolò Zeno, during the very crucial period of the Turk con-

46. C.S. Smith and M. Teach Gnudi, *op. cit.*, 214-216.
47. *Ibidem*, 262.
48. *Ibidem*, 265-266.
49. *Ibidem*, 228.
50. H. Kilbride-Jones, *Celtic craftmanship in bronze*, London, 1980, 34.
51. F. Braudel, *La dinamica del capitalismo*, Bologna, Il Mulino, 1988, 32.

flict. Namely, in the year 1591, the Venetian arsenal was provided with 1813 cannons ready to equip 100 light galleys and 10 great galleys[52]. Moreover, the Consiglio dei Dieci that is, the Venetian government tried to set a serious control on the movement of the rough material, namely copper, to be used in the casting-houses which led to the 1535 issue to stop and forbid the practice of the smiths who always tried to exchange the copper they were provided from the government with another of lower quality[53].

The arsenal of Venice gives, thus, evidence of the framework within which works such as that of Biringuccio had grown. As a matter of fact, the main trend of a certain cultural environment pointed clearly to a radical change : the people who were working in arsenals and workshops thought to a completely different scope while operating in respect to what the vast majority of the traditional scholastic philosophers were still regarding as important for the evolution of the society[54]. Cooperation, progress, perfectibility, and invention became all part of the new concepts of science, technology, and culture. Finally, knowledge was regarded as to be something which is able to constantly grow thank to the contributions of many different people. This opinion, as Sir Francis Bacon stated, clearly reacted against the traditional scholastic idea that is, the transformation of the technical inadequacy into an ontology[55] ; moreover, according to Leibniz, to collect technical procedures and then edit them became one of the most relevant duty for the " modern philosophers "[56]. Nevertheless, there was not any sudden event which definitely and rapidly changed the society structure ; on the contrary, for a long period, as it was mentioned in respect to Biringuccio work, two diverging aims were contemporaneously existing together into the same scene : to this situation was tied the difficulty to create a new technological language on the basis of long-lasting workshop traditions. Therefore, from an economic point of view, the fundamental characteristic of the preindustrial economy has, apparently, to be considered as to be the coexistence of the inertial conservatism of a simple and natural economy together with the economic movements typical of the modern period which, however, were still in minority even if lively and incisive. Thus, on the one hand, there is a great portion of the society which lives still within an almost self-sufficiency economy and, on the other, there is another section which supports a market economy and a capitalism in progress which slowly traces the very limits of the modern society[57]. It has, however, to be stressed

52. E. Concina, *L'Arsenale della repubblica di Venezia. Tecniche e istituzioni dal medioevo all'età moderna*, Milan, Electa, 1984, 144.

53. E. Concina, *op. cit.*, 136-137.

54. P. Rossi (ed.), *La rivoluzione scientifica da Copernico a Newton*, Torino, Loescher, 1973, 33, 35.

55. *Ibidem*, *op. cit.*, 36.

56. *Ibidem*, *op. cit.*, 37.

57. F. Braudel, *op. cit.*, 26.

that, between the 15th and the 18th centuries, the dimensions of this economic structure in progress that is, market economy, did not cease to enhance[58]. Actually, during the 15th century, especially after 1450, there was a general economic increase favourable to the urban civilization which was supported by the increase of the industrial prices and the decrease of those of agriculture. Both the cities workshops and markets were the propulsive forces of this economic movement ; during the successive century the core of this propulsion shifted to the international trade[59].

Thus, during this period, especially within the second half of the 16th century and the first half of the 17th century, there is evidence of the first attempt to apply science to industry[60]. It has to be stressed that this is the period during which capitals of various origin were collected and invested in industrial activities. Mines, metallurgical and textile plants, breweries, sugar refineries, soap, glass, salt, and alum factories were rapidly developing. Certainly, the rising capitalism encountered several difficulties to become leading in economy : the written documents of the archives and of the Chambers of Commerce of the given period, yield evidence of a situation in which capitals were uselessly searching for occasions for investments ; thus, the problem was not in collecting capitals. Often, the capitalist was forced to invest in real estate but, occasionally, the capitalist could speculate on flats renting in the cities, or on industrial activity such as mining, as it happened several times during the 15th and 16th centuries. These were the new activities of the emerging economic power of England for there were studied the data expressed in the books of Biringucio, Agricola, and Palissy[61].

These cultivated workshop masters, engineers, physicians and artists had little respect for the authorities that enslaved culture for so long. In so far Biringuccio is concerned, his source of information was almost entirely extracted from his own observation and experience in the workshops where metals were smelted, worked, and cast. The people as Biringuccio found intellectual satisfaction more in accomplishing a desired result than in contemplating the causes of things[62]. Within the works of these technicians, engineers, and artists a new concept of labour arose and, consequently, a new idea was formed about the importance that the technical knowledge would have had for the society ; this very concept did play a relevant role in structuring the development of the principle of technological and scientific progress : " If the ancient Romans and Greeks would have behaved as the these contemporaneous people are doing at the moment (that is, disregarding the new inventions), the world would have

58. *Ibidem*, 49.

59. *Ibidem*, 38.

60. B. Farrington, *Bacone filosofo dell'età industriale*, Torino, Einaudi, 1980, 31.

61. F. Braudel, *op. cit.*, 62 ; B. Farrington, *op. cit.*, 42.

62. Smith and Gnudi, *op. cit.*, xiv.

been still into darkness, without civilization, and it would have never attained any of the present glorious achievements. All the inventions which concern weapons, methods and crafts and which are used in this given period for fighting, had once their beginning and during that period they had been indeed new inventions ; therefore, if they would have been disregarded they could not have been put into practice and, thus, there would have been no profit and progress at all ; in fact, all the inventions and crafts, even if good, are to be considered in themselves useless ; it is indeed necessary that men through their constant work operate and accomplish them with honour "[63].

63. G. Bossi, *op. cit.*, 95-96, the translation from Italian is mine.

SCIENCE AND TECHNIQUE IN THE WORK OF GALILEO

Fernando L. CARNEIRO

GALILEO'S METHOD OF SCIENTIFIC RESEARCH

It was a mode in important circles of epistemologists, during the third and the fourth decades of this century, to disdain or even to deny the role of the experimentation and the technique in the work of Galileo. The researches made during the last forty years, mainly those, — like the Stillman Drake ones, based on the Galileo's manuscripts of the Italian National Library of Florence, showed that this was false. In fact his scientific method consists of a sound combination of observation and experimentation, with mathematics, used as instrument of deductive logics. Taking a number of experimental facts or simply certain suggestions of the technique as a starting point, one builds a first hypothesis or theory to interpret them. From this theory, certain conclusions are drawn, by deduction. Next the validity of these conclusions is subject to observation or experiment. The hypothesis is replaced or improved if the tests do not confirm it. The source of the truth always remains, in the last analysis, experience. Galileo, therefore, was neither a pure " platonic " rationalist, nor a pure " baconian " empiricist.

The most important of the manuscripts of Galileo analysed by Stillman Drake is the so called " folio 116 v " of the " Paduan notes " preserved in the Biblioteca Nazionale Centrale di Firenze, volume 72 (and that I had also the opportunity of handling during my sabbatical term at the *Istituto e Museo di Storia della Scienza* of Florence, in 1989). The figures exhibited in that manuscript describe clearly an experiment made by Galileo in 1608[1] : they are a record of measurements of distances of horizontal projection of a ball which had first descended through specified vertical distances along an inclined plane, placed on a table, and were then deflected horizontally, falling from the

1. *Galileo's Notes on Motion*, arranged by S. Drake, Firenze, Istituto e Museo di Storia della Scienza, 1979, 79 ; P. Thuillier, " Galilée a-t-il experimenté ? ", *La Recherche,* n° 143 (1983), 442-453 ; S. Drake, *Galileo at Work*, Chicago, 1978, 127-132.

table to the floor. In that experiment Galileo combined his laws of conservation of speed (also known as horizontal inertia), of the fall of heavy bodies, and of relativity (independent compounding of an uniform motion resulting from an horizontal projection, with a naturally accelerated downward motion). Galileo showed that the resultant motion is the parabolic motion of projectiles, and the experiment was a verification of his theory.

According to the corresponding calculations the horizontal projections of the motion of the ball after leaving the table, until the floor, should be proportional to the square roots of the heights of fall along the inclined plane. Taking the first of five cases as a reference, he noted, for each of the other four cases the calculated or " should be " (*doveria essere* in Italian) horizontal projection, and the measured one. The differences, corresponding to the errors in the measurements, were small (less then 3,7%) but random, fact that demonstrates the authenticity of the experiment.

Two other experiments of Galileo, many times reproduced by modern researchers, are the famous experiment of the inclined plane[2], and the experiment with a pendulum with two centers of oscillation[3], described in detail in the " Third Day " of his famous book *Two New Sciences*. And it is interesting to observe that Alexandre Koyré was very unhappy in choosing as a proof of his opinion against the veracity of the descriptions of experiments, made by Galileo, the example of the narrative about the ascension of red wine, in rosy streaks, to a closed glass reservoir filled with water, through a small hole at its bottom, without mixing. Koyré probably ignored that this is possible if the flow is sufficiently slow to be laminar, non-turbulent. Many researchers have also reproduced this experiment very easily, including, more recently Pierre Thuillier and myself[4]. It was Koyré, and not Galileo, that has not made the experiment.

Science and technology in the work of Galileo

The continual recourse to the verdict of experience requires a permanent collaboration of technique with pure or abstract science, as Galileo's constant activity in scientific research brings out. But Galileo was also known as an engineer : *mathématicien et ingénieur du Duc de Florence*, as we can see in the title of the French translation of *Two New Sciences*, by Mersenne. Inventor of the " geometric and military compass ", he often concerned himself with problems of machine and ship construction, — especially during the period when he was living in Padua, in connection with the activities of the Venice

2. Galileo, *Two New Sciences*, translation by S. Drake, Wisconsin, 1974, 212-213.
3. Galileo, *Two New Sciences*, 206-207.
4. Galileo, *Two New Sciences*, 115-116.

Arsenal, and with public works, like fortifications, regularization of rivers and water supply.

Galileo is justly considered as the founder of modern science. He approached pure sciences as mechanics and astronomy with the objective of reaching a better knowledge of nature and his laws, without presupposing an immediate application, but naturally expecting that something useful to mankind would be a future consequence of his researches. It was many years after the discovery of the Jupiter satellites in 1610 that he developed the idea of a practical application of the chronology of their eclipses and occultations in view to determining longitudes. The theory of the fall of the bodies and the laws of the compounding of motions, derived from his principle of relativity, lead Galileo to the deduction of the parabolic trajectory of projectiles, later used for the construction of tables for the range of shoots. And a few months before his death he made the suggestion of applying the laws of pendulum to clocks. The researches of Galileo about motion are described in the two last “ Days ” (*Giornate*) of his book *Two New Sciences*, and his astronomic discoveries in the book *Sidereus Nuncius*.

But very different is the nature of the “ New Science Concerning the Resistance of the Solid Bodies to be Broken ” (*Scienza Nuova Prima, intorno alla Resistenza dei Corpi Solidi ad essere spezzati*), described in the two first “ Days ” of *Two New Sciences*. It is not a “ pure science ”, but an applied one. Galileo was lead to that science, nowadays known as the “ Science of the Strength of Materials ”, in view to solve some technical problems occurred in the Venice Arsenal, and inspired in the practical knowledge of the artisans of this famous dockyard, as he emphasizes in the first paragraph of the book : a “ science very necessary to construct machines and all type of buildings ” (*scienza molto necessaria nel fabricare machine ed ogni sorte di edifizio*), explains Galileo in a letter to Antonio de Medici. According to Timoshenko’s *History of the Strength of Materials*, “ Galileo, in the famous book *Two New Sciences*, made “ the first attempts to find the safe dimensions of structural elements analytically, and this represents the beginning of the science of strength of materials ”.

In the following sections of this paper more information is given about the scientific and technical achievements of Galileo.

The researches of Galileo about motion.
The laws of vertical free fall

The researches of Galileo about motion are described in the final part of the first, and mainly in the third and fourth “ Days ” (*Giornate*) of the book *Two New Sciences* : *New Science of Local Motions* (*Scienza Nuova dei Movimenti Locali*), and *On Motion of Projectiles* (*Del Moto dei Proietti*)

In the final part of the first " Day " of *Two New Sciences* Galileo refutes the opinion of Aristotle, generally accepted at his time, that " moveables of different weight are moved through the same medium with speeds proportional to their weights " and then that " a rock of twenty pounds is moved (in falling) ten times as fast as a two-pound rock ". " I say ", speaks Galileo, " this is false "[5]. Galileo, based in observations of the fall of heavy bodies through the air and other media with different densities " came to the opinion that if one were to remove entirely the resistance of the medium, all materials would descend with equal speed "[6].

In order to reduce the retarding effect of the resistance of the medium, Galileo recommends to make experiments with heavy rounded bodies, made from a dense material like metallic balls. He never mentioned the " experiment of the tower of Pisa ", that seems to be a fantasy or an exaggeration of Viviani, but we can find in his writings references to observation of free fall of moveables from towers, like the following : " A lead ball falling from a tower two hundred braccia high will be found to anticipate an ebony ball by less than four inches "[7]. Galileo finally formulated his famous conclusion : " It seems to me that we may believe, by a highly probable guess, that in the void all speeds (of falling moveables of different weights) would be entirely equal "[8]. The equality of the periods of oscillation of pendulums with the same length but different weights was also mentioned by Galileo as an experimental verification of his hypothesis.

As Galileo later discovered that " the naturally accelerated motion of descending heavy bodies is an uniformly accelerated motion ", we may so formulate his law of free vertical fall : the acceleration due to gravity is the same for all bodies, independently of their mass, shape or other properties. That law was later basic for the Newtonian theory of gravitation and the theory of general relativity of Einstein.

Galileo never presented himself as the discoverer of the properties of an accelerated motion, as an abstract motion, defined mathematically, *ex suppositione*. This was done by the medieval scholars of Merton and Paris, but without any reference to free fall. The original finding of Galileo, as he clearly states in the definition of " naturally accelerated motion "[9] is that the " natural motion " of the free vertical falling bodies is an uniformly accelerated motion. " And first it is appropriate to seek out and clarify the definition that best agrees with that which nature employs. Not there is anything wrong with inventing at pleasure some kind of motion and theorizing about its consequent

5. Galileo, *Two New Sciences*, 110.
6. Galileo, *Two New Sciences*, 117.
7. Galileo, *Two New Sciences*, 120.
8. Galileo, *Two New Sciences*, 117.
9. Galileo, *Two New Sciences*, 197.

properties... But since nature does employ a certain kind of acceleration for descending heavy kinds, we decided to look into their properties so that we might be sure that the definition of accelerated motion which we are about to adduce agrees with the essence of naturally accelerate motion ".

In fact this is an approximation valid only for short courses in the neighbourhood of the surface of the earth, because the acceleration is not constant but proportional to the inverse of the squared distance to its center. Galileo could not, in consequence, arrive to that conclusion *a priori*, without recourse to observation or experiment, as certain authors pretend. " In order to make use of motions as slow as possible ", he thought " of making the moveables descend along an inclined plane not much raised above the horizontal "[10]. This famous experiment is meticulously described in the " Third Day " of *Two New Sciences*[11] and was reproduced many times by modern researchers.

GALILEO'S " POSTULATE " OF CONSERVATION OF MECHANICAL ENERGY IN A GRAVITATIONAL FIELD

To justify the adoption of the device. of the inclined plane Galileo formulate a very important " postulate ", " of which the absolute truth will be later established for us seeing that our conclusions, built on this hypothesis, that indeed correspond with and exactly conform to experience "[12]. The " postulate " corresponds to what we know nowadays as " the principle of conservation of the mechanical energy in a gravitational field : the speed acquired by a body falling without friction or other impediments along a curve depends solely of the difference of level between the point of departure and the point of arrival, and is equal to the speed acquired in a vertical free fall from the same height ". " Every speed acquired by descent through a curve equals one which can make the same moveable rise to the same height through the same curve or another curve " (in fact Galileo speaks only of descent or ascent of balls along inclined planes, along arcs of circle, in pendulums, and along polygonals inscribed in those arcs). Galileo was inspired, to formulate the " postulate " by the experiment with a checked pendulum, — a pendulum with two centers of oscillation, described in the " Third Day " of *Two New Sciences*[13]. In this experiment an interfering nail is driven in a wall in the vertical below the nail on which the pendulum is suspended, and the ball descends from the initial position along a circle with center on this nail until the thread is stopped by the interfering nail and constrained to ascend along a circle with center in this second nail, until

10. Galileo, *Two New Sciences*, 128.
11. Galileo, *Two New Sciences*, 212-213.
12. Galileo, *Two New Sciences*, 205, 207-208.
13. Galileo, *Two New Sciences*, 206.

to reach the initial level, " less a very small interval due to the impediment of the air and of the thread " ; and then begins a motion in the opposite direction.

Applying the " postulate " to inclined planes, it is easy to show that if the free vertical fall is an uniformly accelerated motion, the fall along an inclined plane is also an uniform accelerated motion, but with a smaller acceleration : " the times of the movements will be to one another as the heights of the plane and of the vertical ". In fact, in the case of rolling of a perfect round ball the reduction of the acceleration is bigger, due to the rotational inertia, ignored by Galileo, but this does not affect his conclusions.

The hypothesis of an uniformly accelerated motion in the free vertical fall was verified by Galileo by the experiment with the inclined plane. As a consequence, the final speeds are proportional to the square roots of the heights of fall.

Inertia and relativity according to Galileo. The ship of Galileo and the elevator of Einstein

Galileo's concept of inertia is defined in the " Third Day " of *Two New Sciences*[14] : " Whatever degree of speed is found in the moveable, this is by its nature indelibly impressed on it when external causes of acceleration or retardation ace removed, which occurs only on the horizontal plane ", " all impediments being put aside ".

And the principle of the compounding of motions is then formulated : " That equable motion on the plane would be perpetual if the plane were of infinite extent, but if we assume it to be ended, and situated on high, the moveable, driven to the end of this plane and going further, adds to its previous equable and indelible motion that downward tendency which it has from its own heaviness. Thus there emerges a certain motion compounded from equable horizontal and from naturally accelerated downward motion "[15]. The experiment registered in the " folio 116 " of the manuscripts of Galileo was clearly mounted to verify that theory.

This principle of compounding of motions can be deduced from the principle of relativity of Galileo, that was firstly formulated in his polemic book *Dialogue Concerning the Two Chief World Systems —Ptolemaic and Copernican*, and consisting in the conciliation of the relative motions in the interior of a mechanical system with a common uniform motion : it would be impossible in a closed cabin of a ship in uniform motion and " not fluctuating this way and that ", to know if the ship is moving or standing still " The cause " motion and

14. Galileo, *Two New Sciences*, 243.
15. Galileo, *Two New Sciences*, 268.

" of all these correspondences in effects is the fact that the ship's motion is common to all the things contained in it, and to the air also "[16].

But Galileo considered also the case of a common accelerated motion : " A large stone placed in a balance acquires weight with the placement on it of another stone… But if you let the stone fall freely from a height… in free and natural fall the smaller stone does not weigh upon the larger, and hence does not increase the weight as it does at rest ". " We feel weight on our shoulders when we try to oppose the motion that the burdening weight would make ; but if we descend with the same speed with which such a heavy body would naturally fall, how would have it press and weigh on us ? "[17]. This statement of Galileo coincides exactly with the " most happy thought of my life " (*der glücklichste Gedanke meines Lebens*) of Einstein, that lead the great physicist to the general theory of relativity. The " ship of Galileo " in uniform motion, is substituted by the " elevator of Einstein ", in accelerate motion.

The astronomic discoveries of Galileo

Galileo never claimed for himself the invention of the telescope. The instrument with which he observed the moon and the galaxy, and discovered the satellites of Jupiter was a considerable improvement of a spy-glass originally fabricated in Holland, as he relates in the beginning of the book *Sidereus Nuncius*[18], published immediately after that discovery : " About ten months ago a rumor came to our ears that a spyglass was made by a certain Dutchman by means of which visible objects, although far removed from the eye of the observer, were distinctly perceived as though nearby ". " The rumor was confirmed to me a few days later by a letter from Paris from the noble Frenchman Jacques Badovere. This finally caused to apply myself totally to investigate the principles and figuring out the means by which I might arrive the invention of a similar instrument ". With the first spy-glass by him fabricated Galileo " perceived objects three times closer and nine times larger (in area) : with the second, sixty times (in area) and finally " sparing no labor or expense ", he succeeded " in constructing an excellent instrument that the objects seen through it appeared nearly a thousand times larger an over thirty times closer than when regarded with our natural vision only ". This was in fact the true telescope, no more a simple spy-glass.

All the researches that Galileo was doing at Padua on mechanics were then stopped, and replaced by astronomical observations and supporting copernicanism, until the condemnation by the Holy Office, in 1633. The findings made

16. Galileo, *Dialogue Concerning the Two Chief World Systems, translation* by S. Drake, foreword by A. Einstein, Berkeley, 1967, 186.

17. Galileo, *Two New Sciences*, 108.

18. Galileo, *Sidereus Nuncius*, translation by A. van Helden, Chicago, 1989, 16-18.

in the Paduan period were then resumed during his last years, in his domiciliary prison of Arcetri, and registered in the book *Two New Sciences*.

In the book *Sidereus Nuncius* Galileo presents sixty-two sketches of the position of four " starlets ", observed during two months. From these sketches he deduced the " revolutionary " conclusion that the " starlets " were satellites of Jupiter, and this was the beginning of modern astronomy.

The fortuitous contact with the invention of a modest artisan was the most spectacular example of the influence of technics on the work of Galileo. And we can yet join that the manner by which Galileo made the improvement of the spy-glass was clearly artisanal, despite his references to the " doctrine of refraction ". In fact Galileo adopted, for the improvement, the empirical method of trial and error.

Galileo, founder of the science of the Strength of Materials

Galileo was lead to found the " First New Science concerning the Resistance of Solid Bodies to be broken " (*Scienza Nuova Prima intorno alla Resistenza dei Corpi Solidi all essere spezatti*), object of the beginning or the " First Day " and of all the " Second Day " of *Two New Sciences*, in view to solve some technical problems of the Arsenal of Venice, in which he was, as a professor at the University of Padua, a technical adviser[19]. He was inspired, in these researches, by the experience of the artisans of that great shipyard, as he proclaims himself in the opening of the book : " The constant activity which you Venitians display in your famous arsenal suggests to the studious mind a large field for investigation, especially in that part of the work that involves mechanics ; for in this department all types of instruments and machines are constantly being constructed by many artisans, among whom there must be who, partly by inherited experience and partly by their own observations, have become highly expert and clever explanation "[20].

The problem that struck initially Galileo was that of geometrically similar structures of machines or buildings which, having behaved satisfactorily when executed on a certain scale, fail completely when they are executed on a larger scale. He was led to the solution of the problem by an observation made by an old worker (*quel buon vecchio*) that the sustaining apparatus, supports, blocks and other strengthening devices were made much larger around a huge galley than around smaller vessels in order " to avoid the peril of its splitting under the weight of its own vast bulk, a trouble to which smaller boats are not

19. F.L. Carneiro, *Galileo-Founder of the Science of Strength of Materials, RILEM Fifty Years of Evolution of Science and Technology of Building Materials and Structures*, Freiburg, 1997, 7-26.

20. F.L. Carneiro, *Galileo Founder of the Science of Strength of Materials*, 7.

Obs. : the paginations of *Two New Sciences* are those of the Italian " national edition ", shown in the margins of the pages of the English translation.

subject ". Galileo found the correct explanation of the phenomenon : geometrical similitude does not assure physical similitude when the weight of the structure must be considered. His concept of " limit size " of structures of the same type and geometrically similar, popularly known as the theory of the " relative weakness of giants " is logical consequence of that theory.

The theory of flexure of Galileo was based in the " lever principle " of statics, and in a theory about the strength of prismatic bodies to traction. As he relates, the later was inspired to Galileo by the explanation of a worker called by a friend of Galileo to repair a suction pump that he supposed to be worn out. He learned from the worker that " neither with pumps nor with any other device that lift the water by suction is it possible to make this rise more than eighteen braccia (about nine meters), whether pumps are of large bore or small : that is the measure of this absolutely limited height ". Galileo, based on the wrong hypothesis that a column of water, attached at his top by the " force of the vacuum ", breaks by traction, like a wire fixed at its upper end and with a great weight added at the other end. He then deduced that, each material having, as water, a " length of rupture ", the resistance to traction is proportional to the area of the cross section. The text of Galileo describes a test of traction of a copper wire, and gives for the tensile strength of this material a perfectly acceptable figure.

Galileo supposed that in bending an internal resistant force, equal to the resistance to traction of the bar, is applied to the center of gravity of the cross section, and his formula for the bending strength has in consequence an error in a numerical coefficient. Galileo, attracted by his astronomic researches, did not submit his theory of flexure to an experimental verification, as he would normally have done. The experiment was made about forty years after his death by Mariotte, in the presence of Carcavy, Huyghens and Roberval, and demonstrated that the formula of Galileo should be corrected, but that the error did not affect the relative bending strengths of beams with similar cross sections and the achievements of Galileo as a precursor of the theory of physical similitude.

James Watt's Application of Scientific Technology to Industry

Richard Leslie Hills

When James Watt was asked by Professor John Anderson to repair a model of a Newcomen steam engine belonging to the University of Glasgow during the winter of 1763 to 1764, he could not have realised that it would change both his own life as well as making one of the most important contributions to the progress of the Industrial Revolution. It would lead to his inspiration for the separate condenser in the spring of 1765 and his subsequent development of a much more efficient pumping engine and then a successful rotative steam engine. First we will see how this was based on his previous industrial experience and then follow the experiments he carried out with scientific instruments with which he was familiar from his own business.

In the autumn of 1763, Watt styled himself " merchant " for he had opened a shop in Glasgow which sold a wide variety of goods. This was in addition to his business as a maker of various instruments. These included musical instruments such as flutes, violins and even organs but his main line was in mathematical instruments, principally those used for navigation at sea such as the quadrant. At this period, most of these instruments including the scientific ones were made from wood and it was to wood that Watt turned for some of his earliest experiments with steam and a model Newcomen engine. In order to have all these instruments made for him, Watt was, to use a modern term, a managing director of a small industrial manufacturing business which had on its books around sixteen or seventeen craftsmen, although it is suspected that some of these were employed on a sub-contract basis.

Watt had received his basic training as a maker of scientific instruments in London and returned north to establish himself in a room in the College of Glasgow during August 1757[1]. He soon titled himself as " Mathematical

1. Birmingham Central Library, Boulton & Watt Collection, Muirhead III, 3/1, J. Watt's Waste Book, 1757-1763.

Instrument Maker to the University " and it was through his connections with the professors there such as Joseph Black and John Anderson that he received commissions to manufacture and repair their scientific teaching apparatus which would culminate in the model Newcomen engine.

During his attempts to make this engine perform reasonably, Watt said that he " set about repairing it as a mere mechanician "[2]. His friend, John Robison, commented about these first trials, " This model was, at first, a fine play thing to Mr. Watt, and to myself... But, like everything which came into his hands, it soon became an object of most serious study "[3].

This model Newcomen engine would change Watt from the mechanician into a scientist, or to use the term then current, a natural philosopher of the front rank.

Watt was able to solve the questions he had raised about the Newcomen engine by using three scientific instruments, the balance for weighing things, the barometer turned into a manometer for measuring pressures and, most important of all, the thermometer for measuring temperatures. Watt had bought himself a thermometer in the autumn of 1758 and, many years later in 1774, he calibrated and engraved scales on thermometers for the series of experiments on steam which he was carrying out that February[4]. There can be little doubt that, in 1763, Watt would have been capable of making his own thermometers and known about their use from reading possibly George Martine's book, *Essays and Observations on the Construction and Graduation of Thermometers*, parts of which had been published at Edinburgh in the 1740s[5]. Examples of his thermometers are preserved in his Garret Workshop at the Science Museum in London.

Likewise Watt was familiar with barometers. An inventory of his goods dated 3 January 1757 includes " Eight glass tubes for Barometers - 9/- "[6]. In fact the only surviving example of any of Watt's scientific instruments from this period is a fine barometer displayed in the National Museum of Scotland in Edinburgh. There is an incomplete one in his Garret Wokshop as well as engraved plates for printing the scales for barometers. While no early references have been found to balances in Watt's commercial records we will see that he owned one himself. Then Joseph Black's especial triumph in his doctoral thesis consisted in showing that the chemical changes occurring in his series of reactions could, without isolating the " fixed air ", be detected by sub-

2. J.P. Muirhead, *The Origin and Progress of the Mechanical Inventions of James Watt*, London, J. Murray, 1854, Vol. I, lxix.

3. *Ibid.*, Vol. I, xlvii, xlix.

4. B.C.L., B. & W. Col., MI 1/20, 17-19 Feb. 1774.

5. G. Martine, *Essays and Observations on the Construction and Graduation of Thermometers, and on the Heating and Cooling of Bodies*, Edinburgh, A. Donaldson, 1780, 3 ed.

6. B.C.L., B. & W. Col., MIII, 3/1, Waste Book.

jecting them at every stage to the arbitrament of the balance[7]. So Watt would have been well aware of the importance and use of the balance. Therefore it is clear that his industrial experience with manufacturing scientific instruments gave him the basis for his experiments with the Newcomen model steam engine.

THE VOLUME OF STEAM AND WATER

Trying to make the University model work properly raise a host of questions in Watt's mind. He could not drastically alter a model which did not belong to him so he had to construct his own. First he tried one with a wooden cylinder but found that wood became soft with the heat of the steam and so had to revert to metal. Even with his wooden cylinder, Watt discovered that " the steam condenser in filling it still exceeded the proportion of that required for large engines, according to the statements of Desaguliers "[8]. This may be the point where Watt began to realise the importance which heat played in the functioning of the steam engine as well as steam. Joseph Black said in his written evidence for the 1796 trial, " ...after he [Watt] had been thus employed a considerable time, he perceived that by far the greatest waste of heat proceeded from the waste of steam in filling the cylinder with steam. In filling the cylinder with steam, for every stroke of the common engine a great part of the steam is chilled and condenser by the coldness of the cylinder, before this last is heated enough to qualify it for being filled with elastic vapour or perfect steam ; he perceived, therefore, that by preventing this waste of steam, an incomparably greater saving of heat and fuel would be attained than by any other contrivance "[9]. Early experiments showed that three quarters of the steam were wasted during the ascent of the piston and later tests with better apparatus confirmed that the loss was much greater than this[10].

Very early in his experiments, Watt must have developed the concept of a perfect steam engine. In September 1765, he would write to his friend James Lind, " I have tried my small model of my perfect engine... I expect almost totally to prevent waste of steam, and consequently to bring the machine to its ultimatum "[11]. This concept may have been partly derived from John Smeaton's publication of his experiments on wind and watermills which was published by the Royal Society in 1760[12]. Antoine Parent in France in 1704

7. C. Singer, *A Short History of Science to the Nineteenth Century*, Oxford, Clarendon Press, 1941, 286.

8. J.P. Muirhead, *The Life of James Watt,* London, J. Murray, 1858, 76.

9. *Ibid.*, 59.

10. *Ibid.*, 67.

11. B.C.L., James Watt Papers, CI/15, 4 Sept. 1765, Watt to J. Lind.

12. J. Smeaton, " An experimental Enquiry concerning the Natural Powers of Water and Wind to turn Mills, and other Machines, depending on a circular Motion ", *Philosophical Transactions*, Vol. 51 (1759), 100-174, (published 1760).

had suggested that an undershot wheel would generate only 4/27 of the total available power or " effort " of the stream but Smeaton showed this figure could be raised to two thirds[13]. It is probable that Watt considered his perfect engine as one which would use only one cylinder full of steam at each stroke and would condense that steam into a perfect vacuum. The importance of this concept of a perfect engine was that it gave him a standard against which he could judge an engine's performance.

Now he needed to know the actual steam consumption to determine just how many cylinders full of steam his engine was taking. He realised that he could determine the volume of steam from the volume of water provided he knew the ratio of their volumes at boiling point. With this information, he could measure the volume of water evaporated from his boiler which was the easiest way of determining steam consumption. This question would lead him to doubt the figures printed by Desaguliers who repeated many times that " water in boiling is expanded 14000 times to generate steam as strong (i.e. as elastick) as common Air " and stated that this figure had been obtained by a great many observations made by himself and Henry Beighton[14].

At the suggestion of Black[15], Watt took a Florence flask and inserted a glass tube which reached nearly to the bottom and carefully sealed the joint. He poured about an ounce of water into it and placed it in an oven for over an hour to slowly boil the water and drive out both air and steam. It was allowed to cool and then weighed. The flask was carefully dried and weighed again when it was found to be 4 grains lighter. Watt wrote, " Consequently the water had expanded to above 1600 times its bulk "[16]. Watt later discovered that he had used the wrong figure for grains in a drop weight so that the expansion of steam should have been 1849, " a Cubic Inch becoming a Cubic foot "[17]. This experiment was repeated with the Florence flask lying on its side without the glass tube. Watt later stated that he did " not consider these experiments as extremely accurate, the scale beam of a proper size which I had then at my

13. D.S.L. Cardwell, *From Watt to Clausius, the Rise of Thermodynamics in the Early Industrial Age*, London, Heinemann, 1971, 69.

14. J.T. Desaguliers, *A Course of Experimental Philosophy*, London, A. Millar *et al.*, 1763, Vol. II, 312, " That Heat will add Elasticity to Fluids is evident from numberless Experiments, especially from Distilling and Chemistry. But what is needful to consider here is only, that it acts more powerfully on Water than common air ; for the same Heat which rarefies Air only 2/3, will rarefy Water very near 14000 times, changing it into Steam or Vapour as it boils it ". See also p. 338-339.

15. J.P. Muirhead, *The Origin*, Vol. 1, lxxii and A.D.C. Simpson (ed.), *Joseph Black, 1728-1799, A Commemorative Symposium*, Edinburgh, The Royal Scottish Museum, 1982, H. Guerlac, " Joseph Black's Work on Heat ", 19.

16. E. Robinson, D. McKie (eds), *Partners in Science, James Watt & Joseph Black*, London, Constable, 1970, 437, 13v. This important Note Book of Watt's is JWP W 14, but I have referenced it from the more accessible version printed at the end of Robinson & McKie.

17. *Ibid.*, 437, 14v.

command not being very sensible "[18]. Here we see the importance of the balance in Watt's experiments.

Watt had discovered the ratio of the volume of steam to that of water through determining the weight of the steam or water left in the flask. From then on he could determine the steam consumption of an engine by measuring the amount of water evaporated from a boiler through either the weight or the volume of that water. He built a boiler " which showed, by inspection, the quantity of water evaporated in any given time, and thereby ascertained the quantity of steam used in every stroke by the engine, which I found to be several times the full of the cylinder "[19].

Watt does not state how he measured this quantity. One way could have been by weighing the whole boiler before and after his trial. Another could have been by refilling the boiler from a known volume of water to a level marked on the boiler. The volume of water would have been weighed first and then what was left weighed again. Once more this shows the importance of the balance, which of course could have weighed the fuel as well.

HEAT CAPACITY OF DIFFERENT MATERIALS

Having confirmed that his engine was using many cylinders full of steam instead of one, Watt must have asked what was the cause of so much being wasted. We have seen that he attributed one reason to loss of heat through the radiation from the cylinder walls, which must have been one reason why he used wood for his cylinder. But he also realised that the cylinder walls themselves absorbed a great deal of heat, which was another reason for his trials with wooden cylinders mentioned earlier. He most have realised that the cylinder walls became a source of heat when the vacuum was being created. He therefore turned his attention to the heat capacities of different materials because he recognised that he had to revert to a cylinder made of some metal. He must have known about Desaguliers's comments on cylinders of atmospheric engines where he recommended the use of brass rather than cast iron because brass had " the advantage of heating and cooling quick "[20].

While Robison was editing Black's lectures, he wrote that, when Watt was investigating the amount of ineffective condensation, he learnt the quantity by warming the cylinder and other parts to determine their heat capacities.

" He made many experiments on various substances, in order to learn their properties in this respect "[21].

18. J.P. Muirhead, *The Life*, *op. cit.*, 78.

19. *Ibid.*, 78.

20. J.T. Desaguliers, *op. cit.*, Vol. II, 536.

21. J. Robison (ed.), J. Black, *Lectures on the Elements of Chemistry, delivered in the University of Edinburgh*, Edinburgh, Mundell & Son, 1803, Vol. I, 504.

Watt told Magellan that he " was the first that tryed experiments to find the quantifies of heat absorbed by wood and some other bodies in the year 1763 but invented no theories on the subject "[22].

The only surviving record of these experiments is a short passage in Watt's notebook : " a piece of Iron being heated to 120°[F] & then plunged into water at 60°[F] was found to have heated the water as much as an equall bulk of [water] heated to 120°[F]. Copper was found to do the same tin is said by Dr. Black to hold the half of the heat of an equall bulk of water "[23].

Robison said it was Watt who first " considered the subject steadily and in a system " and drew Black's attention to the significance and practical importance of such research. Black carried on with this research, involving his students, and, Robison claimed, " had many experiments on the heats communicated to water by different solid bodies, and had completely established their regular and steady differences ". One result, according to Robison, was that Black recognised Watt's " uncommon genius " and that this " was the beginning of a friendship which lasted through life "[24]. While Black and Watt were close friends before this, it could well be that Watt's experiments on the steam engine showed him in a different light to Black and may have subtly altered the form of their friendship.

From knowing the heat capacity of a metal and knowing the mass or weight of a cylinder, Watt could determine the amount of cold water at a given temperature which would be needed to cool that cylinder from boiling point to whatever temperature he wanted. For these experiments, he would use both the thermometer and the balance. At this period he was assuming that the amount of heat contained in boiling water and in steam was the same. He already knew how much steam the cylinder contained. By the law of mixtures, he could calculate how much cold water should be needed to cool that steam down as well. But when he came to add onto the figure for cooling the cylinder that for cooling the steam contained within it, his theoretical sums would not have given the same result as he obtained in practice. He found that the water coming out of his engine was hotter than he expected. He was " ...astonished at the great quantity of water required for the injection, and the great heat it had acquired from the small quantity of water in the form of steam which had been used in filling the cylinder "[25].

He wrote in his note book, " Injection in fire engines was much greater than I thought was necessary to Cool the Quantity of water contained in the Steam

22. E. Robinson & D. McKie, *op. cit.*, 77, Watt to Magellan, 1 March 1780.
23. *Ibid.*, 438, 17r.
24. J. Robison, *op. cit.*, Vol. I, 504.
25. J.P. Muirhead, *The Life*, *op. cit.*, 78.

down to below boiling point "[26]. He would have to carry out different experiments to find out why.

THE BOILING POINTS OF WATER

Before we examine these other experiments, the heat capacities of metals gave him the answer to another problem. Watt had quickly recognised that any hot water in the cylinder might be turned into steam as the pressure was reduced.

This would be exacerbated by the heat contained in the cylinder and would partially vitiate the vacuum, giving a residual pressure inside the cylinder which would resist the pressure of the atmosphere[27]. He knew of Dr. Cullen's work in this field and probably had read his single paper on heat, published in 1756, which dealt with the phenomenon of cooling by evaporation, particularly under reduced pressures[28]. Watt wrote in 1813-14, " Experiments had been made long before by Dr. Cullen, John Robison, and others, in public classes, which proved that water, when placed in an exhausted receiver, boiled, and was converted into steam at the heat of 70° or 80° of Fahrenheit's thermometer, while it was well known that under the pressure of the atmosphere it required 212°[F] of heat to make it boil, and emit steam capable of displacing the air. It was evident that, under intermediate pressures, intermediate degrees of heat would be equired to make it boil "[29].

The extent of Watt's knowledge of this in 1764 is unknown, particularly about the lower temperatures of water boiling under lower pressures, for he seems to have thought that this ceased at 100°F. Cullen's book has no reference to the temperatures at which water will boil in a vacuum and so could have been of little help. Watt proceeded to carry out his own experiments into the boiling points of water.

In Watt's note book, there is a drawing of a spherical copper vessel which could be partially filled with water. Connecting with this was a smaller container into which were sealed the lower ends of a thermometer and a closed mercurial manometer 34 inches long so that both could be read at the same time. The manometer gave readings above normal atmospheric pressure. From his readings, Watt plotted a graph of pressure against temperature. This must be a very early example of such a method instead of the column or tabular

26. E. Robinson, D. McKie, *op. cit.*, 439, 16v.

27. A.L. Donovan, *Philosophical Chemistry in the Scottish Enlightenment, The Doctrines and Discoveries of William Cullen*, Edinburgh University Press, 1975, 261. Donovan wrongly calls this " back pressure ".

28. William Cullen, *Of the Cold Produced by Evaporating Fluids, and Some Other Means of Producing Cold : Essays and Observations*, Edinburgh, 1756, 11, and D.S.L. Cardwell, *op. cit.*, 34.

29. J.P. Muirhead, *The Life*, *op. cit.*, 84.

arrangement more usual at that time[30]. On a graph preserved at Birmingham, a note added in January 1803 says, " This scale seems to have been drawn in 1765 from the first rude experiments & on the supposition that water boiled in vacuo at 100°[F] which was grossly erroneous "[31]. Watt wrote, " By experiments which I then tried upon the heats at which water boils under several pressures greater than that of the atmosphere, it appeared that when the heats proceeded in an arithmetical, the elasticities proceeded in some geometrical ratio ; and by laying down a curve from my data, I ascertained the particular one near enough for my purpose "[32].

Watt extrapolated the curve downwards to below atmospheric pressure to see the relationship between pressure and the temperature of the boiling point of water in a partial vacuum. His second curve showed him the necessity of cooling the steam down to a much lower temperature than the 100°F he thought originally if he were to achieve a perfect vacuum in his steam engine. The large volume of cold water that he found necessary to do this in his engine was another problem which had to be solved.

Latent heat

In the Newcomen engine, the vacuum was created by Newcomen's crucial discovery of direct injection of cold water into the cylinder[33]. We have seen that Watt was surprised by the amount of water needed which he found " was much greater than I thought necessary to cool the quantity of water contained in the Steam down to below boiling point "[34]. Even when trying with his model with the wooden cylinder, he had " perceived that the water issuing from the sinking pipe was always near boyling hot "[35]. If he had succeeded in reducing the steam consumption with the wooden cylinder, from where did all this heat come ? This did not agree with the results he obtained by the law of mixtures of different amounts of water at different temperatures. He said, " I mixed 1 part of boiling water with 30 parts of cold water I found it only heated to the arithmetical mean betwixt the two heats & that it was scarcely sensibly heated to the finger "[36].

30. E. Robinson, D. McKie, *op. cit.*, 440.

31. B.C.L., B. & W. Col., MI 4/61.

32. J.P. Muirhead, *The Origin*, *op. cit.*, Vol. I, lxx.

33. R.L. Hills, *Power from Steam, A Short History of the Stationary Steam Engine*, Cambridge University Press, 1989, 25.

34. Robinson & McKie, *op. cit.*, 439, 16v.

35. *Ibid.*, 436, 12r.

36. *Ibid.*, 439, 16v. Robinson & McKie are wrong here in their comment on this passage, see p. 427.

Watt then carried out his famous experiment with the tea kettle in which he put a bent glass tube into the spout so that the steam would pass through it into a vessel containing cold water, his " frigeratory ".

" I took a bent Glass tube & inverted it into the nose of a tea kettle the other end being Imersed in Cold water I found on making the kettle boil that tho there was only a small increase of the water in the frigeratory that it was become boiling hot this I was surprized at & on telling it to Dr Black & asking him if it was possible that water under the form of steam could contain, more heat than it did when water it was heated to 212°[F] or boiling hot it would receive no more heat tho the fire & continued the same that the steams that go off do not appear sensibly hotter to the thermometer than 212 & that the quantity of water that is evaporated in a minute is small in proportion to the heat that is certainly carried off or contained more heat than was sensible to the thermometer & that in certain circumstances they might part with that heat to other bodies "[37].

These experiments showed that the water in the frigeratory had gained about one-sixth in volume, in other words, water converted into steam can heat about six times its own weight of water from room temperature to boiling point[38]. Watt had tumbled upon the phenomenon of latent heat. The thermometer had played a crucial role. He mentioned measured volumes of water, suggesting that he had a container of a specific volume rather than weighing the water in this case[39].

Watt himself was always very careful to ascribe the origin of the doctrine of latent heat to Black[40]. He would deny Robison's claim that he had attended Black's lectures as well as being, in Robison's words, " completely master of the subject "[41]. Black had first investigated the latent heat of fusion in the December of 1761 when he carried out experiments on the amount of heat absorbed when ice melted[42]. He obtained his first values for the latent heat of steam through some short experiments made on 4 October, 1762[43]. These results convinced Black that his theory was correct and he " ...was little solicitous of more experiments for my own conviction, or for making the doctrine more clearly comprehensible to others "[44]. He did, however, include the doc-

37. *Ibid.*, 439, 17v & 18v.

38. H.W. Dickinson, *James Watt, Craftsman and Engineer,* Cambridge, University Press, 1935, 35.

39. *Ibid.*, 439, 19r.

40. *Ibid.*, 76, *Watt to Magellan,* 1 March 1780 and J.P. Muirhead, *The Origin,* xlix, " Dr. Robison is mistaken in this. I had not attended to Dr. Black's experiment or theory on latent heat, until I was led to it in the course of experiments upon the engines, when the fact proved a stumbling block which the Doctor assisted me to get over ".

41. J. Robison, *op. cit.*, Vol. I, viii.

42. H. Guerlac, *op. cit.*, 17.

43. *Ibid.*, 17.

44. J. Robison, *op. cit.,* Vol. I, 171.

trine in his lecture courses at least from that time on. It seems to have been Watt's questioning him about the amount of heat in steam that prompted Black to make further experiments to obtain more accurate figures for the vapourization of water. With an improved still and the help of William Irvine, Black started his new investigations on 9 October, 1764[45]. Watt repeated the experiments that November with better apparatus, probably in the company of James Lind, but did not record his results in his note book[46].

Ascertaining the correct value for the latent heat of steam enabled Watt to calculate why, on his model engines, the injection water was always so hot. It would also have told him why so large a fire was necessary to raise steam and why so little steam was needed to heat up the cylinder. The doctrine of latent heat together with knowledge of the specific heats of materials would give scientific explanations for the passage of steam through steam engines. Watt recognised that, while knowledge of latent heat was not necessary for the practical functioning of steam engines, it resolved the questions he had been unable to answer otherwise. Watt confirmed this in a letter to William Brewster in 1814 : " Although Dr. Black's theory of latent heat did not suggest my improvements on the steam-engine, yet the knowledge upon various subjects which he was pleased to communicate to me, and the correct mode of reasoning, and of making experiments of which he set me the example, certainly conduced very much to facilitate the progress of my inventions ; and I still remember with respect and gratitude the notice he took of me when I very little merited it, and which continued throughout my life "[47].

Black's explanations gave Watt the final clues to the steam engine but equally Watt's experiments gave Black an understanding of the steam engine also, that it worked according to the scientific theories which he had promulgated. The Newcomen engine could be explained scientifically for the first time. This would also be true of Watt's separate condenser when he had invented it. We know that Black supported Watt in his first experiments with his separate condenser but did not have sufficient funds to continue and so introduced Watt to Dr. John Roebuck. Was Roebuck partly influenced by Black's confirming that Watt's new engine was based upon scientific principles and did this influence Matthew Boulton as well ?

WATT'S DILEMMA

Watt, having determined figures for the latent heat of steam and the heat capacity of his cylinder, could calculate the steam consumption of his engine accurately. He could " judge the merits of Savarys engine in comparison with

45. H. Guerlac, *op. cit.*, 19.

46. E. Robinson, D. McKie, *op. cit.*, 439, 19r.

47. D. Brewster, *Robison's Mechanics*, Vol. II, p. ix, letter to Dr. Brewster, May, 1814.

Newcomens "[48] but this showed him the impossibility of achieving the concept of his perfect engine. First, some steam would be destroyed, as he put it, to heat up the cylinder to boiling point before it could be filled with steam. Then the heat retained by the cylinder itself as well as the latent heat of the steam in the cylinder would have to be destroyed before a perfect vacuum could be attained. He wrestled with the dilemma that presented itself and laid it down as an axiom : " That to make a perfect steam — engine, it was necessary that the cylinder should always be as hot as the steam which entered it, and that the steam should be cooled down below 100° in order to exert its full powers. The gain by such construction would be double : first, no steam would be condensed on entering the cylinder ; and secondly, the power exerted would be greater as the steam was more cooled. The *postulata*, however, seemed to him incompatible, and he continued to grope in the dark, misled by many an *ignis fatuus* "[49].

THE SEPARATE CONDENSER

Robert Hart claimed that James Watt gave him the following account of how the idea of the famous separate condenser originated : " It was in the Green of Glasgow. I had gone to take a walk on a fine Sabbath afternoon… I was thinking upon the engine (…) when the idea came into my mind, that as steam was an elastic body it would rush into a vacuum, and if a communication was made between the cylinder and an exhausted vessel, it would rush into it, and might be there condensed without cooling the cylinder… I had not walked further than the Golf-house when the whole thing was arranged in my mind "[50].

This flash of inspiration must have occurred before 29 April 1765 when he wrote a letter to James Lind giving details of experiments with his new engine[51]. Through his use of those three simple instruments, the thermometer, the barometer and the balance, Watt had provided scientific answers to the mystery of the steam engine and found a solution to developing one which would function much more efficiently. He would continue to develop a new industry with his steam engines which would provide the power for the next stage of the Industrial Revolution. Through his early industrial experience,

48. E. Robinson, D. McKie, *op. cit.*, 440, 20r.

49. J.P. Muirhead, *The Life*, *op. cit.*, 87.

50. H.W. Dickinson & R. Jenkins, *James Watt & the Steam Engine*, 1927, reprint Moorland, Ashbourne, 1981, 23, note 3. The story was related by Mr. Hart to members of the Glasgow Archaeological Society (*Transactions*, 1859, Series I, I.1). Dickinson and Jenkins found no contemporary record of this event among Watt's papers but correspondence between Watt and John and Robert Hart took place at the beginning of 1816 (see JWP C6/8.71, 11 Jan. 1816, from J. & R. Hart and a letter from Watt, JWP LB 5).

51. B.C.L., JWP CI/15, 29 April 1765, Watt to Dr. J. Lind and Muirhead, *Origin*, 3. H.W. Dickinson, in *James Watt, Craftsman & Engineer*, Cambridge University Press, 1935, 36, wrongly ascribes the invention of the separate condenser to May 1765.

Watt had been able to carry out scientific experiments from which he found the solution to the steam engine which led to further industrial enterprises and indeed a new era in the history of civilisation.

INDUSTRY AND INDUSTRIAL RELATIONS WITHIN THE LABORATORY. THE MATERIAL CONDITIONS OF JOULE-THOMSON EXPERIMENTS

Christian SICHAU

" Science owes more to the existence of the steam engine than the steam engine does to science ". This remark is often made, even when its meaning sometimes varies. In this presentation I will make use of a new way of interpreting this statement : I will simply take it literally. It will be illustrated by the joint experiments of James Prescott Joule and William Thomson (Lord Kelvin), performed at different sites in Manchester during the time from 1852 to 1861. In those experiments the steam-engine was not only the subject of research but it entered the laboratory as a central part of the experimental set-up. Those experiments by Joule and Thomson provide therefore an ideal case to explore the general relationship between instrumentation, scientific experiment, and industrialisation in Victorian Britain.

Quite a lot has been written about both protagonists of this story, Joule and Thomson. Crosbie Smith and Norton Wise have set the standard with their excellent biographical study of William Thomson[1]. However, they did not discuss the Joule-Thomson-Experiments in any detail. The latest biography of Joule is the one by Donald Cardwell[2]. Although we find there a lot on the collaboration between Joule and Thomson the general approach is different from mine[3]. In general, I will try to follow along the line of Smith and Wise. In this

1. C. Smith, N. Wise, *Energy and Empire. A biographical study of Lord Kelvin*, Cambridge University Press, 1989.

2. D. Cardwell, *James Joule. A biography*, Manchester University Press, 1989.

3. The differences between those two biographies have been emphasised by I.R. Morus in an essay review. Whereas Cardwell's biography of Joule is, according to Morus : " *a straightforwardly internalist account of Joule's intellectual development, interspersed with a few anecdotes to enliven the otherwise rather dreary proceeding* ", the work by Smith and Wise " *may be useful here by highlighting the ways in which particular historical figures with their own ideological goals and commitments pursued their careers, utilizing a wide range of cultural resources in that process* ". Morus, " Industrious People : Biography and Nineteenth Century Physics ", *Stud. Hist. Phil. Sci.*, Vol. 21, n° 3, (1990), 519-25.

contribution I will concentrate on the material aspects of the experiments which links it additionally to the work of Andy Pickering and his analysis of experimental practice by the three factors " phenomenal model ", " instrumental model " and " material procedure "[4]. Hence, the aim is : to scrutinise the experiments done by Joule and Thomson with respect to the resources, both cultural and material, needed to do them successfully[5].

To set the stage some remarks about the place and time of our drama are necessary. It is Manchester, in the early 1850s, Britain's second city. A lot could and has be said about this " first industrial city of the world ". In order to brief I will reproduce here only what two contemporary observers said about it. First : " It is essentially a commercial town and, in so far as commerce dominates its productive labor, a factory town. ...The city is also the center of science and culture, which always go hand in hand with commerce and industry "[6] ; and the second one : " The manufactories and machine shops form as it were a girdle around the town... "[7].

THE JOULE-THOMSON-EXPERIMENTS

In their publications Joule and Thomson gave two objects for this research : 1. To test what they called " Mayer's Hypothesis ", that is (in its simplest version) the assumed exact equivalence of work spent and heat evolved in compressing a gas ; 2. To determine the value of " Carnot's function " which played a central role in William Thomson's theory of heat ; its value would allow : " to calculate the amount of mechanical effect of a perfect engine of any kind, whether a steam engine, an air engine, or even a thermo-electric engine "[8].

In order to explore " what science owes to the steam engine " I would like to look now more closely at the experimental set-up and start with a very basic and schematic representation of the experiment[9].

4. A. Pickering, " The Uses of Experiment ", in Gooding, Pinch, Schaffer (eds), *Living in the material world : On Realism and Experimental Practice*, Cambridge University Press, 1989, 275-97. For a more extensive account see : A. Pickering, *The mangle of Practice. Time, Agency & Science*, Chicago, London, University of Chicago Press, 1995.

5. This approach — and its consequences for our view of " science as practice " — originate from a " replication " (reworking) of the first experiments done by Joule and Thomson in 1852. Details of the re-working of the experiments are discussed in my " Diplomarbeit " : *Der Joule-Thomson-Effekt. Der Versuch einer Replikation*, Oldenburg, 1995 (unpublished).

6. F. Tönnies, in R.H. Kargon, *Science in Victorian Manchester*, Manchester University Press, 3.

7. L. Faucher, *Manchester in 1844*, London, 1844.

8. W. Thomson, " On the Dynamical Theory of Heat ", 1851. ß 26. In W. Thomson, *Mathematical and Physical Papers*, Cambridge University Press, 1882-1911, Vol. 1, 174-315.

9. W. Thomson used a similar version in his " Dynamical Theory of Heat " (Part IV, April 1851). Thus, there is a remarkable constancy in representation.

Fig. 1 : Schematic Representation of the Joule-Thomson-Effect

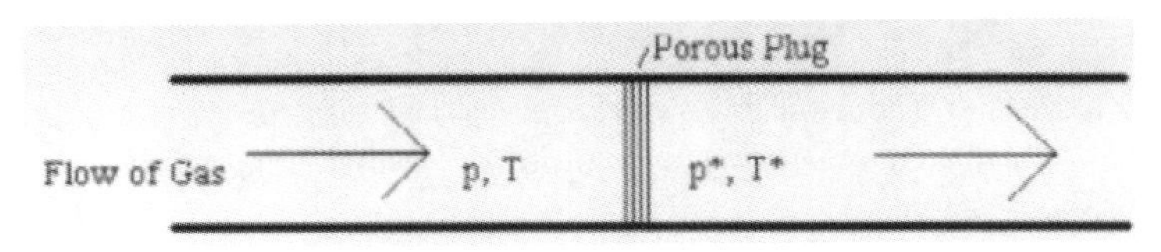

A gas is forced through a porous plug or narrow aperture. Due to the drop of pressure a small cooling effect occurs, in the case of air about a quarter of a degree Celsius per atmosphere pressure difference. From this schematic representation we can already make out some of the essential problems of the experiment : 1. The measurement of small temperature differences ; 2. The measurement of pressures up to several atmospheres ; 3. Very importantly, it is necessary to have a continuous and even flow of the gas[10]. Since Joule's extraordinary thermometric skills have been already discussed by almost all authors who have written about him[11], I will concentrate in the following on the last two aspects.

JOULE'S " SHOPPING LIST "

The experimental set-up used in 1853 was bought by Joule and Thomson with the help of a research grant, out of the so-called " Government Grant ", which was allocated by the Royal Society[12]. A detailed list of the items bought can be seen in Figure 2.

I don't want to go into detail here but with respect to the above mentioned key problems of the experiment several aspects are striking :

1. The amount spent for the thermometers is rather small — compared to the total sum. One reason for this is that both, Joule and Thomson, already possessed some excellent thermometers, but more importantly, thermometers[13] could be bought for a fairly small sum — locally from the Manchester instru-

10. I will ignore in this presentation the problem of heat conduction. I have discussed some aspects of it in the article : " Practising Helps : Thermodynamics, History, and Experiment ", *Science & Education* (forthcoming).

11. See, for example : D. Cardwell, *James Joule, op. cit.,* Manchester University Press, 1989. H.O. Sibum has discussed Joule's thermometric skills and related it to Joule's personal knowledge of the brewing process in several articles. See, for example : Sibum, " Reworking the mechanical value of heat : Instruments of Precision and Gestures of Accuracy ", *Stud. Hist. Phil. Sci.* 26 (1995), 73-106.

12. For details about the funding of the Joule-Thomson-Experiments and its influences see my article : " Ein nationales Experiment und seine Auswirkungen auf einen wissenschaftlichen Versuch ", *Centaurus* (forthcoming).

13. In the following attributes like " small " or " large " are used only with respect to the total sum spent by Joule and Thomson, i.e. roughly £ 200. I don't claim that, for example, the thermometers bought by Joule and Thomson were " cheap " in a general sense. However, the use of those attributes can be justified in so far as Joule or Thomson could have paid a " small " amount of money without a research grant.

ment-maker Dancer, for example, as well as from the Kew Observatory where the manufacture of standard thermometers had just been newly established in the early 1850s.

2. Joule and Thomson paid even a great deal less for " pressure gages ". This indicates that measuring a pressure of several atmospheres was, as it seems, no problem.

3. If we ignore here for the moment the amount spent for iron and copper tubes *etc.*[14] we will find the second largest amount mentioned in the list was used to buy a pump, £ 32 and 7 shillings. But there is no mentioning of a steam-engine to drive the pump — only the costs for " Power " and " Steam " appear at the bottom of the list.

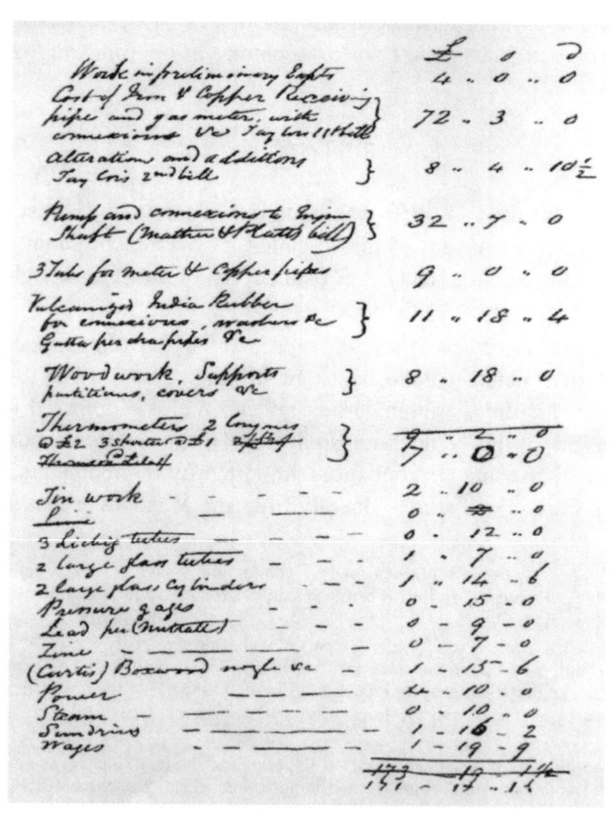

	£ s d
Work in preliminary Expts	4 .. 0 .. 0
Cost of Iron & Copper Receiving pipe and gas meter, with connexions &c Taylor's 1st bill	72 .. 3 .. 0
alterations and additions Taylor's 2nd bill	8 .. 4 .. 10½
Pump and connexions to Engine Shaft (Mather & Platt's bill)	32 .. 7 .. 0
3 Tubs for meter & Copper pipes	9 .. 0 .. 0
Vulcanized India Rubber for connexions, washers &c Gutta percha pipes &c	11 .. 18 .. 4
Woodwork, Supports partitions, covers &c	8 .. 18 .. 0
Thermometers 2 long ones @ £2 3 shorter @ £1 ~~[illegible]~~ ~~[illegible]~~	~~9 .. 2 .. 0~~ 7 .. 0 .. 0
Tin work	2 .. 10 .. 0
[illegible]	0 .. [illegible] .. 0
3 Liebig tubes	0 .. 12 .. 0
2 large glass tubes	1 .. 7 .. 0
2 large glass Cylinders	1 .. 14 .. 6
Pressure gages	0 .. 15 .. 0
Lead pipe (Nuttall)	0 .. 9 .. 0
Zinc	0 .. 7 .. 0
(Curtis) Boxwood nozzle &c	1 .. 15 .. 6
Power	[illegible] .. 10 .. 0
Steam	0 .. 10 .. 0
Sundries	1 .. 16 .. 2
Wages	1 .. 19 .. 9
	~~173 .. 19 .. 1½~~
	171 .. 17 .. 1½

Fig. 2 : List made by Joule in 1853 (From a letter to W. Thomson, Dec. 9, 1853. Cambridge University Library, Kelvin Collection J163)

14. It is important if it is taken into account that in the first experiments by Joule and Thomson lead pipes were employed which were commonly used for water- and gas-pipes (see, for example : W. Gibbs, " Extraction and Production of Metals ", S. 123, in Singer ; Holmyard, Hall, Williams (eds.), *A History of Technology. Vol. IV : The Industrial Revolution, c. 1750 - c. 1850*, Oxford, Clarendon Press, 1958.

THE IMPORTANCE OF THE BREWERY AS A MATERIAL RESOURCE

So, let us now see how the set-up which they build with this equipment looked like (Figure 3) :

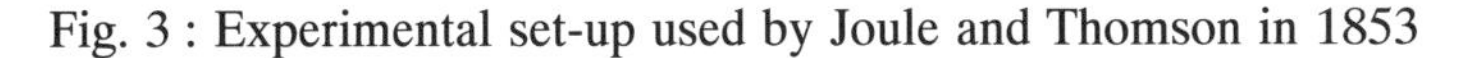
Fig. 3 : Experimental set-up used by Joule and Thomson in 1853

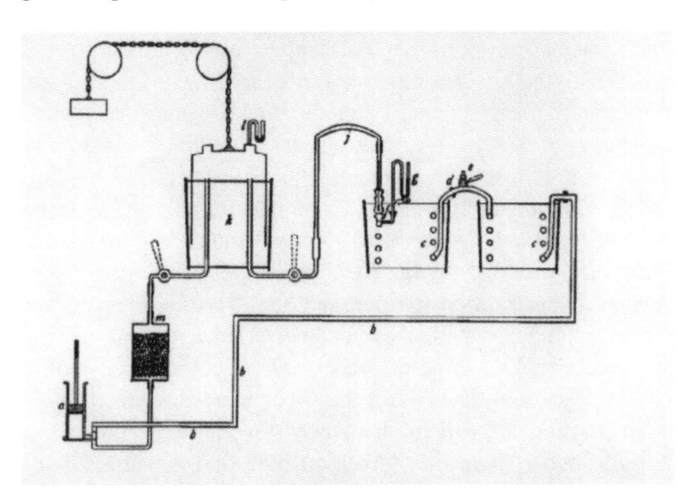

Again, it is remarkable : there is the pump — also mentioned in the list — in the bottom left corner. But there is no machinery to drive the pump. This peculiarity can be explained by a simple fact which does however already lead us right into the middle of the discussion on of the material dimension of experiment. At that time, Joule and Thomson just employed the steam engine of the brewery owned by the Joule-family. The importance of having this steam-engine ready for use becomes evident if we look a little into the future of the collaboration between Joule and Thomson. In February 1853 Joule's father had a " paralytic stroke "[15]. Because neither James nor his brother Benjamin was willing to continue the brewing business the brewery was up for sale. In November 1853 James Joule wrote to William Thomson : " It is now more likely that the brewery will be given up by Xmas "[16]. At this time Joule was busy doing experiments and a sale at this time would have meant an interruption without any clear prospect how and when the experiments could be taken up again : " It would be a pity if the apparatus had to be removed before the last part... is accomplished "[17]. The whole matter hung like the sword of

15. Letter from Joule to Thomson, March 1, 1853 ; Cambridge Univ. Library, Kelvin Collection J137.

16. Letter from Joule to Thomson, Nov. 19, 1853, Cambridge Univ. Library, Kelvin Collection J157.

17. Letter from Joule to Thomson, Nov. 22, 1853 ; Cambridge Univ. Library, Kelvin Collection J158.

Damocles above the experiments[18] and, at the same time, Joule was very much occupied with this the sale of the brewery which restricted his time spent for experiments. In the end, it took nearly a year, until autumn 1854, till the brewery was finally sold.

Two further aspects illustrate the importance of the brewery as an important resource used by Joule. First, it supplied Joule and Thomson with hot and cold water which was necessary to vary the temperature of the gas. So, even when the brewery finally had been sold they wanted to draw on this resource : Joule wanted to put up the new apparatus " on the brewery premises where there will be a good supply of either hot or cold water "[19]. Second, and as it turned out, crucial for the whole research project, they could use carbon dioxide produced in fermenting tuns at the brewery[20]. Up until autumn 1853 only dry air had been used and Joule and Thomson had not yet found any simple law relating the pressure difference and the observed cooling effect. Partly, this was due to the fact that their expectations were based on a " wrong theory " and because of problems encountered with the machinery. However, probably the single most important reason was that the observable effect is very small in the case of dry air — for carbon dioxide on the other hand it is much greater. Thus, shortly after starting to work with carbon dioxide Joule could report some remarkable results to Thomson : " You will find that by subtracting the atmosphere pressure from the pressure of the gas, numbers will be obtained very nearly proportional to the cooling effect, particularly in the case of carbonic acid "[21].This was the breakthrough[22].

Craftsmanship, machinery and experiment

Now, I want to look more closely at how Joule obtained the pump. Similar remarks could be made with respect to the later purchase of a steam engine. Central to the argument is an insight to which the historian Samuel has — already in the 1970s — drawn attention : " In metalwork and engineering — at

18. See, for example : Letter from Joule to Thomson, Feb. 3, 1854 ; Cambridge Univ. Library, Kelvin Collection J168.

19. Letter from Joule to Thomson, March 6, 1854 ; Cambridge Univ. Library, Kelvin Collection J173.

20. See, for example : Letter from Joule to Thomson, Sept. 17, 1853, Cambridge Univ. Library, Kelvin Collection J148.

21. Letter from Joule to Thomson, Nov., 1853, Cambridge Univ. Library, Kelvin Collection J153. On those experiments with carbon dioxide see also the following letters from Joule to Thomson : Oct. 11, 1853 (CUL J150) ; Nov. 8, 1853 (CUL 154) ; Nov. 10, 1853 (Glasgow : Univ. Library J99) ; Dec. 3, 1853 (CUL J161).

22. However, carbon dioxide was not always available as a letter from Joule to Thomson shows : " It being now so far advanced in the summer weather we seldom brew in the fermenting tuns from which the gutta percha pipe leads ; so that it will be best to insert the previous expt. as a preliminary one and a more exact one can be done afterwards in the autumn " (Letter from Joule to Thomson, June 1, 1854, CUL, Kelvin Collection J183).

least until the 1880s — it was the workshop rather than the factory which prevailed "[23] which meant that : " Much of the[ir] work was " bespoke " and carried out according to customer specifications and requirements "[24].

The first plans for the new apparatus were made by Joule in September 1852[25] and within a month everything was fixed. Focusing here solely on the pump we find the following statement made by Joule in a letter to Thomson : " I am getting the piston made at Matthew & Platts manufacturing which is only 100 yards from our Brewery. Their patent piston consists of a spiral spring which opens and contracts with any variation in the diameter of the cylinder. From all I can learn this form of packing is the most perfect known, a fact which is revealed by the increasing demand for their pistons. In the piston I have ordered there is a layer of antifriction metal on the outside or rubbing surface. I hope that by this arrangement it will be possible to work the pump without oil... "[26].

So, here we have :

1. The work was indeed " bespoke ". In order to be able to discuss details of construction it was necessary to " know " the workshop well where it was made ; it was therefore not " by accident " that it was only 100 yards away from Joule's brewery. Only this proximity ensured that the finished product turned out to be " a very beautiful piece of workmanship "[27].

2. The work was of high quality and it needed to be so because of experimental requirements such as the one mentioned in the letter. Besides the patented pistons made by Matthew & Platts other pieces of the apparatus were made, for example, at the well-known workshop of William Fairbairn[28].

It is worth pointing out that Joule relied almost in all cases on local workshops. He was highly selective but found all he wanted in Manchester.

The reason why I emphasise those material aspects of scientific experimental work is simply because it had real importance. This can be seen very clearly when we look at an experiment done by the Cambridge scientist William Hopkins. He investigated experimentally the influence of pressure on the melting

23. R. Samuel, " Workshop of the World : Stean Power and Hand-Technology in mid-Victorian Britain ", S. 8, in *History Workshop* 3 (1974), 6-72.

24. R. Samuel, *op. cit.*, 19, 39.

25. Letter from Joule to Thomson, Sept. 9, 1852 ; Cambridge Univ. Library, Kelvin Collection J124.

26. Letter from Joule to Thomson, Oct. 4, 1852 ; Cambridge Univ. Library, Kelvin Collection J126.

27. Letter from Joule to Thomson, Nov. 11, 1852 ; Cambridge Univ. Library, Kelvin Collection J128. The importance of such a close contact can be seen in another case : The construction of the gasimeter and the pipes. Here Joule wrote to Thomson : " I have just been to the coppersmith to see the progress of our work " ; when later, the gasmeter needed to be repaired it could be done more easily (Letter from Joule to Thomson, Jan. 17, 1854 ; CUL, Kelvin Collection J166).

28. Letters from Joule to Thomson, Oct. 4 and 21, 1852 ; Cambridge Univ. Library, Kelvin Collection J126 and J127. Joule knew Fairbairn well since both were members of the " Manchester Literary and Philosophical Society ".

point of various substances. When he presented his first results he made the following remark right at the beginning : " I am likewise bound to express in the strongest terms my obligations to my friends Mr. Fairbairn and Mr. Joule. Without the aid of the former of these gentlemen I should have been unable even to commence the series of experiments which I have now nearly concluded... "[29]. What was the reason for this gratitude ? The answer is that Hopkins had to do the whole series of experiments at the workshop of Fairbairn. Each time he wanted to experiment he had to make the journey from Cambridge to Manchester — because he did not have the resources to do them at home.

Hence, so far, we can conclude : the local availability of material resources did have an influence on the possibility to carry out experiments of this kind. And it was Manchester where these experiments could be done.

Instruments and their origins

Another instance which illustrates this close link between Joule's experiments and the engineering industry of Manchester is the measurement of pressure. Seemingly, the measurement of pressure was not a problem at all for Joule and Thomson. The pressure gauges cost hardly anything, and Joule and Thomson did not once refer to the exact means of pressure measurement in their publications. So, why bother ? But compare the set-up of 1853 with the one Joule and Thomson used at the end of their series of experiments around 1860 (Figure 4).

Fig. 4 : Experimental set-up used by Joule and Thomson in 1861

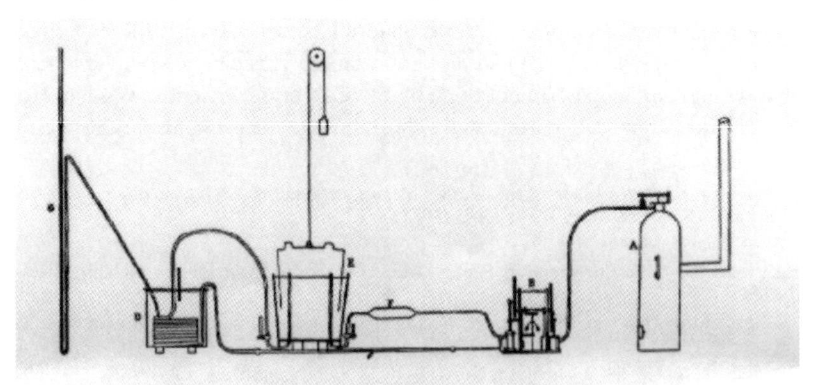

29. William Pole, *The Life of Sir William Fairbairn, Bart.*, London, Longmans, Green, and Co., 1877, 305.

What is striking here is the extraordinary length of the pressure gauge ; the longer leg of it was 17 feet which came nearer to the one used by their rival Victor Regnault who worked at the Collège de France in Paris and who was the leading experimentalist in the field of thermodynamics. In some experiments Regnault employed a manometer which had an approximate height of 23 meters.

If we look now again at the set-up of 1853 we can see that the pressure gauges used had a height of about 10 inches. It's a simple " trick " : In those instruments there is a gas enclosed and the mercury has to compress it instead of rising against a vacuum. In reality, the instrument is a bit more sophisticated in order to ensure that one obtains a reasonable accuracy even at high pressures. What has this to do with the engineering industry of Manchester ? I am not sure yet but instruments of this kind were used to measure the pressure of high-pressure steam boilers (Figure 5). I would suggest, that the problem of measuring high pressure was indeed no problem for Joule because he was familiar with those instruments used probably in many places in Manchester, maybe even at his own brewery.

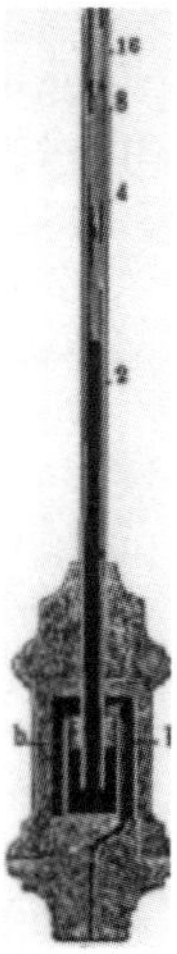

Fig. 5 : Pressure gauge as used for steam-boilers (Müller-Pouillet, Lehrbuch der Physik, Neunte Auflage, 1886. S. 507)

Industrial Relations within the Laboratory

One further aspect needs to be told here. Again, R. Samuel has noted it in this above mentioned article. In mid-Victorian Britain there was still a " gap between expectation and performance... Even if brought " nearly... to perfection " by its inventor, a machine would often prove difficult to operate "[30]. This is also true for the Joule-Thomson-apparatus. As noted in the very beginning, the success of the experiment depends crucially on one issue :

30. R. Samuel, *op. cit.*, 51.

Can you really make sure that you have got an continuous and even flow of gas ? Well, in 1853 Joule and Thomson couldn't — they obtained many dubious results : " owing to a considerable oscillation of pressure from the action of the pump "[31]. It was clear that before they could go on they had to test the machinery. Figure 6 shows some results of this test.

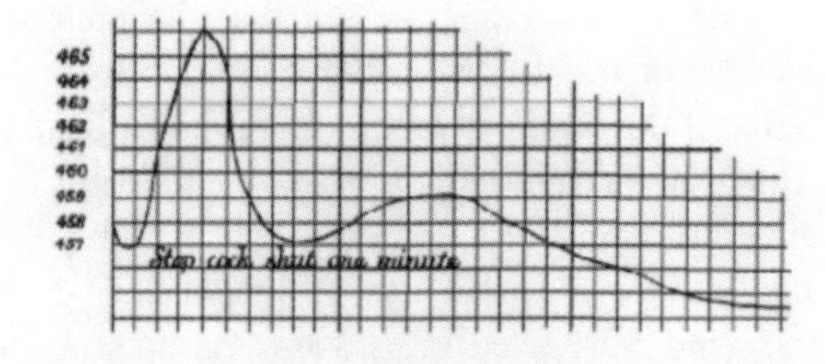

Fig. 6 : Fluctuations of Temperature caused by uneven action of the pump

The conclusion drawn from this test was : " ...simply to render the action of the pump as uniform as possible, and to commence the record of observations only one hour and a half or two hours had elapsed from the starting of the pump "[32]. In 1852, in their first experiments, they did most likely a whole series of measurements or more during such a time — as my reworking of those earlier experiments has shown[33] ; simply because the pumping then was done by hand and not by steam-driven machinery. Then, the pumping was probably done by a workman of Joule's brewery. Joule never mentioned him[34].

So, why did Joule and Thomson opt to work with the complex machinery, expensive and troublesome as it was ? Several causes enforcing each other probably contributed to their decision : 1. On the material side, the steam-driven pump lead to a much more copious flow of air which reduced temperature fluctuations in the issuing stream of gas ; 2. The example was set by their rival Victor Regnault — if he did use this kind of machinery Joule and Thomson had to do the same in order to be credible ; 3. The whole research project should illustrate what good might be achieved in experimental science with the help of a Government willing to support financially science but abstaining from any direct interference ; 4. The widely shared ideal of self-acting machinery had probably strengthened the decision, especially since it has a an additional appeal for scientists : the human body should be eliminated from the experiment as much as possible.

31. Joule & Thomson, " On the Thermal Effects of Fluids in Motion ", *Philosophical Transactions*, 143, 1853.

32. Joule & Thomson, " On the Thermal Effects of Fluids in Motion. Part II ", *Philosophical Transactions*, 144, 1854.

33. Details of the re-working of the experiments are discussed in my " Diplomarbeit ", see footnote 5.

34. For details about the " *unknown helping hand* " see my paper : " Necessary but not mentioned : Helping Hands in Joule's Laboratory Practice ", presented at : EASST/4S : *Signatures of Knowledge Societies*, Bielefeld, 1996.

Conclusions

To sum up : Science does owe at least the Joule-Thomson-Effect to the steam-engine.

Literally, because a steam-engine was used to drive the pump. More generally, on the level of “ ideas ” : because to be able to calculate the efficiency of the steam-engine was one reason why Thomson wanted to do the experiments ; still generally, but on the “ material level ” : because not only the steam-engine but many more pieces of apparatus were provided by the many engineering and metal workshops of Manchester, the “ first industrial city of the world ”.

What does the steam-engine owe to the Joule-Thomson-Effect on the other hand ? Well, nothing — on the practical level. Thomson could use the results to improve his theory and, in the end, they could be used to calculate the theoretical efficiency of a steam-engine. The real practical application of the Joule-Thomson-Effect had to wait for another 40 years until 1895 when Carl Linde used it to liquefy oxygen. So, low temperature research does owe something to the Joule-Thomson-Experiments. But the steam engine probably nothing.

Science industrielle et seconde industrialisation en France. Une proposition d'organisation rationnelle des relations entre la science et l'industrie 1880-1914[1]

Michel Lette

Rationalisation et seconde industrialisation

Au tournant des XIXe et XXe siècles s'opère ce que le sociologue allemand Max Weber qualifie d'accélération du processus de rationalisation des sociétés occidentales. Le concept ambitionne de caractériser, entre autres, la forme moderne de l'activité économique, et ce, suivant toutes ses composantes : idéologique, sociale et technique. Dans ce cas, le mouvement de rationalisation détermine aussi l'extension des domaines de la société soumis aux critères de décision rationnelle, validés par la science et la technique. Et aussi bien il qualifie la société occidentale, légitime l'autorité de son principal acteur : l'Etat moderne[2].

Toutefois le présent papier se réfère à une autre dimension — plus restrictive — de la notion weberienne. Celle qui vise un type d'activité rationnelle par rapport à une fin. La fin visée ? L'accroissement de la production et de la productivité de toute activité sociale constitutive de l'économie industrielle, de son efficacité, de sa performance selon des schémas validés par le calcul, la science et la technique. Dès lors, le projet exige des savoirs scientifiques et des techniques qui légitiment tant une idéologie que la mise en oeuvre d'une organisation pratique de la société.

Ainsi présenté, le point de vue conforte la conviction que la rationalisation de l'économie industrielle a valeur de définition de la seconde industrialisation. Telle est en effet la position adoptée. En d'autres termes, ce qui est ici désigné

1. La présente contribution a bénéficié d'une bourse versée par la Société Cockerill Sambre.

2. M. Weber, *Economie et Société*, Paris, Plon, 1995 (2 tomes). Sur cet aspect de la rationalisation selon Weber voir J. Habermas, *La technique et la science comme " idéologie "*, (trad. française de Ladmiral), Paris, Gallimard, 1993, 1-74 et B. Badie, *Sociologie de l'Etat*, Paris, Grasset, 1979, 38-50.

par “ seconde industrialisation ” se définit non seulement par un type particulier d'évolution économique entre les environs de 1880 et 1914, mais aussi comme le projet explicite de réorganisation rationnelle des liens entre la science et l'industrie, l'activité scientifique et la production, le savant et l'industriel, le laboratoire et l'usine, l'Etat et l'entreprise, et de façon générale entre les acteurs de l'économie.

L'ORGANISATION DES LIENS SCIENCE/INDUSTRIE : DU DISCOURS CONVENU...

Au milieu des années 1870, encore humiliée par la défaite de Sedan et déjà confrontée à la crise de la fin du siècle, la France cherche un second souffle. Les élites et la République s'efforcent de maintenir le cap de l'industrialisation et de porter l'économie française au niveau de celle d'outre-Rhin, d'outre-Manche et bientôt d'outre-Atlantique. Mais comment doper l'outil de production sans innover ? Autrement dit, comment assurer l'exploitation efficace des savoirs scientifiques récents produits en abondance par la chimie, l'électricité ou la métallurgie ? Comment les convertir en exportation de produits compétitifs ? Nécessairement en innovant.

Les discours rhétoriques abondent, relayés par des slogans et des formules incantatoires, et suggèrent même une solution : il faut revoir les modes traditionnels d'association de la science et de l'industrie qui ne correspondent plus aux besoins générés par la restructuration de l'économie à l'échelle des nations industrielles. Discours convenus sans cesse répétés ; la répétition vaut alors démonstration.

Tout en considérant que la science et l'industrie constituent deux ordres d'activité dont les buts divergent — le “ savoir ” et le “ faire ” — savants et producteurs admettent, en effet, que l'une et l'autre entretiennent depuis toujours des rapports étroits, contribuent à leurs progrès respectifs et cultivent des relations de type symbiotique. Ainsi peut-on considérer que tel professeur de la Sorbonne traduit une opinion partagée quand il écrit, en 1908, que “ la science et l'industrie sont si étroitement solidaires qu'aucun progrès de l'une n'est absolument sans influence sur la marche de l'autre ”[3]. Plus audacieux, Louis Houllevigue, un autre universitaire, proclame que l'avènement de la société moderne, scientifique et industrielle, passe par l'organisation rationnelle des relations entre la recherche et l'activité productive[4]. La science engendre des savoirs que l'industrie exploite. La production fournit pour sa part des interrogations pragmatiques que la science promet de résoudre. Les relations sont pensées comme la contribution d'une science qui explique, éclaire les pratiques empiriques de l'usine.

3. E. Bouty, *La vérité scientifique, sa poursuite*, Paris, Flammarion, 1908, 61.
4. L. Houllevigue, *Du laboratoire à l'usine*, Paris, A. Colin, 1904.

Il est vrai que ce type de discours est produit par les scientifiques. Les industriels manifestent, quant à eux, plus de réserve. Cela tient à ce que les premiers, socialement et culturellement dominés par les disciplines traditionnelles, cherchent à étendre autant que possible le champ de leurs prérogatives[5]. Quelques universitaires étendent même leurs visées rationalisatrices à l'ensemble de la société moderne. Ce qu'au demeurant elle n'a pas forcément sollicité. Cette volonté, les scientifiques la légitiment, entre autres, par le rôle croissant de la science dans le développement de l'industrie. Leur attachement à la notion de science appliquée est donc compréhensible : elle garantit le maintien d'une hiérarchie des pratiques en fonction du prestige des savoirs associés. Tout au long du XIXe siècle, la science appliquée aux arts industriels assure le rôle de fournisseur officiel des progrès techniques et industriels[6].

...À LA MISE EN OEUVRE DE SCHÉMAS ORGANISATIONNELS

Un tel contexte incite à la mise en oeuvre pratique. La seconde industrialisation est en effet riche de transformations du genre. Les liens réels, de toute nature et multiformes, entre la science et l'industrie s'amplifient, se structurent avec l'apparence du systématique et du rationnel. En témoignent la floraison des revues technico-scientifiques, la volonté de l'Etat de penser une politique nationale de recherche ou la multiplication des enseignements techniques supérieurs au caractère franchement pratique. Les rapports science/industrie accélèrent leur propre rythme d'institutionnalisation. La tendance et les initiatives s'engagent d'ailleurs de part et d'autre. Depuis la manufacture de type familial du début du siècle, l'entreprise a adopté progressivement le schéma de la firme dirigée par des cadres salariés, dotés de compétences scientifiques et techniques, disposant de capacités organisationnelles solides[7]. Avec l'objectif d'atteindre la performance optimale de ses modes de fabrication, de maîtriser la qualité de ses produits, de maintenir sa présence ou de s'imposer sur des marchés concurrentiels, l'industriel admet une activité, sinon de recherche et de développement, au moins de contrôle. Cette activité acquiert le statut d'unité opérationnelle autonome, c'est-à-dire inscrite dans une structure globale, mais disposant de sa fonction propre, de son personnel et de son budget. Elle devient un élément essentiel d'une stratégie au service de l'entreprise. De fait, s'instaurent, au laboratoire surtout, des liens d'une nature différente entre des savoirs traditionnellement détenus par la communauté scientifique et une visée utilitariste. Avec le souci ferme de contribuer aux performances écono-

5. C. Charle, *La République des universitaires, 1870-1940*, Paris, Seuil, 1994, 135-188.

6. B. Bensaude-Vincent, " De la chimie appliquée à la recherche industrielle : profils du XIXe siècle ", *Actes et communications, Economie et Sociologie rurales*, 6 (1991), (Histoire des techniques et compréhension de l'innovation, Séminaire de recherche (mars 1989-février 1990)), 83-98.

7. A. Chandler, *Strategy and Structure : Chapters in the History of the industrial enterprise*, Cambridge (Mass.), MIT Press, 1990.

miques du pays, d'étendre leur influence et leur prestige social, mais aussi d'élargir les opportunités de rémunération, chercheurs et professeurs mettent, sous des formes les plus diverses, leurs compétences au service de la production. La révision de l'alliance entre le savant et le producteur de richesses exige d'autres façons de concevoir leurs rapports sur le plan institutionnel, et par suite la professionnalisation d'une activité scientifique mise au service de l'entreprise, adaptée à ses besoins.

Dans le détail, des facteurs multiples et complexes déterminent les modes de pensée et d'action auxquels se réfèrent (parfois plus ou moins tacitement) les acteurs de l'économie industrielle pour organiser, institutionnaliser et professionnaliser les liens de la production avec la science et la technique. La voie française de seconde industrialisation et les façons de penser l'organisation ne ressemblent pas tout a fait à celles de ses voisines. Au demeurant, on doit distinguer les espaces nationaux et les espaces régionaux, ou bien encore considérer chaque secteur d'activité. A n'en point douter il y a place pour des chronologies décalées, des voies qui, si elles se ressemblent, restent cependant différentes. La complexité du genre impose d'élaborer quelques grilles d'analyse ; des schémas explicatifs s'avèrent indispensables pour comparer et comprendre l'originalité de chaque processus d'industrialisation.

L'objectif est de concevoir un tel outil d'analyse, pour la France, durant la seconde industrialisation : proposer une grille de lecture d'un schéma suggéré, à la fin du siècle dernier, par l'une des figures emblématiques parmi les promoteurs de l'organisation rationnelle des liens entre l'activité scientifique et industrielle.

Henry Le Chatelier, spécialiste des relations science/industrie[8]

Né en 1850, Henry Le Chatelier est une figure importante de la chimie en France au tournant des XIXe et XXe siècles. Inutile toutefois de le chercher parmi les savants mythiques de la IIIe République. Loin du Panthéon, il repose parmi les célébrités scientifiques méconnues. A l'opposé d'un Marcellin Berthelot ou d'un Louis Pasteur dont la notoriété publique s'affiche sur le portail des établissements scolaires, Le Chatelier est un personnage discret dans la mémoire française[9]. Comme bien des chimistes de sa génération, Le Chatelier a investi des domaines nombreux de la recherche, mais aussi amplement alimenté la réflexion sur le rôle de la science dans l'édification de la société industrielle : thermodynamique chimique, explosifs, grisou, métallurgie, synthèse de l'ammoniaque, mesure des températures élevées, liants hydrauliques,

8. M. Lette, *Henry Le Chatelier (1850-1936) et la constitution d'une science industrielle. Un modèle pour l'organisation rationnelle des relations entre la science et l'industrie au tournant des XIXe et XXe siècles, 1880-1914*, Paris, Thèse de doctorat de l'EHESS, 1998.

9. A notre connaissance, seul un lycée d'enseignement professionnel à Marseille porte son nom.

enseignement scientifique et technique, taylorisme et organisation scientifique du travail. La liste pourrait et devrait être plus longue. Et pourtant, à l'exception de la loi dite " de Le Chatelier " sur le déplacement des équilibres chimiques, aucune réalisation précise, aucune découverte retentissante, n'a garanti sa postérité.

L'attrait du personnage réside en fait ailleurs. Incontestablement, Le Chatelier bénéficie d'une popularité solide quand il s'agit de questionner les relations science/industrie aux environs de 1900[10]. Quiconque s'intéresse aux discours produits sur la science mise au service du progrès industriel fréquentera à coup sûr les siens. En ce sens, la publication en 1923 de son ouvrage *Science et industrie* parachève son oeuvre de défenseur officiel d'une liaison entre les deux. Pour ses contemporains, et a fortiori pour le chercheur, Le Chatelier incarne l'action militante en faveur d'une alliance nouvelle de la science et de l'industrie. De la fin des années 1870, début de sa carrière, jusqu'au conflit mondial à l'issue duquel une impulsion déterminante est donnée à l'organisation des relations entre la science et l'industrie, cet ingénieur des mines, professeur de l'enseignement supérieur, élabore ce qu'il nomme la " science industrielle ". C'est ce projet ambitieux d'un modèle pour l'organisation théorique et pratique des rapports science/industrie qui motive notre intérêt.

Comment comprendre son attitude militante aux frontières de la science et de l'industrie ? Est-ce même une exclusivité ? Ne vient-on pas de considérer comme une évidence commune et caractéristique de la seconde industrialisation l'opinion que le progrès industriel résulte forcément d'une extension de la science aux domaines de la production ? Sa banalité apparente mérite d'être creusée. Si bien des savants du XIXe siècle — physiciens et chimistes — proclament haut et fort la nécessité d'une liaison entre la science et l'industrie, la plupart s'en tiennent à des formules rhétoriques. L'inertie domine dans la communauté scientifique française[11]. L'originalité de Le Chatelier est de dépasser les discours convenus. Il ne se contente pas de répéter inlassablement l'impératif de liaisons, certes sur un ton plus combatif que tout autre : il transforme ses imprécations en système de pensée, et surtout en impératif d'action. D'où son statut de figure emblématique.

10. En voici une illustration. Le colloque organisé en 1994 à l'occasion du bicentenaire de l'Ecole polytechnique n'entendait-il pas à de multiples reprises citer le nom de cet ancien élève ? Parmi d'autres interventions : son rôle dans le tournant taylorien de la société française était évoqué par Aimée Moutet, ou Yves Cohen rappelait la spécificité d'une pensée qui tente de concilier les intérêts propres de la cité scientifique et le pragmatisme gestionnaire de l'entreprise.

11. Comme l'a souligné par exemple Dominique Pestre à propos des physiciens : D. Pestre, *Physique et physiciens en France, 1918-1940,* Ed. des Archives Contemporaines, 1984.

SCIENCE INDUSTRIELLE, UNE NOTION DÉJÀ ANCIENNE

Avant de poursuivre, précisons l'expression " science industrielle ". Elle apparaît peu en dehors des discours de Le Chatelier et même pour ainsi dire a disparu avec son promoteur en 1936. Une histoire de la science industrielle serait-elle l'histoire d'une impasse, d'un projet avorté ? A un niveau superficiel, il semblerait qu'elle permette simplement de formuler autrement la question de l'organisation des relations science/industrie à un moment où s'impose la révision de schémas traditionnels. Mais cela ne signifie pas que la science industrielle n'a pas une histoire, et encore moins que Le Chatelier soit son unique penseur. L'expression était déjà employée pour désigner, entre autres, l'enseignement de l'Ecole centrale lors de sa création en 1829 : des contenus et un mode d'exposition destinés aux ingénieurs civils, situés entre ceux de l'Ecole polytechnique et des Ecoles des arts et métiers[12]. L'ambition de Le Chatelier prolonge assurément la préoccupation des fondateurs de l'Ecole centrale. Il faut seulement l'inscrire dans le cadre de la seconde industrialisation et de ses problématiques spécifiques. Pour Le Chatelier, il s'agit non seulement de développer des connaissances utiles à l'industrie, mais de constituer une zone intermédiaire entre " savoir " et " produire " qu'il dote d'un schéma intellectuel spécifique et qu'il conçoit comme un guide opérationnel.

Il faut insister. La science industrielle constitue un objet d'histoire, c'est-à-dire une entité dont la signification varie en fonction du temps. Le Chatelier ne l'explicite d'abord que timidement à partir des années 1890, n'érige la science industrielle en modèle qu'à la veille du conflit mondial, après une longue maturation. On doit en fait considérer que sa pensée arrive à son terme en 1914, la période de formation se situant entre 1895 et 1905. Les deux dates associées à des initiatives précises fournissent les repères chronologiques de son évolution. En outre, à l'âge de 69 ans, juste après la Première Guerre mondiale, Le Chatelier entame une retraite. Carrière et production scientifiques s'arrêtent pour ainsi dire à cet instant. Commence ensuite la vie d'un autre personnage. Le Chatelier se consacre à ses élèves et à ses disciples, fonde une école, mais surtout développe ses réflexions qui prolongent, au delà de la chimie et de la science, les principes de la science industrielle. Il ne s'impose plus seulement comme un chercheur et un professeur, mais comme une autorité consacrée. Et c'est fort de sa réputation qu'il s'engage dans le débat sur

12. Voir Lavallée, Dumas, Olivier,..., " Ecole centrale des arts et manufactures. Prospectus ", *Annales de l'Industrie Française et Etrangère*, 2 (1828), 379-397 (surtout), où le pluriel " sciences industrielles " l'emporte sur le singulier. Ce qui lui confère un sens pas tout à fait identique ou témoigne d'une confusion. Dumas lui donnera plus tard une signification plus nette : J.B. Dumas, " Projet de rapport à présenter au Ministre de l'Agriculture et du Commerce ", *AAS*, fonds Dumas, carton 16 (1873). On doit enfin voir J.H. Weiss, *The Technological Man. The Social Origins of French Engineering Education*, Cambridge (Mass.), MIT Press, 1982, 89-122. Dernière remarque, le chimiste Chaptal employait ces termes pour spécifier sa conception de la science appliquée. On le voit, l'expression " science industrielle " employée par Le Chatelier et les options qui la soustendent émergent sur fond d'une tradition.

l'extension d'une application de la science à l'ensemble de la société. Il devient propagandiste et caution scientifique du taylorisme qu'il s'efforce d'adapter au contexte français[13]. Il considère les théories de l'ingénieur américain Taylor comme l'application juste de la science industrielle au cas de la production et de l'économie suivant toutes ses composantes. Aussi inscrit-il son nom sur la liste des théoriciens de l'organisation scientifique du travail. Plus généralement, il contribue au mouvement de rationalisation de la production qui se précise dans l'entre-deux-guerres[14]. Pour autant ses ambitions ne se limitent pas à cela. Les prétentions scientistes de Le Chatelier s'étendent jusque et y compris au domaine des relations sociales, de la formation, de l'économie politique et même de la morale.

Quoi qu'il en soit, Henry Le Chatelier tente d'édifier un champ nouveau de la connaissance et de l'action. En d'autres termes, il se veut fondateur d'une discipline. La constitution de la science industrielle et sa promotion mettent en oeuvre des stratégies. En sa qualité de membre d'une élite, Le Chatelier met à profit l'ensemble des opportunités pour imposer ses convictions. Tous les espaces où peut s'élaborer et s'exposer la science industrielle sont investis en vue d'un contrôle : lieux de la sociabilité savante et industrielle, institutions publiques ou privées, chaires de l'enseignement supérieur et revues scientifiques et techniques.

La science industrielle prise comme grille d'analyse

Où situer la science industrielle par rapport aux autres schémas organisationnels ? Quel type d'association originale Le Chatelier propose-t-il ? Pour répondre, il faut concevoir la science industrielle en vue de désigner clairement ses éléments constitutifs, de les identifier et de les organiser. Le projet d'approcher la science industrielle comme une grille de lecture des relations science/industrie tente d'y pourvoir. Ceci devrait contribuer à la compréhension d'une composante essentielle de l'industrialisation en examinant la signification de la science industrielle, par comparaison notamment avec la notion encore dominante, durant la seconde industrialisation, de science appliquée.

Au-delà d'une réflexion sur la nature des rapports entre la science et la production, c'est la rapidité de leurs échanges que l'on considère comme une caractéristique de l'industrialisation. A tel point qu'elle engendre une confusion entre deux domaines que l'on tenait pourtant à distinguer nettement. Mau-

13. Voir par exemple P. Fridenson, " La circulation internationale des modes managériales ", dans J.P. Bouilloud & B.P. Lecuyer (dir.), *L'invention de la gestion. Histoire et pratiques*, Paris, L'Harmattan, 1994, 81-89 et " Un tournant taylorien de la société française (1904-1918) ", *Annales ESC*, 42 (1987), 1031-1060.

14. Sur le thème de la rationalisation, le rôle des ingénieurs et de Le Chatelier, il faut se reporter à A. Moutet, *Les logiques de l'entreprise. La rationalisation dans l'industrie française de l'entre-deux-guerres*, Paris, EHESS, 1997 (en particulier le premier chapitre).

rice Daumas pouvait écrire, en 1962, que la science et la technologie, tissées dans un réseau de liens complexes, sont si proches que l'on a parfois du mal à les distinguer[15]. Le terme de " technologie " auquel Daumas fait allusion, carrefour entre la science, la technique, l'industrie et l'économie, mérite une attention particulière. Il renvoie à une zone complexe de réflexion sur les liens science/industrie qui dépasse l'idée réductrice d'une science appliquée. Avec d'autres, Daumas propose de partir de cette notion ancienne pour désigner un domaine d'activité créatrice où science et technique sont si étroitement associées qu'il est difficile de faire la part de ce qui revient à l'une et à l'autre. La suggestion vise alors à dépasser l'opposition entre deux genres que les historiens s'attachent à cultiver séparément[16].

La proposition d'établir l'unité et l'identité de la technologie, de lui assurer le statut de territoire intermédiaire, de champ autonome à la fois de la connaissance scientifique et de l'activité industrielle, bute cependant sur des limites plus insaisissables au fur et à mesure que l'on tente de les atteindre. Cette voie n'a jusqu'à présent abouti qu'à une indétermination : les débats conduisent irrémédiablement à la discussion indéfinie d'une collection de définitions, sans pour autant aboutir à une modélisation, à une grille de lecture des rapports entre la science et la production[17]. Des travaux sur le concept se dégage l'évidence d'un monstre polymorphe qui ne se laisse pas circonscrire, même réduit à sa seule composante scientifique[18]. Pourtant, les propositions de schémas explicatifs via la technologie, continuent de recueillir une faveur étonnante. Et c'est tant mieux : ces schémas cultivent la vision plurielle d'une alliance entre le fondamental et l'appliqué. Surtout, ils discréditent le modèle unique qui limite la dimension d'une technologie médiatrice à une simple application des produits de la science pure aux problèmes industriels particuliers. En outre, ces schémas démontrent que les rapports science/industrie sont variables en fonction du temps et suivant les contextes, c'est-à-dire qu'ils sont précisément des objets d'histoire[19]. Enfin, ils sont surtout utiles en ce qu'ils attirent l'attention sur la nécessité de comprendre, au-delà de ce qu'est la nature d'une connexion

15. M. Daumas, " Rapport entre sciences et techniques : étude générale du point de vue de l'histoire des sciences et des techniques ", *Revue de Synthèse*, 25 (1962), 15-37.

16. M. Daumas, " L'histoire des techniques : son objet, ses limites, ses méthodes ", *Revue d'Histoire des Sciences et de leur Applications*, 22 (1969), 5-32.

17. J. Guillerme & J. Sebestik, " Les commencements de la technologie ", *Thalès*, 12 (1966), 1-72 ; E. Layton, " Technology as Knowledge ", *Technology and Culture,* 15 (1974), 31-41 ; J. Salomon, " What is Technology ? The Issue of its Origins and Definitions ", *History and Technology*, 1 (1984), 113-156 ; F. Sigaut, " More (and enough) on Technology ! ", *History and Technology,* 2 (1985), 115-132.

18. J.C. Beaune, *La technologie introuvable. Recherche sur la définition et l'unité de la Technologie à partir de quelques modèles du XVIIIe et XIXe siècles*, Paris, J. Vrin.

19. R. Kline, " Construing " Technology " as " Applied Science ". " Public Rhetoric of Scientists and Engineers in the United States, 1880-1945 ", *Isis*, 86 (1995), 194-221.

science/industrie à un instant donné et pour un domaine déterminé, comment les acteurs eux-mêmes les perçoivent[20].

Parmi les schémas explicatifs des relations science/industrie, il en est un qui force l'attention. C'est une proposition qui rebondit sur l'échec d'une délimitation de la technologie : celle de l'historien américain Edwin Layton. Il affirme que la technologie se transforme aux environs de 1900 pour donner naissance à une technologie scientifique[21]. Son existence avait, selon lui, été occultée par la perception classique d'une science seule créatrice, fournissant à la technologie les matériaux de sa construction. Renversant le propos, il suggère de considérer qu'une technologie se développe à la charnière des XIXe et XXe siècles, non seulement de manière autonome, mais encore qu'elle se constitue en parallèle de la science, à l'identique, comme son image dans un miroir. Il entend par là qu'elle structure sa propre communauté, à partir d'établissements spécifiques d'enseignement et de diffusion, de ses lieux académiques, ses institutions de sociabilité, son langage et une production littéraire. Mais tout cela reste calqué sur le modèle hégémonique de la science. Leurs relations sont dès lors logiquement symétriques. En somme, produit de l'industrialisation, la technologie serait un équivalent industriel d'une organisation de la communauté scientifique. Il reste que chacune évolue, l'une à côté de l'autre, mais avec des objectifs et des systèmes de valeurs spécifiques.

Si cette interprétation suscite l'attention, c'est que l'ambition, à l'instar de la science industrielle, est de parvenir à un modèle explicatif. Elle fournit même une perspective d'examen intéressante : la technologie scientifique mobilise les mêmes types d'approche que l'étude de la science et de ses objets : savoirs, communauté et pratiques. Cerner les contours d'un schéma, tel que la technologie scientifique selon Layton, ou la science industrielle d'après Le Chatelier, revient à identifier leurs éléments constitutifs et les mécanismes de leur organisation pour aboutir à un système intelligible.

La science industrielle serait-elle une sorte d'équivalent de la technologie scientifique selon Layton ? En fait, les modèles diffèrent. Si leur organisation intellectuelle et pratique présente quelques analogies, ils ne sont en aucun cas semblables. Le propos mérite une explication.

L'élément central auquel Layton fait appel pour spécifier la technologie scientifique est l'engineering science. Le concept, même s'il n'est pas d'usage courant sous le vocable de science de l'ingénieur, ni d'une signification tout à fait identique en France, s'y développe néanmoins. Au pluriel, les sciences de l'ingénieur forment un système complexe qui s'étend des connaissances scien-

20. O. Mayr, " The Science-Technology Relationship as a Historiographic Problem ", *Technology and Culture*, 17 (1976), 663-673 ; R. Laudan, " Natural Alliance or Forced Marriage ? Changing Relations between the Histories of Science and Technology ", *Technology and Culture*, 36 (1995), 17-28.

21. E. Layton, " Mirror-Image Twins : The Communities of Science and Technology in 19th Century America ", *Technology and Culture*, 12 (1971), 562-580.

tifiques les plus méthodiques aux protocoles parfaitement organisés du savoir-faire de l'ingénieur. Pour autant, les sciences de l'ingénieur possèdent une unité. A l'instar des sciences académiques, elles se réfèrent à une méthode, prétendent à la rigueur, recourent à l'organisation systématique et proclament leur confiance dans l'expérimentation et les mathématiques[22]. Un exemple probant invoqué par Layton est celui de la résistance des matériaux, une science de l'ingénieur qui trouve ses origines sur le vieux continent au XIXe siècle[23].

Toutefois, et à la différence de la science commise par les savants officiels, les sciences de l'ingénieur privilégient l'efficacité et les résultats pratiques. La maîtrise de la production, l'amélioration des rendements et de l'économie en général, déterminent ses objets, mais aussi ses méthodes, notamment le recours à des approximations qu'imposent les facteurs liés à la fabrication. C'est un renversement de priorité : les sciences à finalité cognitive façonnent la démarche analytique, l'abstraction et le théorique. Les sciences de l'ingénieur ont, elles, plutôt une tendance à la synthèse des résultats en vue d'un usage pratique matériellement performant.

Enfin, Layton désigne celui qui possède les dispositions pour produire, pratiquer et diffuser au mieux la science de l'ingénieur. Membre des deux communautés, scientifique et industrielle, il doit être capable de comprendre le fonctionnement, les enjeux et les besoins de l'une et de l'autre. Ce qui fait de lui un intermédiaire qu'il qualifie " d'ingénieur-scientifique " ou de " scientifique-ingénieur ". Peu importe en fait la formation initiale, seule compte la conviction d'appartenir à un domaine intermédiaire de savoirs et de pratiques répondant aux besoins de l'industrialisation. Ainsi Layton a-t-il identifié un certain nombre d'éléments constitutifs de ce qui doit être considéré comme un modèle : savoirs, savoir-faire, méthodes et acteurs essentiellement, mais d'autres sur lesquels il faudrait insister[24].

L'objectif de modéliser la science industrielle à partir de l'examen des discours et des pratiques de Le Chatelier ne diffère pas fondamentalement de la démarche de Layton. Seulement on pousse la logique jusqu'au bout afin de parvenir à une délimitation de tous les éléments constitutifs d'un système, du discours théorique à leurs applications concrètes, mais également de leur orga-

22. E. Layton, *art. cit.*, 567-568.

23. S. Timoshenko, *History of Strength of Materials*, New York, Dover, 1983. Layton livre d'autres exemples convaincants : E. Layton, " Millwrights and Engineers, Science, Social Roles, and the Evolution of the Turbine in America ", dans W. Krohn, E. Layton & P. Weingart (dir.), *The Dynamics of Science and Technology*, Boston, 1978, 61-88. Voir aussi E. Layton, " Scientific Technology, 1845-1900 : The Hydraulic and the Origins of American Industrial Research ", *Technology and Culture*, 20 (1979), 64-89.

24. La dimension sociale de l'objet " sciences de l'ingénieur " a été passée sous silence. Elle est toutefois une préoccupation majeure de Layton. La stratégie des ingénieurs en charge de la technologie scientifique vise à bousculer la hiérarchie des savoirs et des pratiques, mais aussi la stratification sociale. Voir E. Layton, *The Revolt of the Engineers,* Cleveland, Press of Case Western Reserve Univ., 1971 ; et " Science as a form of Action : The Role of the Engineering Sciences ", *Technology and Culture*, 29 (1988), 82-97.

nisation. C'est ici que ressort l'avantage d'envisager la problématique à partir d'un individu. Là où Layton considère un phénomène global associé à une communauté, la science industrielle livre un modèle d'inspiration personnelle. D'un niveau de généralité moindre et d'un degré d'empirisme supérieur, la science industrielle et Le Chatelier se laissent plus aisément saisir comme un modèle aux contours mieux définis.

Il existe une autre différence entre la science industrielle et la technologie scientifique. Le contexte de l'industrialisation aux Etats-Unis n'est pas celui de la France sous la IIIe République. Saisir la science industrielle, c'est aussi concevoir la représentation que forge son promoteur d'une industrialisation où les rapports entre la communauté scientifique, l'Etat et l'entreprise relèvent d'une appréhension propre à la société française. Autrement dit, le modèle est perçu comme l'expression d'une conception hexagonale des liens entre la science et la production. La figure de l'ingénieur d'Etat, acteur de l'industrialisation, avec les convictions qui l'animent, incarnée ici par Le Chatelier, est un élément sans lequel il ne serait pas raisonnable d'envisager le processus en cours[25].

Les difficultés d'acclimatation en France de la notion de science de l'ingénieur sont symptomatiques d'une spécificité de conception des relations science/industrie[26]. Ceux qui détiennent le monopole du discours ont sans doute quelque réticence à désigner clairement un domaine qui les subordonnerait à la " science pure " et les rapprocherait de l'application, moins valorisante du point de vue académique. Dans quelle mesure donc la science industrielle doit-elle être comprise comme la traduction d'un modèle d'organisation, " à la française ", de la science et de l'industrie ?

SCIENCE INDUSTRIELLE ET RATIONALISATION SOCIALE

La science industrielle placée sous les auspices d'une représentation sociale peut être assimilée à une adhésion à ce que l'on peut envisager comme le paradigme d'un groupe qui impose ses normes, son système de valeurs, son idéologie.

Quel est ce groupe ? Celui des ingénieurs sélectionnés sur le mode méritocratique et investis de la mission de servir l'Etat, la collectivité. Ils forment la

25. Quelques historiens ont fort justement consacré la figure de l'ingénieur d'Etat comme un acteur essentiel de la structuration des relations entre la science et la production, de l'industrialisation dans le contexte français. Voir en particulier A. Picon, *L'invention de l'ingénieur moderne. L'Ecole des Ponts & Chaussées, 1747-1851*, Paris, Presses de l'ENCP, 1992 ; et A. Thépot, *Les ingénieurs des Mines du XIXe siècle. Histoire d'un corps technique d'Etat*, tome 1 : 1810-1914, Paris, Ed. Eska, 1998.

26. Elle livre une signification dont la géométrie est variable quand il s'agit de spécifier savoirs et savoir-faire de l'ingénieur, mais qui ne s'est pour autant pas inscrite durablement. Aujourd'hui encore elle cherche son identité et son unité reste, sinon à démontrer, au moins à établir précisément : G. Ramunni, *Les sciences pour l'ingénieur. Histoire du rendez-vous des sciences et de la société*, Paris, CNRS, 1995.

partie visible de l'élite scientifique, technique et administrative. A ce titre, Le Chatelier se considère comme le dépositaire de l'intérêt général. Il entend jouer un rôle déterminant, tant dans la structuration des savoirs récents que dans leur diffusion auprès des acteurs de l'économie. Autrement dit, il est pour lui question d'une contribution à l'éducation industrielle de la nation. L'Etat lui délègue un rôle de stimulation, d'animation, d'organisation et parfois de conduite directe de la novation, de la recherche scientifique et industrielle en vue d'une coopération efficiente entre savants et producteurs. L'histoire de la science industrielle présente par conséquent un intérêt certain pour qui cherche à mieux cerner les visées technocratiques de la classe des administrateurs scientifiques et techniques de l'Etat, par suite de leur influence sur le processus d'industrialisation. En quoi cependant participe-t-elle de la représentation d'une certaine science, de l'industrie et du rôle de l'ingénieur d'Etat dans le progrès industriel ?

La science industrielle implique un cadre organisationnel pratique, de contrôle et d'orientation des conduites. Telle est au demeurant la fonction d'un modèle. On ne doit donc pas seulement considérer la science industrielle comme le reflet d'une réalité objectivée, mais au-delà comme un guide pour l'action, un repère pour prendre position. Cette perspective invite à regarder la science industrielle comme un ensemble organisé d'opinions, de convictions, de règles, de normes et de comportements, qui dans l'exercice concret se traduit entre autres par l'attribution d'un rôle précis à des acteurs désignés.

La science industrielle est explicitement associée à l'idée d'une organisation rationnelle des relations entre la science et la production. Elle traduit la conviction d'une subordination nécessaire de la pensée et de l'action à des critères rationnels. Le modèle supporte dès lors le projet d'une transformation de l'environnement, tant intellectuel que physique, et ce en vue d'un rendement optimal de l'activité productive. Autrement dit, la science industrielle sous-tend un projet de rationalisation. Les efforts de Le Chatelier pour institutionnaliser les échanges entre le savant et le producteur illustrent le propos. Le promoteur de la science industrielle est un représentant zélé de ce mouvement. En témoigne la mission d'éducation scientifique de la nation dont il se considère investi, ou sa conviction de devoir garantir le progrès à coup d'organisation rationnelle. Là encore pointe la figure du serviteur de l'Etat aux visées technocratiques, sans laquelle l'industrialisation à la française ne serait pas tout à fait compréhensible.

L'étude des relations science/industrie, à la charnière des XIXe et XXe siècles, ne peut dès lors se limiter à l'examen d'un transfert des savoirs, des techniques et d'autres objets issus de la science vers l'usine. Elle doit tenir compte d'une volonté ferme de standardisation des connaissances, des savoir-faire et des produits à des fins économiques. La standardisation doit être considérée comme un élément essentiel de la science industrielle, et par conséquent de la seconde industrialisation.

Aussi la science industrielle incarne-t-elle un type de rationalité, rattachée aux critères de la science et de l'économie, mais qui en définitive débouche sur la démonstration que la science industrielle se définit elle-même comme science. L'adhésion à un système de critères rationnels appuyés sur une science orientée vers l'économie autorise une sorte de perspective de domination légitime, l'exercice d'un contrôle. à la fois du monde scientifique et industriel, mais également de l'espace social. Au-delà des divergences individuelles sur la liaison science/industrie, la science industrielle recouvre une série d'attitudes et de comportements communs. Le collectif se situant au sommet d'une hiérarchie, il impose ses analyses du processus de production scientifique et industrielle, et la manière dont il faut les structurer.

En conséquence, l'organisation rationnelle selon Henry Le Chatelier prend sa vraie dimension à travers la hiérarchie qui l'accompagne : quelques individus dotés des plus hautes compétences scientifiques dirigent une majorité d'exécutants pour la production d'objets standardisés d'une nature déterminée et imposent la méthode pour y parvenir. Tel est l'enjeu : refondre les fonctions du savant, de l'industriel et de leur intermédiaire privilégié, l'ingénieur, dans le cadre d'une programmation méthodique du progrès.

En effet, le projet permanent de rationalisation auquel invite la science industrielle précède l'entreprise de Le Chatelier. Antoine Picon recourt à la notion de “ rationalité technique ”, avancée par les ingénieurs des Ponts et Chaussées, pour caractériser au tournant du siècle précédent une dynamique de transformation des structures productives[27]. Elle se rapporte dans ce cas à une conduite des acteurs de la production, conformément à des règles dictées par des critères rationnels parés de l'objectivité. Au delà, elle est assimilée à un paradigme, compris comme la somme des convictions, valeurs et techniques partagées par la communauté des ingénieurs. Enfin, Picon émettait le voeu de voir explorer plus avant une histoire de la rationalité technique. Une définition de la science industrielle voudrait répondre à ce voeu pour la seconde industrialisation.

Définition de la science industrielle

Il faut donc désormais définir la science industrielle telle qu'elle a été élaborée par Le Chatelier et ainsi garantir sa signification comme mode d'organisation rationnelle des relations entre la science et l'industrie. Conformément aux vues de départ, il faut délimiter en priorité ses éléments constitutifs, ceux qui assurent la cohérence d'une grille de lecture, c'est-à-dire explicitent un

27. A. Picon, “ Towards a History of Technological Thought ”, dans R. Fox (dir.), *Technological Change. Methods and Terms in the History of Technology*, Amsterdam, Harwood Academic Publishers, 1992 et 1996, 37-49.

schéma organisationnel lisible qui puisse être comparé à d'autres formalisations contemporaines des liens science/industrie.

Cinq éléments constitutifs sont à retenir. Le premier, celui qui force l'attention avant que ne soit envisagée tout autre dimension plus complexe, est la série d'énoncés sur ce qu'est la science, la production industrielle et la science industrielle. Ce point de départ est à la fois simple et le plus essentiel. La limitation des principes à quelques axiomes figés dans les discours rend compte de la simplicité. C'est le plus important parce qu'il constitue la base sur laquelle s'érige l'édifice : la signification des autres éléments en dépend. En dernière instance, les énoncés aboutissent à la définition, mais d'abord théorique, de la science industrielle comme champ autonome de la pensée et du savoir. Ils révèlent dans le même temps son cadre épistémologique.

Un second élément résulte de l'identification des objets manipulés : ceux nécessaires à la construction concrète de la science industrielle et de sa mise en oeuvre. Ils se déduisent en fait logiquement des définitions, préceptes et énoncés précédents.

Dans le prolongement de ces premiers éléments, Le Chatelier revendique l'autonomie d'une étendue disciplinaire. Il ne pouvait dès lors que valider ses énoncés en proposant un enseignement type de la science industrielle. C'est le troisième élément constitutif du modèle.

Par qui la science industrielle est-elle manipulée ? A qui cet enseignement est-il destiné ? La réponse fournit l'avant dernier élément : les acteurs chargés de convertir les principes théoriques en mode de pensée, et surtout d'action.

Enfin, la science industrielle se produit et s'exerce quelque part. Elle dispose de ses lieux privilégiés de production et d'exercice. Ce sont ces espaces aménagés en vue de la production des objets et de l'exercice de la science industrielle qui constituent le cinquième élément. Bien entendu, il faut tenir compte de l'organisation entre eux des entités identifiées : énoncés, objets, acteurs, enseignement et lieux.

Enoncés et principes

La science, telle que la conçoit ou se la représente Le Chatelier, se définit d'abord par une croyance inébranlable au déterminisme : celui qui affirme que le passé détermine la connaissance objective du présent et fixe le futur. Tous les phénomènes naturels, sans exception, sont liés suivant des rapports nécessaires entre des facteurs d'influence déterminés ; ils dépendent tous d'un nombre limité d'entre eux. Pour résumer, retenons l'expression favorite de Le Chatelier : “ Nier le déterminisme, c'est nier la science ”.

Mais ces relations entre les facteurs ne sont pas seulement qualitatives : elles s'expriment quantitativement sous la forme d'une équation reliant ces différents paramètres traduits par des grandeurs mesurées. Il est donc question d'une science explicative des phénomènes qui vise la mise en équation expérimentale. Autrement dit une science érigée sur l'observation et la prise en considération exclusive des grandeurs mesurables, accessibles à l'expérience.

Le but ultime de la science est d'établir ces lois numériques. Ce sont elles qui expliquent et prévoient, et permettent finalement la maîtrise des phénomènes. Et si la plupart sont encore peu étudiées, complexes, ou ne peuvent être traduites sous la forme d'une expression mathématique simple, ce n'est que le résultat d'un développement insuffisant des sciences particulières et de données expérimentales en nombre encore trop faible.

La science se définit aussi par sa méthode qualifiée de scientifique, celle mise en oeuvre pour la recherche des lois déterministes. Elle est unique et fonde tant la science que la science industrielle. Elle consiste à mobiliser la démarche analytique qui observe et détermine les facteurs d'influence. Ensuite, elle les convertit en grandeurs quantifiables précisément mesurées. C'est une première étape. Au delà, le raisonnement et la synthèse établissent la loi expérimentale de variation d'un des facteurs en fonction des autres. Une méthode que Le Chatelier énonce selon : " La science, telle que nous venons de la définir, peut être représentée par un tétraèdre reposant sur un plan horizontal ; les trois sommets de base, que l'on doit traverser avant d'arriver au sommet, sont l'observation, les mesures et le raisonnement : le sommet supérieur représente la loi, but définitif de la Science. Pour atteindre ce but final, il ne suffit pas cependant d'observer n'importe quoi, de mesurer autre chose, puis de raisonner sur une hypothèse quelconque ; il est nécessaire de relier les différentes étapes de la science en se conformant à des règles précises, qui constituent ce que l'on appelle la méthode scientifique. Dans la représentation géométrique proposée ici cette méthode serait figurée par les arêtes de tétraèdre "[28]. Enoncé que l'on peut figurer ainsi :

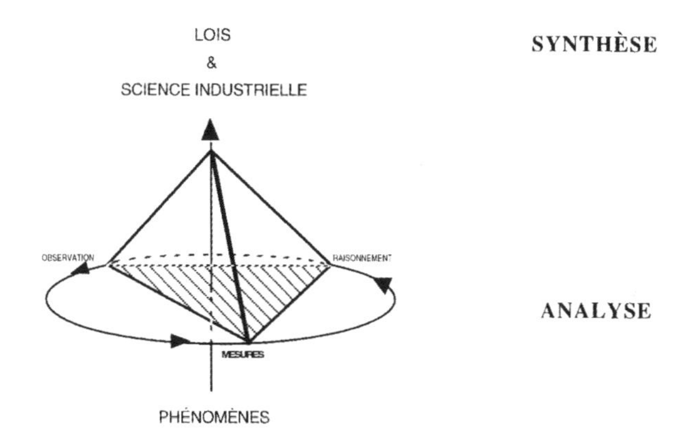

28. H. Le Chatelier, *Science et industrie*, Paris, E. Flammarion, 1925, 21-22.

Ce mode d'appréhension s'applique rigoureusement à tous les ordres de phénomènes, y compris aux cas industriels les plus complexes. Ils dépendent tout autant des lois déterministes. Leur étude doit être envisagée par une méthode scientifique en tout point identique. La recherche de ces lois permet seule, selon Le Chatelier, d'augmenter la puissance d'action du travailleur scientifique et industriel. C'est en somme l'unique voie de développement rationnel de la production.

Si donc la science industrielle a pour vocation d'établir les lois expérimentales déterministes qui régissent les phénomènes, il lui faut, pour s'élaborer, la contribution de toutes les sciences arrivées à un degré suffisant de développement. Toutes les sciences particulières ont, dans ce cadre conceptuel, pour fonction de contribuer à l'édification de la science industrielle, qui exprime dès lors le vieux rêve de réunifier les savoirs.

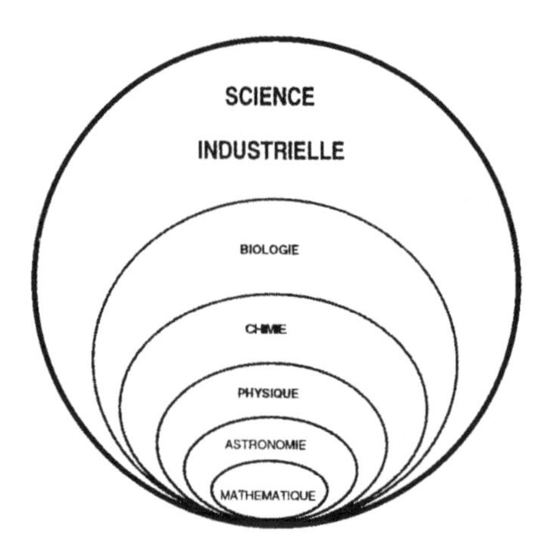

Le Chatelier adapte ainsi la hiérarchie des sciences proposée par Auguste Comte, au sommet de laquelle il substitue la science industrielle à la sociologie, cette dernière devenant l'un de ses éléments. La construction de la science industrielle comme ultime science positive ne se limite cependant pas à reprendre une théorie de la connaissance d'inspiration comtienne. Le Chatelier s'attache également à condamner l'hypothèse qui ne serait pas établie sur des bases expérimentales, refuse d'envisager l'examen des phénomènes autrement que par l'expérience et l'observation, rejette les explications établies à partir d'entités abstraites sur la nature et l'origine des phénomènes ou proscrit du champ de la science industrielle la statistique et le hasard. La constitution d'une mécanique chimique de type phénoménologique en est l'expression la plus convaincante. Ces énoncés, principes ou préceptes de la science industrielle guident l'activité de Le Chatelier dès ses premiers travaux. Ils ne sont cependant formulés de façon explicite que plus tard, lorsque la science industrielle aura acquis la conscience de son originalité et la certitude de sa légitimité. Ils deviennent ensuite des slogans sans cesse scandés.

Les objets de la science industrielle

Il découle des énoncés qui précèdent que les objets exclusifs de la science industrielle sont les éléments utiles à l'élaboration des lois expérimentales. C'est-à-dire les grandeurs quantifiables associées aux facteurs d'influence, mais aussi les lois intermédiaires établies par chacune des sciences particulières, celles qui servent à l'établissement des lois définitives de la science industrielle.

Les données à atteindre sont immédiatement visibles après l'analyse. La production des données numériques peut dès lors faire l'objet d'un programme. Pour cette raison, la science industrielle implique une organisation méthodique où les exécutants produisent de manière systématique les éléments de base indispensables à l'édification des lois générales. C'est pourquoi Le Chatelier cultive le souci permanent de l'accumulation patiente des données expérimentales nombreuses. Ainsi a-t-il pour priorité, quand il milite pour la promotion de la science industrielle, de diriger des programmes qui visent la production de mesures systématiques sur les propriétés physiques, et non de confier à l'inventeur le soin d'une hypothétique découverte.

L'enseignement de la science industrielle

Il comporte forcément les certitudes de Le Chatelier sur le déterminisme et son application aux phénomènes industriels. Sa finalité est de convaincre de l'intérêt d'une adhésion à cette conviction et à la méthode qui lui est associée pour la production tant scientifique qu'industrielle. Ensuite, il faut assurer la transmission des résultats acquis : les lois générales connues ou la synthèse provisoire des quelques données encore incomplètes.

L'enseignement de la science industrielle a donc pour finalité la synthèse, c'est-à-dire l'exposé des lois synthétiques en nombre limité.

Par ailleurs, compte tenu de la définition stricte de ses objets, l'enseignement de la science industrielle impose l'acquisition préalable des méthodes expérimentales de mesure. Après plusieurs tentatives durant les premières années de professorat, cet enseignement est arrivé à maturité au Collège de France et à l'Ecole des mines après 1900. Le Chatelier propose dès lors d'exposer la science industrielle en prenant prétexte d'un thème général sur lequel convergent les connaissances acquises, empruntant ainsi à toutes les sciences. Aucune description technologique ne vient accompagner le cours. Ces considérations sont en effet étrangères aux principes théoriques, énoncés et objets de la science industrielle. Cet aspect est confié aux stages dans les usines ou cantonné aux débuts de la carrière professionnelle.

Les attributs de la science industrielle enseignée sont donc l'adhésion au déterminisme et la conviction que tous les phénomènes industriels sont régis par des lois à dévoiler par la méthode expérimentale. L'enseignement de la science industrielle comprend nécessairement l'apprentissage des procédés de mesure, l'exposé synthétique des lois générales et le rejet des descriptions technologiques, mais également le refus de la spécialisation.

LES ACTEURS

La maîtrise des méthodes de mesure et l'utilisation des lois générales à des fins de contrôle rationnel de la production nécessitent des compétences précises. Un individu possédant par sa formation une connaissance suffisante des sciences, mais également de la production, doit être l'acteur naturellement désigné pour accueillir l'enseignement de la science industrielle et l'appliquer au cas concret de l'usine. Le Chatelier désigne l'ingénieur qui s'est initié à la recherche dans le cadre de la préparation d'un doctorat. Il incarne une sorte d'idéal-typique. On reconnaît l'ingénieur, civil ou d'Etat, issu des grandes Ecoles qui a contribué par ses travaux au progrès scientifique, tout en occupant une fonction liée à la production. Ce sont effectivement eux que l'on croise tout au long des programmes de recherche de science industrielle dirigés par Le Chatelier, mais également à la Sorbonne, au laboratoire du Collège de France.

Détenteur de savoirs synthétiques, d'une méthode jugée la plus efficace pour l'examen et la maîtrise totale de la production, l'ingénieur acteur de la science industrielle est appelé à occuper une position privilégiée dans l'organisation pratique de l'usine. Le Chatelier lui confie d'emblée le rôle d'un dirigeant, d'un penseur de la production. Le scientifique industriel programme et organise l'activité ; les autres travailleurs, scientifiques ou industriel, s'exécutent sous sa direction.

LES LIEUX D'ÉLABORATION ET DE PRATIQUE

Au centre des relations science/industrie : le laboratoire. On y trouve en effet, tout au long du parcours de Le Chatelier, les principes en action de la science industrielle, l'instrumentation utile à la production des données numériques et un ingénieur scientifique maître des lieux. Un laboratoire de science industrielle ne se spécifie pour autant pas seulement par ces traits. Plus encore, sa vocation le désigne comme le dernier élément constitutif d'un modèle.

Le Chatelier lui attribue un rôle identique à celui de l'ingénieur : celui de diriger, de penser pour le producteur ce qui convient le mieux à l'organisation performante de son activité. Attitude qui débouche sur la maîtrise, le contrôle strict de toutes les étapes du processus de production. Le laboratoire de science industrielle dirige donc. Il prend la production comme source d'inspiration et l'usine comme champ d'application.

L'ORGANISATION RATIONNELLE DES RELATIONS SCIENCE/INDUSTRIE

La science industrielle, science qui fédère toutes les sciences, structure savoirs et savoir-faire, avec désormais une seule justification : contribuer à la rationalisation de la production. De là doit découler naturellement le progrès tant scientifique qu'industriel. Puisque la science industrielle est située au sommet d'une hiérarchie de la connaissance, toute contribution devient nécessairement une participation directe au progrès de la science.

Cette structuration des savoirs génère une organisation conséquente du travail, mais aussi du social. Puisque tous les éléments de la science industrielle sont par avance déterminés, normalisés, standardisés, qu'ils font l'objet de programmes de production systématique, la division des tâches imposée est aussi strictement méthodique et déterminée. Quelques individus, détenteurs des plus hautes compétences que leur assure la science industrielle, détiennent le monopole de la direction des projets de rationalisation de la production et de l'innovation. La majorité des travailleurs — scientifiques, ingénieurs et techniciens — sont chargés d'exécuter les plans de recherche, de développement et de contrôle, décidés et préétablis suivant des critères considérés comme objectifs parce qu'issus de la science et de la technique.

C'est finalement le modèle technocratique de gestion des relations entre la science et la production, avec pour acteur la figure de l'ingénieur scientifique, organisateur et rationalisateur, qui ressort de la délimitation de la science industrielle. L'étude des implications pratiques d'un tel schéma montre que si les résultats ne sont pas toujours mesurables, voire visibles en terme d'application directe de la science à l'industrie, ils n'en sont pas moins réels. La science industrielle selon Le Chatelier démontre qu'au delà des brevets, des inventions, des découvertes ou des laboratoires voués explicitement à l'innovation, les projets de standardisation, de normalisation des gestes, pratiques, produits et procédés sont aussi la traduction convaincante d'une relation réelle entre la science et l'industrie, l'expression d'un apport significatif au progrès tant scientifique qu'industriel. Ce sont là des objets qu'il faut expliciter si l'on veut spécifier une voie particulière de seconde industrialisation.

Changes in the Concepts of Flight in the First Decade of the 20th Century : The Pioneering Work of Alberto Santos-Dumont

Henrique Lins de Barros - Mauro Lins de Barros

Introduction

The conquest of sustained and controlled flight is a landmark of our century. Several inventors tackled the problem since Montgolfier's free balloons first lifted off the ground at the end of the 18th century. However, the lack of a light-weight, safe and reliable power plant hindered progress for a long time. Only by the late 19th century, the invention of the internal combustion engine promised a solution, even though it still needed important improvements.

An interesting aspect in the history of heavier-than-air flight is the path linking Sir George Cayley's revolutionary model gliders of ca. 1804 to the first aircraft built at the beginning of the 20th century. In spite of Cayley's gliders incorporating the essential elements for an auto-stable configuration, early 20th century pioneers generally adopted an intrinsically unstable configuration in regard to one of the 3 axes.

In the career of the Brazilian Alberto Santos-Dumont (1873-1932)[1] we follow a process that leads from an unstable configuration to a successful airplane. At the same time, we show how the dominant empiricism of the times suggested detours requiring enormous energies to reject an inappropriate mental model largely based on the idea that force is associated to speed. Interestingly, this concept is paralleled in the history of the science by the physics of impetus. We focus on the brief period from mid-1906, when Santos-Dumont directed his attention to heavier-than-the air flight, to November 1907 when he

1. H. Lins de Barros, *Santos Dumont*, Ed. Index/API. RJ, 1986. C. Dollfus, " Alberto Santos Dumont né le 20 juillet 1873 ", *Ícare*, n° 64 bis, 1973, 95-145. H.D. Villares, *Quem deu asas ao homem*, Ed. Autor, SP, 1953.

attained what could be considered as the true physical precursor of the modern airplane.

Conceptual Precursors and Physical Precursors

The airplane is a complex system, " an unique system composed of compatible parts each of which interacts with others and contributes to the basic function, so combined that the removal of one part prevents the efficient working of the whole "[2]. To trace the evolution of the airplane it is therefore convenient to borrow some concepts from the study of complex systems. Our purpose is not to treat mechanical systems as if they were subjected to the rules of an evolutionist theory in Darwin's manner or to reduce the question to a mere formalism, which at worst can take a purely mathematical nature. Rather, our intention is to put the discussion in perspective with the help concepts that allow the distinction of nuances in various designs, which reflect changes in the mental models required for a successful heavier-than-air machine. The history of aviation is already too rich in conflicting claims and counter claims and we do not want to add our share of controversy.

Two notions — conceptual and physical precursors — will be useful in our discussion. A conceptual precursor is an artefact that is capable to accomplish a certain function. It is not necessarily tied up to any given technical solution. In our case, one of the concepts involved is that of controlled flight. In this respect, one conceptual precursor is, for example, the dirigible-balloon that accomplishes the same function as the airplane (flying under control).

The physical precursor, on the other hand, is a link in an unbroken chain that leads to the final artefact. It may precede it by one or many stages in development. Thus, it must be possible to arrive at the final artefact, starting from the precursor, through incremental transformations and adaptations.

Finally, we understand that the minimum function is the capacity to accomplish a given task (in our case, controlled flight by a heavier-than-air machine) in physically realistic conditions. The minimum function concept allows to discriminate inventions that, although seemingly adequate, are not capable to accomplish the required task, be it for the lack of appropriate materials or because of the incorrect use of some other element. In the case of the aviation we will see that the minimum function will explicitly appear in the test criteria instituted by the inventors themselves.

2. M. Behe, *A caixa preta de Darwin*, Jorge Zahar ed., 1997 (portuguese translation *Darwin's black box*, NY, The Free Press, 1996).

METHOD AND 19th CENTURY INVENTORS

One important aspect in the invention of the airplane is the almost total absence of a scientific body of work. Inventors were, in the vast majority of cases, mechanically gifted laymen with little knowledge of well-established basic principles[3]. This can be clearly seen, for instance, in the absence of considerations on the torque of the several forces involved, while the balance of forces received its due attention[4].

This inadequate training in physics led the 19th century inventors to do experimental work. Raw experiments drove the path to discovery without the elaboration of theoretical models that could aid and accelerate the progress. The road to success was much longer than necessary because the mental models suggested by experimentation led away from the optimum solution. It is an ironic aspect, in the history of aviation, that the main characteristics of a stable configuration were well known for a long time, having been proposed by Cayley and tested by Otto Lilienthal during the 19th century.

In this sense, the invention of the airplane is similar to that of the marine chronometer, made by Harrison in the 18th century[5]. In both cases, the contribution of important names in contemporary science was negligible, or null in the case of aviation, frequently even to point of trying to prove the impossibility of the invention. This disbelief about flight in general is referred to in an article published in 1885 : " The conclusion is reached that art of aerostation is much nearer the practically applicable state than scientific men generally suppose. The objects now sought are the attainment of better and stable forms, the more effective arrangement of parts, the invention of lighter motors, invariable in weight, and convenient of operation, and securing of higher efficiency of propelling instruments. Even now, with the experience of the past, it is possible to build a machine of this class capable of making at least ten meters per second through the surrounding medium "[6].

Scientific development requires progressing towards abstraction and generalization : experiments, as the basis for reasoning, must be depurated of extraneous influences and should be able to supply the elements for the construction of a model in which the several variables can be manipulated or understood.

3. G. Voisin, *Mes 10000 cerfs-volants*, Paris, La Table Ronde, 1960. E. Petit, *Nouvelle histoire mondiale de l'aviation*, Hachette Réalités, 3e éd., 1977. T.D. Crouch, *A dream of wings : Americans and the airplane 1875-1905*, Smithsonian Institution Press, 1989. *Les Cahiers de Science et Vie : Les Grandes Controverses Scientifiques*, n° 1, février 1991 : Naissance de l'aviation.

4. H.S. Wolko, *The Wright Flyer : An Engineering Perspective*, NASM, Smithsonian Institution Press, 1987.

5. D. Sobel, *Longitude*, Ediouro, 1996 (portuguese translation *Longitude*, Walker Publishing Co., 1995).

6. R.H. Thurston, *The Status of Aeronautics in 1884*. *Science, 5* (April 10, 1885), 295.

In the case of technological development, the process is reversed in that it seeks progress from the general to the specific. This requires other forms of thinking in which abstraction has little use. The driving force is the desired goal, be it the measurement of time in the high seas or the construction of a machine capable to fly. This difference in thinking and in approach made it impossible to eminent scientists like Galileo, Newton and Huygens, among others, to solve the problem of the measurement of time in ships, although they contributed with the notion of a linear time. By the same token, Lord Kelvin in the 19th century questioned the possibility of flying[7] with a heavier-than-air machine. It also made people with a better academic background like Samuel Pierpont Langley to fail, while bold and creative laymen succeeded.

Without developmental work based in fluid mechanics or systematic experiments designed with appropriate criteria, the aeronautical pioneers were left to exercise their creativity through naïve experiments which came to follow two trend lines. The first was the development of scale models that, although achieving some successes, did not always produce meaningful data. This occurred because in fluid dynamics as well as in resistance of materials changes in scale require special treatment. The other trend was the construction and testing of prototypes.

It is also interesting to note that no inventors were able to exploit the results obtained by Bernoulli in fluid mechanics in the 18th century. That could have simplified things enormously, but the equations cannot be directly applied to bodies of complex geometry. Another attitude was required : a mathematical approach with a high level of abstraction to establish the limits of application of Bernoulli's equations and to direct experimental conclusions.

However, an airplane is not an easily modelled object mathematically and the flight regime is not sufficiently regular to be treated in a mathematical way. Of course, after the success of an invention, it becomes much easier to re-examine it under a theoretical light and accelerate development with mathematical models, as the means of calibrating them and evaluating the validity of conclusions are at hand.

Anyhow, the experimentalism of the 19th century produced useful common knowledge relating to the essential characteristics of flight. Rudimentary wind tunnels and whirling arms helped to show that the lift generated by a surface moving through air grows with the angle of incidence until a maximum value is reached at the stall. Lilienthal produced tables ; Horatio Phillips and Langley studied airfoils and F.H. Wenham and J. Browning in 1871, and the Wright brothers in 1901 studied the lift of surfaces in wind tunnels and compared their results with Lilienthal's. Experiments with scale models demonstrated auto-stabilized flight. In 1871 Alphonse Pénaud flew a model airplane with built-in

7. T.D. Crouch, *Op. cit.*

stability and Langley flew successfully his unmanned Aerodrome over the Potomac in 1896. And so forth[8].

The flights by Lilienthal, accomplished in the short period of 1893-96 proved that the Cayley configuration supplied a good solution to the stability requirement. However, on August 9, 1896 his glider stalled and spun into the ground, fatally injuring him. His death cast a long shadow. The fear of the stall, produced by the excessive angle of incidence of the wing, seemed to require a new approach. The road to success was about to enter a long detour.

In the United States, Orville and Wilbur Wright introduced a horizontal front surface (elevator) intended to avoid the perceived danger of the stall[9]. This surface created instability during flight, but it avoided that the glider dropped its nose if the wing stalled due to excessive incidence or loss of speed. The successful flights of the Wright brothers set a new paradigm.

TESTING CRITERIA AND THE QUESTION OF TAKE-OFF

Another fundamental issue in the development of the aeronautics, the capacity of the machine to take-off under its own means, appeared prominently as one of the requirements of the testing criteria set to judge success. And, as in the case of the marine chronometer, the public test, scheduled in advance and judged by a previously nominated commission constituted the only generally accepted proof of success, in opposition to reports by witnesses.

Langley, however, resorted to the report of a credible witness to claim that his Aerodrome no. 5 demonstrated the possibility of flight[10] :

" No one could have witnessed these experiments without being convinced that the practicability of mechanical flight had been demonstrated.

Yours very truly,

Alexander Graham Bell

1331 Connecticut Avenue,

Washington, AD, May 12, 1896 ".

Differently from scientific progresses, where the main success criterion is acceptance by peers, testing criteria in aeronautics required the airing, in advance and for the knowledge of all interested parties, of the requirements to be met and which would form the basis for the judgement. Voisin synthesizes the most important points that should be satisfied[11] :

8. E. Petit, *Op. cit.* T.D. Crouch, *Op. cit. Les Cahiers de Science et Vie*, *op. cit.* M.J.H. Taylor and D. Mondey, *Milestones of Flight*, Jane's ed., 1983.

9. T.D. Crouche, *he Bishop's Boys : a Life of Wilburn and Orville Wright*, W.W. Norton and Co., 1989. H. Combs, *Kill Devil Hill. Discovering the Secret of the Wright brothers*, TernStyle Press Ltd., 1979. M.J.H. Taylor and D. Mondey, *Op. cit.*

10. S.P. Langley, *A successful trial of the Aerodrome*, *Science, 3*, n° 73 (1896), 753-754.

11. G. Voisin, *Op. cit.*

“ The first of all these demonstrations was the following : (…that the test must be accomplished…)

a) in the presence of an official body enabled by an homologating authority,

b) in calm weather and over a measured and controlled horizontal terrain,

c) leaving the ground, under its own means, at a previously determined point, with a man on board,

d) carrying on board the necessary sources of energy,

e) manoeuvring in altitude (straight line),

f) manoeuvring in direction (curves and circle),

g) returning to the starting point, everything without accident.

We accepted that neither height nor the duration of the tests would enter into consideration ”.

In 1903 the Aéro-Club de France instituted the Archdeacon Cup in the value of 1.500 francs, incorporating all the requirements above, except (f), to the first machine to fly 50 m. Flying in circles was considered as a stage to be reached later and would be the object of another prize.

It is worth noting that the requirement for take-off over level ground, in calm weather and under its own power represented the greatest technical challenge of the competition. It is not a detail but an essential point that involves the comprehension of what is meant by flight and how it takes place, and it demands a solution that cannot be considered intuitive.

In fact, to take-off an airplane is necessary to provide an adequate angle of incidence for the wing during the take-off run. In most of the contemporary designs, this required a torque to rotate the airplane in the traverse (pitch) axis. An upward force in front of the airplane or a downward force in the rear of the airplane can provide this torque. The first solution seems to have been preferred for two reasons : in first place, because it made the torque producing force to collaborate in the generation of lift.

In second place, this solution is coherent with a mental model[12] in which forces are parallel to the speed : the upward torque producing force moves the nose up. It is worth reminding that this reasoning does not find theoretical back up — being in fact a serious misconception — but it seems to be the one that prevailed at the beginning of the century.

Another important aspect is tied up to the subject of the stability. The adoption of a front elevator by all the pioneers may minimize the perceived danger of nosing down in a stall but it introduces inherent stability in the pitch axis that demands the pilot’s continuous action. As Santos-Dumont wrote in

12. C. Franco, H. Lins de Barros, D. Colinvaux, S. Krapas and G. Queiroz, “ Towards a model of mental models. Perspective on Models and Modelling ”, C.J. Convernors Boultier and J.K. Gilbert, *First Conference of European Science Education Research Association* (Rome, Sept. 2-6, 1997), 1-8.

1918[13] : " I struggled, at first, with the greatest difficulty to gain the complete obedience of the airplane ; in my first machine (14 bis) I placed the elevator ahead because, at that time, there was general faith in the need to do so. The reason was that, if it were placed behind, it would be necessary to force the tail down so that the machine could climb. There was some truth in that, but the control difficulties were so great that I had to abandon this disposition. It was the same as trying to throw an arrow with the tail first ".

The second alternative, placing the elevator behind, demands a down force at the rear contributing negatively to lift. On the other hand, it allows for stability in pitch and better control.

The conceptual evolution from a tail-first machine back to the original Cayley configuration cannot be accomplished through incremental development. Rather, the two conceptions are antagonistic and, in that sense, it is unreasonable to think that such a change could happen gradually. The canard (or tail-first) airplane was a fundamentally different conception from what we call today a conventional airplane[14]. It is a conceptual precursor but not a physical precursor, although both can perform the same minimum function.

ALBERTO SANTOS-DUMONT

Alberto Santos-Dumont was born in Brazil in 1873. Already in 1898, he was active in aeronautics, designing, building and flying his prototypes. As with other inventors, he did not possess a solid academic background having had some private classes for four years and attended, as a listener, a few other courses. However, he soon attracted attention when he flew the smallest spherical balloon until then built in Brazil[15].

He immediately devoted himself to the problem of the aerial navigation of balloons. As early as 1898, he built his first dirigible prototype powered by an internal combustion engine. Although it was not really successful, it did demonstrate the use internal combustion engines in hydrogen balloons without the risk of explosions, which had claimed already a number of lives. In the next 3 years he built four more dirigibles improving gradually their performance.

In 1901, he won the Deutsch de la Meurthe prize for aerial navigation by flying his dirigible n° 6 round the Eiffel tower and returning to the take-off point in less than thirty minutes. With that flight he won the acclaim of the aeronautical world : the aerial navigation of balloons was firmly established.

13. Early canards had no stabilizer that made them inherently unstable, requiring constant control inputs from the pilot. Successful tailless aircraft and flying wings depend on wing torsion and sweep so that wing tips counteract the nose-down torque generated by the airfoil, effectively functioning as stabilizers. There were few successful canards before computers and fly-by-wire technology opened the way to artificial stability. Therefore, it is most unreasonable consider early canards, in any way, as remote physical precursors of the Eurofighter !

14. *Ibidem.*

15. H. Lins de Barros, *Op. cit.* C. Dollfus, *Op. cit.* H.D. Villares, *Op. cit.*

From 1901 to 1905 he designed and built several other dirigibles. In 1903 he demonstrated the practical use of a non-rigid dirigible (n° 9), flying and demonstrating it countless times around Paris.

It seems that only in beginning of 1906 he turned to the problem of heavier-than-air flight. He was, it is true, informed of trials made in France with gliders and he was present at the tests made by Gabriel Voisin and Louis Blériot at the Seine with a glider towed by a boat, which suggested that 50 CV was the minimum power required for heavier-than-air flight. He knew the results found by Lawrence Hargrave in Australia, showing that box-kite structures could provide stability in two axes (rolling and direction). He actively interacted with the members of the Aéro-Club de France. He was a friend of important names of the period like Ferber, Voisin, Blériot…

He was certainly informed of the flight by Wilbur Wright at Kitty Hawk on December 17, 1903, as well as of the unsuccessful attempt made by Langley with his Aerodrome, only nine days earlier.

From 1906 to 1909, as we shall see, Santos-Dumont pursued the ideal of a practical and safe airplane, turning then to discussions on the future of aviation. Until his death by suicide in 1932, Santos-Dumont played a distinguished part in the history of aviation. But, not having patented any of his inventions, for thinking that the inventor's role is invention itself without any consideration of profits, his name has been fading slowly, inexorably, from the annals of aeronautics.

From June 1906 to November 1907

Santos-Dumont designated his projects with a simple sequential numbering system (n° 6, n° 14), in one case modified with a suffix (n° 14 bis). One difficulty when charting his progress is that he introduced a great number of changes after each test without changing the designations. It is therefore impossible to use his numbering to represent the learning curve of an invention. For example, n° 16 will appear now with an engine driving one propeller, now with two engines set up in another structure and driving two propellers. We deliberately simplified the description of each project keeping its essential elements and discarding smaller details[16].

Santos-Dumont n° 14 bis (Fig. 1)

Probably by mid 1906, Santos-Dumont began working in n° 14 bis. It was a biplane of canard configuration powered by a 50 CV Levaseur engine (the 25 CV engine initially installed was soon abandoned). Hargrave cells formed the wings as well as the elevator. The wings had a hefty 10° dihedral angle to guarantee stability in the longitudinal (rolling) axis. It possessed, in its initial

16. A. Santos Dumont, *O que eu vi o que nós veremos*, Ed. Autor, Petrópolis, 1918.

configuration, a landing gear with three wheels, soon simplified for two. Weight was about 300 kg (cell 115 kg, power plant group 125 kg, pilot 50 kg, gasoline and water 10 kg)[17].

Stability tests of n° 14 bis were performed during July 18-23, 1906, while hanging from dirigible n° 14 (incidentally, explaining the break in the numbering system). From July 26 to August 22 he suspended it by cables or placed it on top of a chariot for further testing. On September 3 he installed the 50 CV engine. During September 4-13 more trials were performed.

FIGURE 1

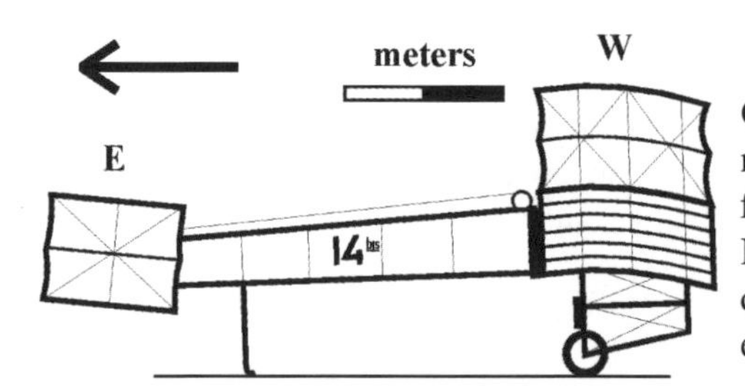

W=wings, E=elevator
Canard biplane Santos Dumont n° 14 bis. First homologated flight in 12 November 1906, in Bagatelle, Paris, France. It is a conceptual precursor of modern airplane.

Finally, on October 23, in the field of Bagatelle, Paris, he took off and flew for a distance of 60m at a height of about 3 m, winning the Archdeacon Coup. On November 12, 1906, again in Bagatelle, he flew a distance of 220 m at a height of 6 m in 21.2 s to win the Aéro-Club de France prize as the first homologated flight in the history of the aeronautics. On April 4 he flew n° 14 bis for the last time. After about 50 m the airplane entered into oscillations and dropped, being destroyed in the crash[18].

Santos-Dumont n° 15 (fig. 2)

FIGURE 2

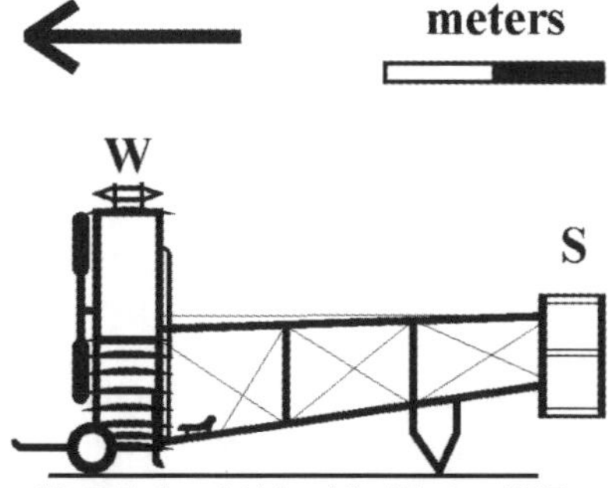

W=wings, S=stabilizer
A physical precursor of airplane, the n° 15's landing gear was the reason it could not take-off. First tried on 21 March 1907.

17. F. Ferber, *L'Aérophile*, Février, 1907.

18. R. Vitrotto, *Histoire des techniques : Une Étude Systématique sur le XIV bis de Santos Dumont. Pégase*, n° 29-30 (juin 1993), 12-28. P. Lissarrague, *Histoire des Techniques : Une Étude Systématique sur le XIVbis de Santos Dumont. Pegase,* n° 31 (septembre 1983), 4-30. F. Peyrey, *Les Oiseaux Artificiels*, H. Dumont et E. Pinat éditeurs, 1909, 265-449.

Keeping the structure of Hargrave cells for the wings and the elevator, n° 15 had an inverted configuration in regard to n° 14 bis : the wings were now set ahead of the elevator which took on the role of a stabilizer. The narrow chord wings were built in lacquered wood with steel fittings and piano wires for rigidity. The aluminium and steel propeller attached directly to the 50 CV Levavasseur engine as in n° 14 bis. It used only one wheel as landing gear. Tests, starting from March 21, 1907, failed to achieve flight[19].

The analysis of n° 15 shows that Santos-Dumont understood the stability problem of n° 14 bis and adopted the appropriate correction of inverting the configuration. The project failed because the shallow ground angle prevented the wings to achieve the required incidence during the take-off run. However, Santos-Dumont did not diagnose correctly the problem and reacted with a retrograde step, back to the balloons and dirigibles that he was familiar with. The n° 15 was a physical precursor of the modern airplane because the only restriction for success was the inadequate design of its landing gear.

Santos-Dumont n° 16 (Fig. 3)

FIGURE 3

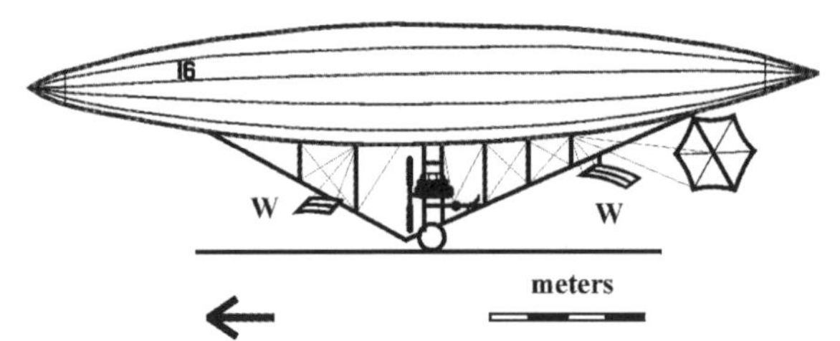

W=wings
n° 16 hybrid tried to ease take-off with the help of a hydrogen balloon. Tested in June 1907.

Worried with the failure of n° 15 Santos-Dumont tried a hybrid solution, combining a 99 m^3 hydrogen balloon with a lift of 110 kg with a structure containing wings, elevators and a 50 CV engine. Total weight was 190 kg[20]. The craft was heavier-than-air and should not be mistaken for a dirigible. Santos-Dumont had already resorted, in at least two other occasions, to hybrid attempting to improve lift : balloon n° 13 used an apparatus for heating the gas and his balloon *Les Deux Amériques* of 1906 used a group of horizontal propellers. In none of these instances did he achieve success.

He was looking at the wrong direction. The take-off of an aircraft does not relate lift to weight in a simple way. The question revolves on how to accelerate sufficiently and then to gain lift by increasing the angle of attack. n° 16 was a test-bed that only showed the degree of its inventor's ignorance in regard to heavier-than-air flight. It was a conceptual precursor that did not reach its minimum function. It was certainly not a physical precursor.

19. P. Lissarrague, *Op. cit.* F. Peyrey, *Op. cit.*
20. *L'Illustration*, n° 3.355 (15 juin 1907), 400.

5.4. Santos-Dumont n° 17 (fig. 4)

FIGURE 4

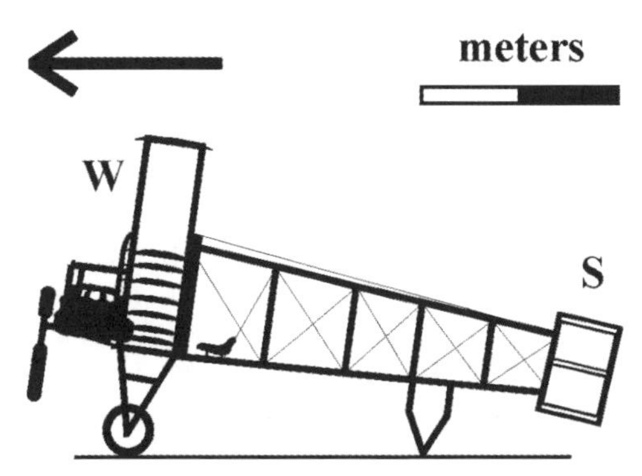

W=Wings, S=stabilizer
As nearly as can be ascertained, this was the appearance of n° 17, an improvement of n° 15. It is a physical precursor of the airplane, with adequate landing gear but frail wing structure. It did not satisfy the basic function of an airplane.

After the failure of n° 16, Santos-Dumont went back to his previous project. Scarce information is available on n° 17, perhaps Santos-Dumont's least documented design. It was a modification of n° 15 that attempted to increase lift by doubling engine power (50 to 100 CV) and increasing take-off speed. Santos-Dumont was also forced to design a new and stalkier landing gear because the lower positioning of the engine required greater propeller clearance[21].

Unfortunately, tests failed miserably. The narrow chord wings did not have adequate torsional rigidity to absorb the power and the airplane simply disintegrated. That was doubly regrettable as, perhaps unintentionally, the new undercarriage gave a built-in ground angle that allowed adequate wing incidence during take-off. With n° 17 Santos-Dumont arrived at the optimum configuration for successful controlled flight.

Although it did not accomplish its basic function, n° 17 can also be considered as belonging to the line of physical precursors of the airplane. Better knowledge of the dynamics of flight would have led, without a doubt, to improving the structural rigidity. Instead of doing that, for the moment, Santos-Dumont took yet another detour.

21. H. Lins de Barros, *Op. cit.*

Santos-Dumont n° 18, the " Hidroglisseur " (fig. 5)

FIGURE 5

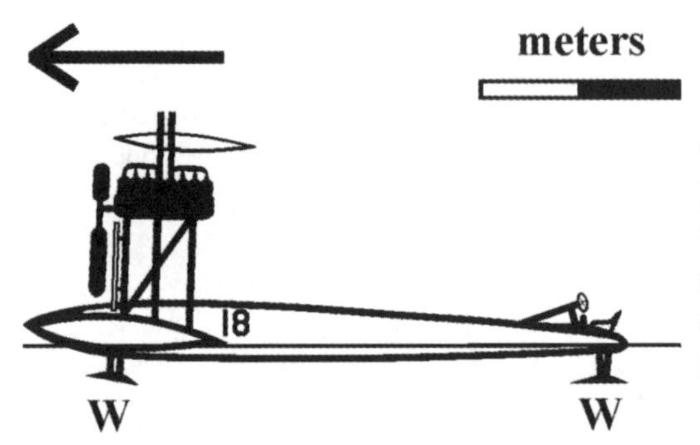

W=submerged wings (rear one movable) The *Hidroglisseur*, Santos-Dumont n° 18, a study of " flight " in water. Note the seat, mounted at the extreme rear, and the out-rigger pontoons.

In n° 18 Santos-Dumont seems to have abandoned, for the moment, the quest for flight. The *Hidroglisseur* was a boat with the same engine as n° 17. Tandem wings were submerged, in the way of a modern hydrofoil[22]. The rear one was movable. In this way, he intended to further investigate lift and stability in a fluid with higher viscosity than air.

A boat towed n° 18 on the Seine under the watchful eyes of Blériot, Ferber, Delagrange, Cody, Voisin and others. Apparently, it was never run under power. Of course, the real issue that had prevented success with n° 15 was not lift but lack of incidence in take-off. However Santos-Dumont must have profited from the tests for his next design was a winner.

Santos-Dumont n° 19, the first " Demoiselle " (fig. 6)

FIGURE 6

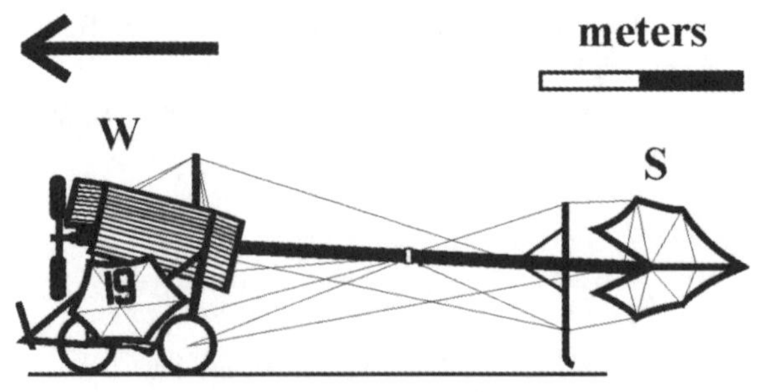

W=wings, S=stabilizer The first Demoiselle, Santos Dumont n° 19. Span 5m, length 8m, Dutheil et Chalmer 20 hp two cylinder horizontally opposed engine.

22. *Ibidem.*

First flight in November 16, 1907. The first invention that met all the requirements of a modern airplane.

Combining the knowledge gained in his five previous designs, Santos-Dumont now built the world's first really practical airplane. The n° 19, called the *Demoiselle*[23], finally realized the configuration proposed by Cayley back in the early 18th century : a monoplane with cruciform tail at the rear of a long tubular fuselage. The engine was installed in a cut-out in the leading edge of the wings, which had generous dihedral to ensure longitudinal stability. The pilot sat below the wings on a 3-wheeled trolley. The landing gear allowed for ample propeller clearance and adequate wing incidence in take-off. Initially, Santos-Dumont placed auxiliary horizontal and vertical surfaces ahead of the wings, a gesture towards current mental models. They were soon deleted as he found that they contributed neither to stability or lift.

The n° 19 *Demoiselle* first flew on November 16, 1907. Small, fast, stable and safe, looking very much like a modern ultra-light, n° 19 performed flawlessly. From it, the 1908 n° 21 *Demoiselle* was developed and copied throughout Europe up to 1910. The n° 19 should be regarded as the first modern airplane.

CONCLUSIONS

The precedence for the first successful flight has been the object of passionate discussions, more governed by patriotic considerations than by objective scrutiny. Several pioneers, in particular Santos-Dumont, have challenged the claim for the Wright Flyer's flight on December 17, 1903, based on the absence of public proof and the failure to meet established criteria (unassisted take-of, calm air). In fact, all Wright airplanes until 1910 required some form of assisted take-off and their configuration was inherently unstable in the pitch and rolling axes (no stabilizer, negative dihedral).

Santos-Dumont's n° 14 bis also lacked a stabilizer and was unstable in pitch. While both the Flyer and n° 14 bis flew under control and satisfied the minimum function, they cannot be considered as physical precursors of the modern airplane because of their configuration.

But Santos-Dumont's continued determinedly to pursue practical and his designs show consistent progress towards his goal. Both n° 15 and 17, which had stabilizers but did not fly, were clearly physical precursors that did not meet the minimum function. In n° 16, the balloon-airplane hybrid, Santos-Dumont tried unsuccessfully to reformulate his mental model based on his former experience in dirigibles. The n° 18 *Hidroglisseur* immersed wings in water to understand lift and stability. Finally, all problems were solved in n° 19

23. *Ibidem*. F. Peyrey, *Op. cit.*

Demoiselle. After the success of n° 19, Santos-Dumont launched his second generation *Demoiselles* in 1908, sturdier and more developed designs that would leave their mark on the first years of aviation.

An oft-repeated question remains : who invented the airplane ? There may more than one valid answer, depending on how one understands flight, or in other words, how the minimum function is defined. Nowadays, after airplanes became commonplace, we tend to drop the requirements for public proof and unassisted, calm air take-off and we consider flight, simply, as moving under control through the air. Precedence for the first flight, in this sense, stays firmly with the Wrights.

If public proof and take-off criteria are required, as contemporaries of the pioneers did, then the first airplane to fly was n° 14 bis on November 12, 1906, as attested by the first speed and distance records registered with the FAI — *Fédération Aéronautique Internationale*[24].

However, we have tried to demonstrate that neither of these airplanes can be considered as the first modern airplane[25]. Other answers may be ventured. Santos-Dumont's nos. 15 and 17 had the right configuration but did not fly (so they can claim to be physical precursors that did not meet the minimum function).

On October 26, 1907, Voisin's n° 1 piloted by Farman gained the speed and closed circuit records from the *FAI*. It represented yet another solution for the stability problem : movable forward elevators and rear mounted fixed stabilizer. It is somewhat of a hybrid and it is questionable if it can be considered as a physical precursor. But fly it did.

The Santos-Dumont n° 19 *Demoiselle* which first flew on November 16, 1907 is unquestionably representative of the configuration of the modern airplane. We claim it is the best candidate for the title of the first modern airplane.

24. K. Munson, *Jane's Pocket Book of Record-breaking Aircraft*, J.W.R. Taylor editor, Collier books, 1981.

25. *Ibidem*.

A PRE-STUDY FOR THE HISTORIOGRAPHY OF THE PHILIPS RESEARCH LABORATORIES

Marc J. DE VRIES

INTRODUCTION

AIM OF THE HISTORIOGRAPHY OF THE PHILIPS RESEARCH LABORATORIES

Historiography has a well-established place within Philips. Four volumes of the company historiography have already been published by Heerding and Blanken ; various jubilee years were a reason for bringing out more popular descriptions of periods in the history of the company. Some attention has been devoted to the Research Laboratories in those publications, but a comprehensive historiography of the Laboratories is not available yet. Such a historiography does exist for other, similar research laboratories, like the Bell Labs and the GE Labs. The Philips Research Laboratories has its own particular position within the company, just like other research laboratories have that in their own companies. For an insight into the role of scientific research within such companies it is useful to have a separate historiography of the research laboratories, in particular as the position of the research facilities is debated seriously in our time.

It is also of importance for the historiography of technology in the 20th century that a historiography of the Philips Research Laboratories is written because of the importance of the emergence of the industrial research laboratories in this century. In the TIN-20 project technological developments of the 20th century in the Netherlands are described. The historiography of the Philips Research Laboratories will contribute as a resource for that project. The project as a whole yields information about the economical and social context for technological developments in the Netherlands. Hence, the project as a whole is a useful resource for the historiography of the Philips Research Laboratories. This way mutual relationship can be seen between the two projects.

Early in the 20th century various industrial research laboratories are initiated. This is partly caused by the fact that the new technological developments in that time depend on knowledge of microscopic phenomena more strongly than the technological developments before that time. Thus a need emerges for research into such phenomena in order to be able to use scientific knowledge in the development of new technologies. Noble in his book *America by Design* mentions the years 1880-1920 as those in which the " wedding of science to useful arts " takes place. In the title of one of his chapters he uses the term " the rise of science-based industry ". In particular in the electrical and chemical industries this happens. For the Netherlands, Hutter has shown that particularly in the period 1900-1930 industrial research laboratories are started. In those labs, research is not limited to microscopic phenomena only, but also deals with the more classical scientific theories (classical mechanics, classical thermodynamics and classical electrodynamics).

The industrial research laboratories through the years have struggled to find a place of their own in the scientific, technological and social developments. On the one hand they were places where new scientific knowledge was developed, for which interaction with other research institutions such as universities was important. On the other hand their scientific activities had to be embedded in the commercial interests of the company of which they were a part and where technological and economical factors were of importance too. Another complicating factor is the fact that the development of new technologies often required a combination of various disciplines and fields of research.

A historiography of the Philips Research Laboratories can illustrate how these labs dealt with these problems and how changes in the scientific, technological and social context influenced the policy, programme and practice of the labs. If that is to be the case a condition is that such a historiography pays attention to :

- the contact of the labs with the scientific world outside the labs and the way this was used to optimise the knowledge level of the labs,
- the relationships with the other industrial activities within the company aimed at making the developed knowledge fruitful for developing new products and processes,
- the influence of changes in the economical and social context of the company on the activities of the research labs,
- the way the research policy was defined based on considerations about those relationships and influences,
- the transformation of that policy into concrete research programmes and activities, including the co-operation of various disciplines and research fields. That implies that a description of the population of the labs (what disciplines are represented and how are they distributed over the labs) should be part of the historiography too.

Because of the two motives that were mentioned (the relevance of the Research Labs within Philips and the relevance of Philips for the technological developments in the Netherlands) a historiography will be written that has as its main research question : what was the contribution of the Philips Research Laboratories in the course of its history ? In the current mission statement of the Philips Research Laboratories we find a number of contributions of the research to the company :

- to generate options for successful industrial innovation,
- to take care of timely transfer of technical results to the Product Divisions and to initiate new businesses within the scope of the company,
- to help establish a strong patent position,
- to develop a portfolio of key technologies for future products,
- to add value through synergy,
- to serve as a source of highly skilled scientists and engineers for the company,
- to be the objective technical conscience of the company,
- to maintain effective interfaces with the Product Divisions and the scientific world and other industrial research laboratories.

These missions will be used as a checklist for possible contributions of the Philips Research Laboratories, whereby the historical research will show how the role and relevance of these missions changes through time.

PHILIPS AND PHILIPS RESEARCH

Philips started in 1891 in Eindhoven, the Netherlands, as a light bulb producing company. Since then its products range has widened substantially to many fields. From light bulbs it was not a big step to various sorts of tubes (e.g. amplifier tubes, X-ray tubes) and from tubes the next logical step was to develop whole systems, of which tubes were a device (e.g. radio, television, X-ray equipment) and thus Philips became involved in the whole process of processing (transmitting and receiving telecommunication signals). Devices remained an important area for Philips, especially in terms of transistors and later ICs. Once involved in sound and sight, Philips also started developing various technologies for recording such signals (gramophones, magnetic tapes, optical recording). Philips is now a multi-national company with a turnover of about 70 billion Dutch guilders (approx. 35 billion Am. dollars) and about 250 thousand employees (Annual Report 1996).

Philips in 1914 decided to create a research programme that was formally independent from the industry groups within this company. In practice this meant that the laboratory could set its own priorities for doing research and determine its preferences for different types of research (" phenomenon-oriented " or " fundamental " versus " application-oriented ", " long-term " versus " short-term "). This is essentially different from a situation where research is done by a variety of laboratories that are embedded within the industry groups. By the end of the period we describe here, in 1990, the situation was as follows. The corporate research programme was carried out by a laboratory in Eindhoven, the Netherlands (the oldest one and still the largest, the so-called " Nat. Lab. ") and other laboratories in Belgium, France, Germany, the UK and the USA. Outside the corporate research programme, development laboratories within the various industry groups are focused on the " D " of " R&D ", although the separation between " R " and " D " in the company can not be sharply defined. The Lighting group is the only product division that has its own large research facility, which caused a different relationship with corporate research than in the cases of the other industry groups. The total amount that was spent on R&D, both in the corporate research programme and in the development laboratories within the industry groups, through the years varied from about 7 to 8 percent of the company's total turnover (about 1 percent of the total turnover is allocated to corporate research). The research topics range from phenomena oriented materials research to research into mechanics, all sorts of electronic devices, IC-design and -technology, systems, and software.

In the early years of the laboratories, ferrite's and other magnetic materials were an important research field and in particular the work on ferrite's has yielded very positive outcomes, because they were applied successfully in all sorts of products as magnetic materials (e.g. loudspeakers, antennae, magnetic coils in televisions). Several Main Industry Groups have gained from the transfer of this research to them and for a long time there was continued research support for working on applications. Later on the attention shifted to semiconducting materials. In the beginning of the period we describe here, this was already seen as the main issue in the " materials " area.

For the " devices " area there was again a " natural " partner in the industry groups, namely Elcoma (ELectical COmponents and MAterials). The fact that devices were produced by a separate Main Industry Group sometimes caused specific problems for transferring research outputs in this area. On the other hand, in the development of ICs contacts on the worker's level often was good. It could be expected that the devices research would draw from the materials research. In fact the possibility of transferring knowledge from different disciplines is one of the most important reasons for having a central, corporate research programme. This transfer indeed did take place, although the two fields remained separated and the relationships were not always clear even to the researchers.

In the " systems " area the systems research by its very nature had to have close contacts with industry groups, because systems are directly related to users and therefore technology transfer could be expected to be easier in this area. There was also contact with the " devices " and " materials " areas. As in the contact between " materials " and " devices ", here too we find examples of " walls " between the areas. Thus it is told that in the devices group started working on the Video Long Play (VLP) even though the systems people (the " natural " place because the VLP of course was a system) had already decided not to start working on this topic. When the ICs became very complex and in fact were systems in themselves, as well as parts of a larger system, the boundaries between the " devices " and " systems " areas became fuzzy.

It can be questioned if the (quasi-)disciplinary division into " materials ", " devices " and " systems " research creates a barrier for transfer to industry groups, that are not disciplinary, but product oriented. Other industrial research laboratories have made different choices here. The programme of the GEC Labs for example was changed in 1960 from a discipline-oriented (physics, telecommunication, chemistry/engineering, mark the similarity with the Philips Research programme structure) to a more product-oriented structure. At the same time we see that there are several topics (of which ICs is the most striking example) for which the practical meaning of the division into " materials ", " devices " and " systems " is hard to see and hence the difficulties in connecting research to development, caused by the seemingly discipline oriented structure of the research programme, should not be over-estimated.

Existing historiographies of industrial research laboratories

As there are only few research laboratories of the same size and scope as the Philips Research Laboratories, there are only few historiographies that can serve as a standard for the historiography of the Philips labs. The existing historiographies are quite various in nature.

The most extensive one probably is the one about the Bell Labs, that consists of seven impressive volumes that came out in the years 1975-1985 and were edited by M.D. Fagan. They focus strongly on individual research projects and not much is told about the research policy behind those projects. They also lack a clear periodisation, nor is there a description of the people that worked in the various projects and their background. The description is mainly based on publications of the Bell people and interviews.

Quite different in nature are the historiographies of the GE Labs by G. Wise (entitled *Willis R. Whitney, General Electric and the origin of US industrial research*) that came out in 1985, and L.S. Reich's *The making of American industrial research. Science and business at GE and Bell, 1876-1926*, that came out in the same year. The first book focuses on the person of Willis R. Whitney, but also gives some insight into the activities of the research lab as a whole. The second book describes the role of the research labs of both companies (GE and Bell) in their respective companies and in American industry in general. There is also a historiography of Du Pont's research labs by Hounshell and Smith, that came out in 1988. This book also pays a lot of attention to the general industrial policy of the company and the research policy of the labs as related to that.

In summary : there are a number of historiographies of comparible industrial research labs, that vary quite much in nature.

Sources for a historiography of the Philips Research Laboratories

In general the sources of the historiographies of industrial research labs are : the scientific publications that came out in the course of time, archive material (like letters, reports of meetings, other reports about specific issues), and interviews. By combining data from several of such sources a reliable description is strived for. As secondary sources we often find later publications about parts of the history of the labs or about parts of the research programme (like survey articles, previous historiographies, biographies of researchers or specific case studies of research topics or products).

For the historiography of the Philips Research Laboratories the following sources are available :

- all issues of the Philips Technical Review (1936-1943 and 1946-1989) with extensive indexes both author oriented and subject oriented,
- all issues of the Philips Research Reports and the Philips Journal of Research (1945-now),
- registers of publications in other journals and registers of internal reports (1914-now),
- some special survey publications for the occasion of a jubilee year,
- the archive of the Philips Research Laboratories that is a separate part of the Philips Company Archives,
- a collection of interviews about the pre-WWII years,
- the Nat. Lab. (=Dutch abbreviation for the main lab in Eindhoven) Technical Notes on various research topics, that are published since about 1970,
- reports of the Concern Research Conferences and the Research Directors Conferences (since 1950),
- reports of the meetings between the directors and representatives of the lab population,
- catalogues of the Concern Research Exhibitions, in which research activities were displayed to the Main Industry Groups (later called Product Divisions)
- the Concern Research Programmes (the so-called " Blue Books ") that contain the names of all research groups, their members and activities (since about 1960).

Other, secondary sources are :

- the historiography of the company as a whole by Heerding and Blanken (now covering the history of Philips until about 1950),
- autobiographies of Frits Philips and Hendrik Casimir in which several chapters deal with research within Philips,
- an internal historiography of the period 1914-1950 by Garrath that was finished in 1976 but never published (it did not contain any references to sources,

although evidently a lot of archive material and interviews must have been used),

- publications on case studies : gas discharge lamps (by Hutter), the Compact Disc (by Ketteringham and Nayak in the book *Breakthroughs*), the Stirling engine (by Hargreaves), the ferrite's (by Hoitzing), the transistor (by Verbong), FM (by Tielen), the Plumbicon (Sarlemijn and De Vries).

Besides these sources there is a lot of material that has been used for displaying in a historical exhibition about the labs in 1989 when the labs existed 75 years.

For the historiography a new series of interviews will be held with : three of the representatives of Research in the Philips Board of Management (Casimir, Pannenborg and Van Houten ; Holst has died already), thirteen former directors and co-directors, two researchers with special responsibilities (for human resource management and for the Philips Concern Research Bureau), three research assistants, five others from the Main Industry Groups that have been in contact with research as external parties.

In summary, a variety of sources is available for writing the history of the Philips Research Laboratories.

THE OUTLINE OF THE HISTORIOGRAPHY

a. Periodisation

The history of the Philips Research Laboratories will be described according to the following periods :

- 1914-1946 : this is a period in which the research programme to a large extent was directly linked to Philips' industrial activities and in many cases was the origin of new activities. In this period Gilles Holst was in charge of the research lab,
- 1946-1972 : in this period the research scope was extended quite substantially, new labs were started or bought (in Belgium, France, Germany, the UK and the USA) and there was a strong interest in taking up research into new phenomena and the relationship with the Main Industry Groups that had been created in the post-WWII years was rather weak. The economical circumstances were very favourable for industry and there were a lot of new business opportunities so that research got a lot of freedom to look for such new opportunities. In this period Hendrik Casimir first served as a director of the lab and later represented the research programme in the newly formed Board of Management,
- 1972-1990 : in this period gradually the relationship between Research and the Main Industry Groups was strengthened, mainly because of the less favourable economical circumstances that forced Philips to make more careful choices in the development of new products. Gradually certain coherence

between the various labs (the main lab in Eindhoven, the Netherlands and the labs in the other countries) grows. In 1989 there is an important change in the way of funding for the labs (introduction of contract research) that serves as a " natural " point to finish the historiography, because then a new period starts. In this period Eduard Pannenborg and George van Houten are the Board of Management members with Research in their portfolio.

Thus the periodisation combines the change of social and economical circumstances with a change of highest representative for the research laboratories. Evidently the persons that have served as such (Holst, Casimir, Pannenborg and Van Houten) each quite nicely fitted with these circumstances (Holst was very much interested in practical applications for research, Casimir was a famous physicist, Pannenborg was an engineer with an entrepreneurial attitude and Van Houten came from outside research and thus was able to take the dramatic step towards contract research that made Research partly dependant from the Product Divisions).

b. Themes in the description

In the introduction some remarks have already been made with respect to themes that should be included in the historiography. A more comprehensive list of themes to be described, is the following :

I. the content of the research programme (in terms of diversity, scientific content, projects, multidisciplinarity and capability portfolio). The range of research topics that have been dealt with in the course of time is quite large and comprises light sources, biology, X-ray, telephone, radio/radar, Stirling engines and refrigerators, magnetic materials, television and displays, sound, magnetic and optical recording, lasers, glass fibres, passive components, ICs, electron microscopy, nuclear physics and production techniques and automation ; in the last decades of the history, this programme was structured according to three main groups : materials (mostly chemists), devices (mostly physicists), and systems (mostly electrical and mechanical engineers),

II. the relationship between the various labs (the main lab in Eindhoven, the Netherlands, and the lab in Belgium, in France, the two labs in Germany, the lab in the UK and the two labs in the USA),

III. the input from outside (scientific input from universities and other research labs, and to a small extent market input from Main Industry Groups),

IV. the relationships with the Main Industry Groups (later : Product Divisions) in terms of transfer of new products,

V. the relationship with the Main Industry Groups in terms of transfer of improvement of existing products and processes,

VI. contribution to the patent position of the company,

VII. transfer of people from Research to Main Industry Groups,

VIII. practice of the work as experienced by the people.

Themes I, II and VIII are process-oriented, theme III is input-oriented, and themes IV, V, VI and VII are output-oriented. The themes have been derived from the current mission statement of Philips Research (see section 1 of this paper). These are not dealt with as an ideal but as a checklist of possible missions, of which the role changes in the course of time.

So far the impression is that for all these themes the periodisation, that has been chosen, is a relevant one.

c. Case studies

To illustrate the main characteristics of the policy, the programme and the practice of the Research Labs in each of the periods, a number of case studies will be carried out. In making the choice for the case studies the following criteria will be used :

- the case studies should represent the main parts of the research programme,
- they should cover both successes and failures,
- they should be related to a variety of product categories (both professional and consumer products),
- they should be illustrative for the main characteristics of the period in which they belong,
- sufficient sources should be available to enable a reliable description.

So far the following options for case studies have been identified :

- 1914-1946 : gas discharge lamps, X-ray tubes, ferrite's,
- 1946-1972 : the Plumbicon television pickup tube, the Stirling engine, LOCal Oxidation of Silicon (LOCOS, a process for producing ICs),
- 1972-1990 : optical recording (Video LP and CD), optical communication system for the city of Berlin, new type of projector lamp.

PLANNING OF THE STUDY

The study is carried out according to the following schedule.

In the period December 1996 to April 1997 the main themes for the description of the history of the Philips Research Laboratories are defined by studying the reports of the general management meetings and by interviewing a number of former research directors.

In the period May 1997 to March 1988 the data are gathered for the actual description of the main themes of this historical research, as they have been defined in the previous months. For this purpose reports of other types of meetings, e.g. meetings between the research directors and the Main Industry Group directs, will be studies and more interviews will be held, not only with people from Research, but also from Main Industry Groups.

In the period April 1988 to December 1988 the data are gathered for the description of the case studies. Here we will make use of archive material about the specific products of the case studies and have some additional interviews.

In the period January 1999 to December 1999 the book of the history of the Philips Research Laboratories is written. The book is to come out in the year 2000.

The book, however, should be seen as only one of the outcomes of the project. In fact the first aim is to build up an archive with relevant materials about the history of the Philips Research Laboratories, that can serve as a resource for later research studies as well as for writing the historiography of Philips Research in the mentioned book. This archive will remain at the facilities of the Philips Research Laboratories during the historiography project, but later probably will be transferred to the central Philips Company Archives.

The research is carried out by a senior researcher, a junior researcher and a Ph.D. student. The tasks are divided as follows :

- the senior researcher will be responsible for the development of the main themes of the historiography, as well as the description of those themes and all case studies, and the data gathering for the last two periods in the historiography (the periods " Casimir " and " Pannenborg/Van Houten "), as well as for three case studies,
- the junior researcher will gather the data for three case studies,
- the Ph.D. student will do a research of his own for his doctoral dissertation, with a main research question that can differ from the themes in the total historiography, but in the meantime he will gather the data for describing the main themes in the first period of the historiography (the period of " Holst ") and for the three case studies of that period.

During the whole project a project team will guide the project. This team consists of :

- prof. dr. Harry Lintsen, professor of technology history at the universities of Delft and Eindhoven,
- dr. Feye Meijer, former of the Philips Corporate Research Bureau, who after his retirement also plays a vital role in collecting the data,
- prof. dr. Emile Aarts, group leader in Philips Research and professor at the Eindhoven University of Technology,
- dr. Marianne Vincken, responsible for the public relations of Philips Research,
- drs. H. van Bruggen, from the Philips Company Archives.

The team meets with the researchers about every two months to discuss the progress of the research.

In addition there are contacts with the Dutch project TIN-20, that will describe the history of technology in the Netherlands in the Twentieth Century.

In this project there is a sub-theme called : knowledge infrastructure. This sub-theme will deal with the emergence of industrial and other research laboratories in the Netherlands. The historiography of the Philips Research Laboratories of course fits quite well into this sub-theme and therefore mutual transfer of outcomes of the TIN-20 project and the Philips Research historiography project is expected. An informal group of researchers, most of whom are involved in the TIN-20 projects, regularly meets to discuss both projects.

REFERENCES

G. Bekooy, *Philips honderd*, Zaltbommel, Europese Bibliotheek, 1991.

H.B.G. Casimir, *Haphazard reality ; half a century of science*, New York, Harper and Row, 1993.

R. Chandler, *The visible hand*, S.l., Belknap Press, 1977.

M.D. Fagen, *A History of Engineering and Science in the Bell System*, Indianapolis, AT&T Bell Laboratories, 1975-1985 (7 volumes).

A.J. Garrath, *The story of the Philips Laboratory at Eindhoven 1914-1946*, Confidentieel intern Philips-rapport, 1976 (2 volumes).

S. Gradstein en H.B.G. Casimir, *An Anthology of Philips Research*, Eindhoven, Philips Gloeilampenfabriek, 1966.

C.M. Hargreaves, *The Philips stirling engine*, Amsterdam, Elsevier, 1991.

A. Heerding en I.J. Blanken, *Geschiedenis van de N.V. Philips Gloeilampenfabrieken*, Leiden, Martinus Nijhoff, 1980-1997 (4 volumes).

A.H. Hoitzing, *Ferrietonderzoek op het Philips Natuurkundig Laboratorium* (unclassified Report nr. 006/92), Eindhoven, Philips Natuurkundig Laboratorium, 1992.

D.A. Hounshell en J.K. Smith, *Science and Corporate Strategy : Du Pont R&D 1902-1980,* Cambridge, Cambridge University Press, 1988.

J.J. Hutter, *Toepassingsgericht onderzoek in de industrie. De ontwikkeling van kwikdamplampen bij Philips 1900-1940,* Eindhoven, Technische Universiteit Eindhoven, 1988.

J.M. Ketteringham en P.R. Nayak, *Breakthroughs !* , New York, Rawson Associates, 1986.

D.F. Noble, *America by Design. Science, Technology and the Rise of Corporate Capitalism*, Oxford, Oxford University Press, 1979.

F. Philips, *45 jaar met Philips*, Rotterdam, Donker, 1976.

L.S. Reich, *Science and Business at GE and Bell, 1876-1926,* Cambridge, Cambridge University Press, 1985.

A. Sarlemijn, *Tussen academie en industrie. Casimir's visie op wetenschap en researchmanagement*, Amsterdam, Meulenhoff Informatief, 1984.

A. Sarlemijn and M.J. de Vries, " The piecemeal rationality of application oriented research. An analysis of the R&D history leading to the invention of the Philips Plumbicon in the Philips Research Laboratories ", in P.A. Kroes en M. Bakker (eds), *Technological development and science in the industrial age*, Dordrecht, Kluwer Academic Publ., 1992.

G. Verbong, *De ontwikkeling van de transistor bij Philips* (doctoraalscriptie), Eindhoven, Technische Hogeschool Eindhoven, 1981.

M.J. de Vries, “ The Philips Stirling engine development : a historical-methodological case study into design process dynamics ”, *Methodology and Science*, Vol. 26 (1993), 74-86.

G. Wise, *Willis R. Witney, General Electric and the Origin of US Industrial Research*, New York, Columbia University Press, 1985.

Histoire, interaction et union de la technique électronique et de l'optique physiologique

V.M. Tchesnov - M.G. Yarochevsky - M.A. Ostrovsky

Logique du développement de la science moderne. Biologie et technique

L'observation rétrospective nous permet aujourd'hui, grâce au recul dont nous bénéficions, de repérer des unions imprévues entre sciences et techniques. La rencontre de l'optique physiologique et de la technique électronique constitue de ce point de vue un exemple très intéressant. Cette union de deux branches assez différentes de l'activité intellectuelle humaine apparaît comme seulement un des résultats possibles de leur interaction. Nous pouvons déterminer les caractéristiques essentielles et les directions principales de ce processus.

Le principe essentiel qui a déterminé les formes d'interaction entre la technique électronique et la biologie est le suivant. L'homme qui en est le promoteur réel se rend compte, soit très clairement, soit intuitivement, que le système biologique compris comme partie de la Nature est toujours meilleur qu'un objet artificiel. Il est plus perfectionné du point de vue du transport de l'énergie, du traitement de l'information, et de la prise des décisions, et il est plus complexe et achevé d'un point de vue " technologique ". Cette situation a pour résultat que le processus d'interaction s'est orienté vers le perfectionnement de techniques basées sur les observations effectuées lors de l'étude des systèmes biologiques (naturels). L'objet biologique joue toujours le rôle de la roue motrice et la technique peut être comprise comme une roue commandée dans le processus d'interaction entre la biologie et les techniques.

Cependant, le processus n'est pas totalement unidirectionnel et des rétroactions se sont aussi produites. En particulier, ce sont des systèmes techniques de plus en plus perfectionnés qui ont permis de comprendre de mieux en mieux l'organisme vivant.

Une telle conception nous permet de distinguer trois niveaux dans l'interaction de la biologie et de la technique.

1. Interaction au niveau auxiliaire. Des installations électroniques exercent les fonctions de service lors des expérimentations des physiologistes. Parfois leur rôle devient déterminant dans le succès d'une recherche biologique. C'est par exemple le cas à la fin des années 1960, début des années 1970, quand l'expérimentation biologique fut révolutionnée par l'équipement électronique.

2. Interaction au niveau de système. Un objet biologique ou un de ses sous-systèmes sont considérés comme système indépendant. Les principes de son fonctionnement et de son organisation forment le fondement pour l'élaboration de techniques électroniques.

3. Interaction au niveau technologique. Elle peut suivre deux voies : 1) utiliser en technique des composants biologiques " vivants " ou, au contraire 2) construire des installations techniques analogues d'après leur destination aux " organes " optiques biologiques.

La première voie s'est révélée la plus intéressante parce qu'elle a permis d'obtenir des succès tout à fait inattendus. Elle a ouvert la voie à des avancées technologiques extraordinaires. D'habitude, le point terminal et le résultat de l'interaction à ce niveau ne peuvent pas être déterminés d'avance. Il est tout de même possible de les pronostiquer en prenant pour base l'analyse historique du développement de la biologie et de la technique.

INTERACTION DES BRANCHES DE LA SCIENCE ET DE LA TECHNIQUE. PHYSIOLOGIE DE LA VISION ET TECHNIQUE ÉLECTRONIQUE

La physiologie de la vision étudie les processus de la vision, c'est-à-dire les processus de réception de l'information sur le monde environnant par les organismes vivants au moyen de l'enregistrement des rayonnements électromagnétiques de gammes d'ondes de 300 à 800 nm réfléchis ou émis par des objets différents. Les structures d'organisme qui participent à ce processus de reconnaissance forment le système de vision dit " analyseur visuel ". A son tour ce dernier est le sous-système du complexe supérieur, c'est-à-dire de l'organisme vivant.

Le processus de la vision reste toujours l'objet d'une grande attention des biologistes. En même temps, il est très intéressant du point de vue du développement de la technique électronique. Les premiers systèmes électroniques visuels ont été construits au milieu des années 1950. Ces appareils étaient destinés à aider les aveugles à lire des textes ou à reconnaître l'intensité de la lumière. Dès que les ordinateurs sont devenus assez perfectionnés pour permettre de résoudre le problème du traitement des données des senseurs électroniques, les expérimentations sur la vision " technique " se sont constituées en une branche spécialisée de la science et de la technique, la *computer vision*.

Mais le problème théorique principal reste à résoudre. Il faut présenter les processus biologiques à travers les abstractions et les symboles de la cybernétique et des mathématiques. Ce problème touche également les scientifiques qui s'occupent de la neurophysiologie de la vision. La formation d'une théorie unifiée biologique et technico-mathématique de la vision permettrait de changer toute l'idéologie de la construction des ordinateurs.

Depuis les années 1960, la technique électronique restait l'instrument le plus puissant, irremplaçable en fait pour les expérimentations des physiologistes de la vision. Une petite révolution dans cette branche de la biologie fut provoquée par l'introduction du microscope bioélectronique qui permit d'identifier les structures de l'oeil. Comme analyseur visuel, l'oeil a des éléments photosensibles de dimensions moindres que la longueur d'onde de la lumière, dont l'étude était impossible sans le microscope optique. Ce point d'interaction est assez intéressant, mais son histoire relève plutôt de l'histoire de l'équipement électronique.

Le moment le plus important du développement parallèle et de l'influence mutuelle de la physiologie de la vision et de la technique électronique est leur interaction au niveau des technologies. Dans ce cas, la contribution principale est « biologique ». Elle provient de l'optique physiologique, particulièrement de la branche de la physiologie de la vision qui étudie le fonctionnement de l'oeil. Les résultats obtenus pendant les recherches sur la photo-réception, c'est-à-dire pendant l'étude des processus photochimiques primaires et de certains processus bio-physiologiques secondaires de la vision, permettent de parler de l'élaboration d'une nouvelle classe de convertisseurs de l'énergie de la lumière.

L'autre point d'interaction est apparu assez spontanément mais il a acquis rapidement une grande popularité parmi les spécialistes de la technique électronique. Les progrès de cette dernière étaient toujours déterminés par la perfection des éléments proprement électroniques : tubes, transistors, circuits intégrés et autres. Le « moteur » qui remplacera la jonction semi-conductrice n'est pas encore connu. Mais il est bien probable que son rôle peut être assumé par la molécule photosensible d'albumen. Les réalisations de la biologie physique et chimique proposent les molécules de rhodopsine ou de bactériorhodopsine qui peuvent se trouver dans deux états stables en fonction de la fréquence du faisceau lumineux de commande. Cet effet a la plus grande importance pour la technique binaire d'ordinateurs.

HISTOIRE DE LA SCIENCE MODERNE, BASE ET PERSPECTIVE. OPTIQUE PHYSIOLOGIQUE ET ÉLECTRONIQUE

Le moment de fusion de la science biologique et de la technique est précédé par l'histoire semiséculaire des ordinateurs électroniques, par un siècle de recherches sur les albumens de la rétine et par deux décennies d'étude des propriétés de la bactériorhodopsine, albumen de photosynthèse sans chlorophylle.

La rhodopsine ou " pourpre visuelle " a été découvert par le savant autrichien F. Boll en 1876. Il a examiné la rétine de l'oeil et s'est aperçu qu'elle perdait sa couleur rose quand elle était éclairée par la lumière et devenait alors blanchâtre. Boll a supposé que ce phénomène devait être lié au processus de la vision. En réalité, le mécanisme du fonctionnement de l'albumen découvert s'est révélé beaucoup plus compliqué.

L'autre albumen semblable à la rhodopsine a été découvert par deux Américains V. Stokkenius et D. Osterheld de l'université de Californie un siècle plus tard, en 1971. Il se trouve dans les bactéries pourpres et est " responsable " du procédé de photosynthèse sans chlorophylle le plus simple et le plus ancien. C'est pourquoi cet albumen a été baptisé par les biologistes bactériorhodopsine. Il est plus stable que la rhodopsine et possède des caractéristiques optiques satisfaisantes.

L'étude de la rhodopsine visuelle a donné naissance à une nouvelle branche de la biologie, la physiologie moléculaire de la vision. Les recherches sur la rhodopsine bactérienne ont eu comme résultat la découverte du mécanisme biologique le plus simple d'utilisation de l'énergie solaire. Cet ensemble de deux albumens a permis de commencer l'élaboration de dispositifs électroniques bio-optiques à la fin des années 80 de notre siècle. Bien que ce soient d'autres spécialistes qui ont remporté la palme dans ce domaine de la science et de la technique, le fondement théorique de ces recherches biologiques, biophysiques et biochimiques sur l'optique physiologique a été élaboré par des savants russes.

Les premiers travaux furent conduits dans les années 1920 et 1930. C'étaient les académiciens de l'Académie des Science de l'U.R.S.S. L.A. Orbeli, disciple du célèbre I.P. Pavlov, et le physicien S.I. Vavilov qui sont devenus les " pères " de la physiologie moléculaire de la vision en Russie. En 1934, ces deux savants ont été les organisateurs de la première conférence sur l'optique physiologique en Union Soviétique. Pendant la cérémonie de clôture, Orbeli a tenu des propos prophétiques sur la mise en valeur des résultats obtenus en optique physiologique non seulement par les biologistes, mais également par les spécialistes d'autres branches de la science, et surtout des techniques. A la fin des années 30, une commission spécialisée a été formée auprès de l'Académie des Sciences de l'Union Soviétique et une publication périodique a été éditée.

Les premières recherches sur la physique de la vision ont commencé à l'Institut d'Optique d'Etat (GOI) où un laboratoire spécialisé a été créé en 1923. Dès 1932 et jusqu'en 1941, Vavilov y a réalisé une série de travaux devenus classiques sur l'étude des fluctuations quantiques de la lumière. Le physicien avait besoin d'un récepteur sensible de flux de lumière très faible qui ne contenaient que quelques dizaines de quantums. Vavilov a décidé d'utiliser dans ces buts l'oeil humain qu'il estimait être le plus puissant, le plus universel et le plus fin des organes des sens. Ses recherches de physique ont permis à Vavilov d'arriver à des conclusions physiologiques très intéressantes. Il a découvert qu'une molécule de rhodopsine était un récepteur assez perfectionné pour réagir à un seul quantum de la lumière. Vavilov a aussi montré que l'oeil pouvait enregistrer une impulsion lumineuse qui ne contenait que huit quantums en 0.1 seconde.

Les recherches de Vavilov ont suscité un énorme intérêt dans plusieurs pays et sont devenus la base des expérimentations en ce domaine. En 1941, des physiologistes américains, S. Heht, S. Shler et M. Pairren, ont effectué des expérimentations inspirées par la méthodologie proposée par Vavilov, mais sans mentionner son nom. Douze ans après les travaux de Vavilov, le physicien hollandais H. Van der Velden a prétendu découvrir la méthode des fluctuations quantiques. Il est intéressant de noter que les données obtenues plus tard par son collègue M. Bouman ne concordaient ni avec les résultats de Vavilov, ni avec les résultats de Heht.

Dans les années 1930, le biochimiste américain G. Wold a publié les résultats de ses recherches restées classiques. Il a expliqué le processus de changement des couleurs de la rhodopsine dû à la lumière. Lauréat du prix Nobel, Wold a ouvert le " monde " de la photo-réception à plusieurs chercheurs. L'académicien russe V.A. Engelgardt, notamment, s'est beaucoup intéressé aux travaux de Wold dans les années 40 quand il étudiait le problème de la rhodopsine.

Toutes les recherches des spécialistes russes ont été arrêtées par le commencement de la Grande guerre nationale. Mais la guerre finie, à la fin des années 1940, la biologie physique et chimique a enregistré d'importants progrès. La logique de développement de cette branche de la biologie a impliqué la réunion de spécialistes de différents domaines de la science. La première union de cette sorte a été formée au laboratoire de cinétique des processus chimiques et biologiques a la fin des années 50. Au commencement des années 1970, le laboratoire de l'étude des principes de la réception a pris le relais comme centre de recherche interdisciplinaire.

Au commencement des années 70, les recherches dans ce domaine se sont intensifiées grâce à une puissante impulsion en provenance d'un programme d'Etat spécial Rhodopsin, dirigé par l'Académicien Yu.A. Ovtchinnikov. Lui et ses collègues découvrirent la structure primaire de la bactériorhodopsine en 1978 et, trois ans plus tard, déchiffrèrent la structure de la rhodopsine optique. Les premiers films photosensibles qui contenaient des molécules de bactériorhodopsine furent reçus. Ce succès permit d'utiliser les molécules en technique électronique. Ce ne fut cependant pas en Russie-Union Soviétique, mais bien aux Etats-Unis, en Allemagne et au Japon que, dans les années 1970 et 1980, se déroulèrent les premiers travaux pratiques et les expériences qui maintenant, nous offrent la perspective de créer des ordinateurs " biologiques ".

PART TWO

Technology as a Bridge between Economy and Society

IMPRIMERIE ET TRANSFERT DE TECHNOLOGIE À LA RENAISSANCE

Marie-Claude DÉPREZ-MASSON

L'utilisation de l'imprimerie et du livre illustré pour la diffusion des idées et des connaissances apparaît comme l'un des multiples aspects fascinants de la Renaissance. Je n'évoque que pour mémoire son impact dans le domaine religieux de la Réforme et de la Contre-Réforme, si bien étudié dans *La Réforme et le livre*[1]. A la suite des recherches d'Elizabeth Eisenstein[2] sur les conséquences sociales de l'imprimerie et du questionnement de Pamela Long[3] sur une " ouverture du savoir " à cette époque, je souhaite ici m'attacher à l'impact de l'imprimerie dans le domaine minier et métallurgique, en essayant de préciser d'une part le public visé par les diverses oeuvres, et d'autre part, les conséquences de cette diffusion exotérique d'un savoir hors des milieux techniques au sens strict. Il est donc question de recouper les modélisations récentes sur la technologie et ses contingences sociales, avec l'étude des relations de pouvoir dans ce domaine ; le pouvoir provenant autant de la détention des savoirs techniques spécialisés que de la possession des moyens de production[4]. L'accent sera mis surtout sur le *De re metallica* d'Agricola, et, à titre de comparaison éclairante, sur Biringuccio et Ercker, les auteurs des ouvrages originaux les plus importants imprimés relatifs à la métallurgie à la Renaissance.

La Renaissance, en effet, voit la publication d'un certain nombre de textes techniques exposant divers aspects des opérations minières et métallurgiques. Plutôt que de les présenter par ordre chronologique, il semble plus éclairant de

1. J.-F. Gilmont, *La Réforme et le livre. L'Europe de l'imprimé (1517-v. 1570)*, Paris, 1990.

2. E. Eisenstein, *The Printing as an Agent of Change. Communications and Cultural Transformations in Early-Modern Europe*, 2 vol., Cambridge, 1979.

3. P. Long, " The Openness of Knowledge : An Ideal and its Context in 16th-Century Writings on Mining and Metallurgy ", dans *Technology and Culture* (1991), 318-355.

4. Voir, par exemple, P. Lemonnier (ed.), *Technological Choices*, London, New York, 1993, ou B. Pfaffenberger, *Democratizing Information on line : Databases and the Rise of End-User Searching*, Boston, 1990.

les regrouper en fonction du domaine technique particulier auquel ils sont consacrés.

Sur la géologie minière paraît le livret du médecin Rülein von Calw, le *Bergbüchlein*[5] rédigé en allemand, vers 1500. Ce livret de 48 pages connut 10 réimpressions jusqu'en 1616[6]. Les techniques du fer et celles des essayeurs de minerai, surtout des métaux précieux, voient publier nombre d'imprimés, relevant de la littérature des recettes : le *Von Stahel und Eysen*[7], 11 pages, paru en 1532, qui connut 6 éditions en 7 ans, et fut intégré aux *Kunstbüchlein*, dès 1532, puis au *Libro de Secreti* d'Alexis de Piémont, en 1555 ; et les *Kunstbüchlein*[8] à partir de 1532, à Leipzig, aux nombreuses rééditions en allemand et en néerlandais pendant deux siècles dans l'espace germanophone et néerlandophone. Plus spécifiquement consacrés aux techniques des essais, on rencontre : les *Probierbüchlein*[9], anonymes, à partir de 1510 ou 1520, et qui comptent au moins 15 réimpressions identifiées au XVIe siècle et d'autres au XVIIe siècle. Paraissent aussi le *Probier Büchlein* de Modestin Fachs[10], écrit vers 1569, celui de Samuel Zimmerman[11] (Augsbourg, 1573), et celui de Ciriacus Schreittmann[12] (Francfort, 1578). Cet ouvrage fut rédigé dans les années 1550 par un essayeur, jalousement conservé par son patron pendant plus de 20 ans, et publié seulement en 1578, après la mort de ce dernier, par son fils, Valentin Abel. Ces trois derniers livrets comportent un certain nombre de rééditions. Enfin, existe le *Beschreibung der Allerfürnemsten mineralischen ertz und Berc-*

5. *Ulrich Rülein Von Calw und sein Bergbüchlein. Mit Urtext-Faksimile und Übertragung des Bergbüchlein von etwa 1500*, W. Pieper (ed.), Berlin, 1955. Traduction anglaise de A.G. Sisco et C.S. Smith, *Bergwerk- und Probierbüchlein. A Translation from the German of the Bergbüchlein and of the Probierbüchlein with Technical Annotations and Historical Notes*, New York, 1949.

6. Le nombre des réimpressions des divers ouvrages cités provient de la compilation des renseignements donnés par les éditeurs et traducteurs de ces textes, recoupée par le dépouillement du *Dictionary of Scientific Biography*, C.C. Gillispie (ed.), New York, 1970-1980, et celui de l'*Index Aureliensis, catalogus librorum sedecimo saeculo impressorum*, Baden-Baden, 1965. Comme il s'agit là d'éditions encore attestées par les catalogues de bibliothèques, le nombre réel d'éditions a pu, éventuellement se montrer plus élevé.

7. *Von Stahel und Eysen*, Nurenberg, 1532 ; traduction A. Sisco, " On Steel and Iron ", dans C.S. Smith (ed.), *Sources for the History of Steel, 1532-1786*, Cambridge (Ma), 1968.

8. Voir E. Darmstaedter, *Berg-, Probier- und Kunstbüchlein*, Munich, 1926.

9. Traduction anglaise de A.G. Sisco et C.S. Smith, *Bergwerk- und Probierbüchlein. A Translation from the German of the Bergbüchlein and of the Probierbüchlein with Technical Annotations and Historical Notes*, New York, 1949.

10. Modestin Fachs était maître de la monnaie à Leipzig, lorsqu'il écrivit le *Probier Büchlein/ Darinne Gründlicher bericht vormedel/vie man alle Metall/ und derselben zugehörenden Metallischen Ertzen und getöchten ein jedes auff seine eigenschafft und Metall recht Probieren sol.* Ce livre fut édité à Leipzig, en 1595, après sa mort, par les soins de son fils, et dédié au duc d'Anhalt.

11. Samuel Zimmermann, *Probierbüch: Auff alle Metall Müntz/ Ertz/ und Berckwerck/ Dessgleichen auff Edel Gestain/perlen/ Corallen/ und andern dingen mehr...* Augsbourg, 1573. Son ouvrage est décrit dans E. Darmstaedter (*op. cit.*, n. 8) et dans P. Long (art. cit. n. 3), 345-346.

12. Il en existe une édition de 1580 : Ciriacus Schreittman, *Probierbüchlin, Frembde und subtile Künstvormals im Truck nie gesehen...*, Frankfurt am Main, 1580. On trouvera des renseignements et une étude le concernant dans E. Darmstaedter (*op. cit.* n. 8), 189, dans C.S. Smith, " A Sixteenth-Century Decimal System of Weights ", dans *Isis*, 46 (1955), 354-357, et dans P. Long (art. cité n. 3), 342-343.

kwerkcs arten[13], gros ouvrage de 135 pages, de Lazare Ercker, paru en 1574 à Augsbourg et réédité huit fois jusqu'en 1736 ; il fut traduit en anglais en 1683, avec trois éditions rapprochées, puis en hollandais en 1745, et probablement en latin vers 1700.

Seuls, toutefois, trois traités s'attachent à l'ensemble des opérations des mines et de métallurgie extractive. Le premier, également rédigé en langue vernaculaire, l'italien, par le siennois Biringuccio, en 1540 est la *Pirotechnia*[14]. Il connaît 10 éditions attestées, en italien et en français, plus deux autres probables (en latin et en français) jusqu'en 1678, sans compter les plagiats et emprunts (en espagnol[15] et en anglais[16]). Le second, rédigé en latin, publié en 1556 par l'éditeur d'Erasme, le *De re metallica*[17], est l'oeuvre du saxon Georg Agricola, et connaît 10 éditions en un siècle, en latin, en allemand et une en italien, dédiée à la reine Elisabeth d'Angleterre. Et enfin le *De re metalica* de Perez de Vargas, rédigé en espagnol, en 1569. Ce dernier, n'étant qu'une compilation des ouvrages de Biringuccio et Agricola, ne sera pas particulièrement examiné ici.

Cette énumération permet déjà de situer plusieurs aspects de la diffusion des savoirs métallurgiques par le biais de l'imprimé, au XVIe siècle. D'abord, l'origine géographique des textes : l'Italie et, massivement, l'espace allemand ; puis des pays comme l'Espagne, la France, l'Angleterre et les Pays-Bas, en ce qui a trait aux traductions, avouées ou non. Ce sont donc les pays miniers qui donnent d'abord naissance aux auteurs des textes originaux ; cependant que dans les pays européens où les mines présentent une certaine importance économique se composent surtout des traductions.

Parmi les auteurs de ces ouvrages, deux grandes catégories de rédacteurs se distinguent. D'une part des spécialistes des opérations décrites ; cela est particulièrement net dans les cas de Ercker, de Biringuccio et de Schreittman. D'autre part, des intellectuels, et plus précisément, des médecins ; il vaudrait,

13. Lazarus Ercker, *Beschreibung der Allerfürnemsten mineralischen ertz und Berckwerkcs arten vom Jahre1580,* P. R. Beierlein ed., Berlin, 1960 ; traduction anglaise par A.G. Sisco et C.S. Smith, *Lazarus Ercker's Treatise on Ores and Assaying Translated from the German Edition of 1580*, Chicago, 1951.

14. V. Biringuccio, *De la pirot/echnia. / Libri X dove ampiamen/te si tratta non solo di ogni sorte & di/versita di miniere...*, Venise, 1540. Traduction anglaise par C.S. Smith et M.T. Gnudi, *The Pirotechnia of Vanoccio Biringuccio. Translated from the Italian with an Introduction and Notes*, New York, 1959 (1942).

15. En particulier dans l'ouvrage de Perez de Vargas, *De re metalica,* Madrid, 1569.

16. Par Richard Eden (*The Decades of the newe worlde...*, 1555) et par Peter Whitehorn, (*Certain waies for the ordering of souldiers...*, Londres, 1569).

17. *Georgii Agricolae De re metallica libri XII. Qui/bus Officia, Instrumenta, Machinae, ac omnia denique ad Metalli/cam Spectantia, non modo luculentissime describuntur, sed per/ effigies, suis locis inserta, adiunctis Latinis, Germanicisque appel/lationibus ita ob oculos ponuntur, ut clarius tradi non possint*, Froben, Bâle 1556 (réimpr. anastatique Bruxelles, 1967). Traduction anglaise et notes par H.C. et L.H. Hoover, *Georgius Agricola. De re metallica*, Londres, 1912 (réimpr. New York, 1950). Traduction française par A. France-Lanord, *Agricola. De re metallica*, Thionville, 1987. Les citations du *De re metallica* proviennent toutes de l'édition originale.

du reste, la peine de relever le nombre d'écrits majeurs en sciences ou techniques dus à des médecins européens, au fil des siècles.

En ce qui concerne le nombre d'ouvrages miniers ainsi diffusés au XVIe siècle, il est difficile de donner une estimation précise. La majorité des historiens actuels du livre imprimé[18] donnent une fourchette d'exemplaires variant entre 500 et 2.000 par édition, avec une moyenne probable de 1.000. Si on tient compte des seules éditions citées et attestées au XVIe siècle, on arrive à plus de 60 éditions, soit entre 30.000 et 120.000 livres consacrés au domaine minier ou métallurgique, et à un nombre " moyen " de 60.000 exemplaires probables. Il s'agit là d'un chiffre important pour une littérature aussi étroitement spécialisée.

Les destinataires de ces ouvrages sont généralement répartis en deux groupes distincts : d'une part, les artisans, qui souhaitent connaître les procédés de leurs confrères ; d'autre part les " dirigeants ". Toutefois, l'examen plus précis des textes, en commençant par ceux qui concernent les essais, suggère d'abord de se poser une question très simple : les artisans ont-ils vraiment besoin de ces textes ? Leur formation leur est acquise dans le cadre de l'atelier où s'est effectué leur apprentissage. Quel besoin auraient-ils d'investir dans l'achat d'un livre ? Pour découvrir d'autres tours de main ou procédés ? Telle est la réponse habituelle ; peut-être s'avère-t-elle exacte pour quelques uns d'entre eux. Mais, globalement, n'ont-ils pas déjà en main, et en tête, des méthodes suffisantes pour l'exercice de leur profession ?

Or, au fil des années, on voit s'améliorer la qualité et la précision des informations données dans les divers *Probierbüchlein*. Quel public visent-ils donc réellement ? A cet égard, je reste frappée par deux points : une explication de Biringuccio et la place que tiennent, dans la carrière d'Ercker, ses livrets d'essais, d'abord manuscrits avant la rédaction et la publication de son grand traité. On voit ainsi Biringuccio, dans sa *Pirotechnia*, aviser avec force son interlocuteur, un riche Italien, des précautions à prendre, dans le cas où il recevrait la responsabilité d'un atelier de frappe des monnaies : qu'il effectue lui-même tous les essais aux diverses étapes des opérations du monnayage, pour éviter d'être trompé par des essayeurs peu scrupuleux[19]. Quant à Ercker, la première impression de son livre est immédiatement suivie par sa nomination au poste de responsable administratif des mines pour l'Empire[20]. Ce livre a été

18. Outre l'ouvrage fondamental de H.-J. Martin et L. Febvre, *L'apparition du livre*, Paris, 1958, on peut se référer à la synthèse donnée par des spécialistes de l'édition française dans le livre publié sous la direction de H.-J. Martin et R. Chartier, *Histoire de l'édition française*, t. 1, *Le livre conquérant du Moyen Age au milieu du XVIIIe siècle*, Paris, 1982.

19. Livre IX, ch. 3, 358-360 (dans la traduction citée n. 14). Biringuccio a déjà exprimé cette même nécessité plus haut dans son ouvrage, pour évaluer précisément le pourcentage de métaux précieux contenus dans les minerais à essayer (livre III, ch. 6, 159).

20. Sur la carrière de Ercker, voir l'article de P. Long déjà cité (n. 3), 346-350 ; le *Dictionary of Scientific Biography*, s.v. " Ercker (also Erckner or Erckel) " (cité n. 6) ; et P.R. Beierlein, *Lazarus Ercker : Bergmann, Hüttenmann und Münzmeister im 16 Jahrhudert*, Berlin, 1955.

précédé par la rédaction de trois opuscules manuscrits, réservés aux princes des Etats qu'il servait alors ; or chacune de ces rédactions a entraîné une promotion pour Ercker, comme responsable officiel de mines et d'ateliers de frappe de la monnaie. On remarquera que dans l'un de ces livrets, *Das Münzbuch,* daté de 1563[21], dédié à son souverain et protecteur du moment, le duc de Braunschwig-Wolffenbüttel, Ercker, comme Biringuccio, insiste sur la nécessité pour le prince de savoir surveiller les opérations de ses propres frappes de monnaie. Ce qui implique que le Prince connaisse, par lui-même, les techniques des essais. La même idée sous-tendait déjà un autre de ses manuscrits, dédié au duc Auguste de Saxe en 1556, *Das Kleine Probierbuch*. On la retrouve également dans le livret de Zimmermann. Aussi, la longue utilisation des traités d'essai et leur précision croissante semblent en faire essentiellement des textes destinés à des responsables soucieux de mettre un terme aux multiples fraudes que nous révèle l'étude de l'aloi des pièces, soit en pratiquant directement les techniques des essais, soit en devenant capables de vérifier la compétence des essayeurs.

C'est dans un tel contexte qu'il faut mettre en situation les oeuvres de ces non-techniciens que sont Rülein von Calw et Agricola. Je les oppose à Biringuccio quant à la visée de leurs ouvrages. En effet, dans sa *Pirotechnia*, ce dernier donne juste ce qu'il faut comme renseignements pratiques pour appâter un futur employeur, en particulier sur la coulée des canons et l'obtention du salpêtre, produit de base de la poudre à canon ; mais il ne se protège pas moins, avec une grande adresse, en affirmant que " seule une longue pratique de cet art " qu'est la fonte des canons permet de bien appliquer tout ce qu'il va expliquer en détail[22]. Autrement dit : " Embauchez-moi.... ". Rien de tel chez Rülein et Agricola. Dans sa brièveté, le *Bergbüchlein* veut rendre capable un investisseur novice de repérer les filons prometteurs, afin d'investir sciemment dans le domaine de la prospection minière[23].

Le cas du *De re metallica* d'Agricola pourrait sembler moins évident. Cet ouvrage est le premier à présenter une description exhaustive de toutes les techniques d'obtention du métal : recherche des filons, droit minier, fonçage et boisage des puits et galeries, questions de topographie qui y sont liées. Puis vient l'explication détaillée des méthodes d'essai pour évaluer la teneur en métaux précieux des minerais. Ensuite, Agricola décrit tous les outils nécessaires à ces opérations, fussent-ils aussi simples que des marteaux ou des barres de mines. De même détaille-t-il les machineries utilisées dans les mines :

21. Lazarus Ercker, *Drei Schriften : Das kleine Probierbuch von 1556 ; Vom Rammelsberge, und dessen Bergwerk, ein kurzer Bericht von 1565 ; Das Münzbuch von 1563*, P.R. Beierlein et H. Winkelmann ed., Bochum, 1968.

22. Livre VI, Préface, 216 (*op. cit.* n. 14).

23. Voir la traduction de A.G. Sisco et C.S. Smith (*op. cit.* n. 5), 17 : " Then it will be explained to me logically in this little book and I shall be given a reasonnable understanding of which mines can be worked gainfully so that my investments will not be wasted but will show a profit ".

machines d'exhaure (manuelles ou hydrauliques), systèmes de pompes aspirantes, méthodes d'aérage et d'extraction des minerais. Sont également longuement détaillées les opérations subséquentes : tri et concassage du minerai, grillage ou calcination de celui-ci. Puis vient la réduction dans diverses sortes de fours pour tous les métaux alors produits dans l'espace allemand, quoique le fer et l'acier soient assez brièvement traités. Enfin est exposée la récupération des métaux précieux contenus dans les métaux ainsi produits, sans oublier la fabrication des divers produits et réactifs nécessaires à ces opérations — y compris celle de l'alun et du salpêtre. En quelque 500 pages de texte, Agricola propose ainsi un exposé complet des techniques utilisées de son temps, des plus archaïques aux plus sophistiquées, en signalant qu'il faut toujours adapter ces méthodes au minerai traité, afin d'assurer la rentabilité la meilleure. L'accent, du reste, est mis régulièrement sur cette notion de rentabilité[24].

L'ouvrage présente encore une innovation majeure par rapport à tous les autres textes techniques. Ces derniers étaient rédigés en langues vernaculaires. Or le traité d'Agricola, lui, est écrit en latin pour être mieux compris. Une telle affirmation peut, à première vue, surprendre ! Mais, justement, dans les traités comme les *Bergbüchlein* ou les *Probierbüchlein*, les éditeurs, très vite, ont dû adjoindre au texte de véritables glossaires allemand/allemand, pour que les lecteurs puissent les comprendre[25]. Faut-il rappeler la diversité des parlers germaniques, pour ne rien dire de la variété des vocables désignant de nombreuses réalités minières, ni de cette particularité des “ jargons ” techniques de donner des noms différents à des outils semblables, selon les corps de métier et les régions ? Aussi, pour éviter toute ambiguïté, Agricola crée-t-il quelque 1.200 expressions techniques en latin. Pour cela, il utilise un vocabulaire latin classique très simple : il décrit littéralement les objets et les nomme. On trouvera ainsi, pour traduire les multiples termes saxons qui les désignent, une formule de base pour toutes les machines : toutes seront des “ machines qui font quelque chose ” (machine pour enlever l'eau, machine pour l'air, machine pour broyer le minerai, *etc.*). De même, lorsqu'Agricola décrit tout outil présentant la forme “ marteau ”, il l'appelle *malleus,* le définissant par son emploi (marteau du mineur : *malleus metallicus*) ou par sa taille, lorsqu'un métier en utilise de plusieurs dimensions (petit, moyen ou grand des “ petits marteaux ” : *malleorum minorum minimus, medius* ou *maximus*).

24. Entre autres illustrations probantes, les diminutions de postes d'ouvriers dans le cas des machines d'exhaure, livre VI, 157 (cité n. 17) ; ou la nécessité pour le responsable de la mine de se livrer à des inspections surprises et d'accorder des gratifications aux bons mineurs pour augmenter leur zèle, livre II, 19-20.

25. Voir, par ex., l'édition de 1518 du *Bergbüchlein* éditée par W. Pieper (*op. cit.* n. 5) : les 20 pages de texte sont complétées par un glossaire de 8 pages. Pour le *Probierbüchlein* anonyme de 1500, un certain nombre de réimpressions présentent des glossaires : Strasbourg 1530, une autre, sans date, de la même époque. D'autres éditions présentent des synonymes allemands dans le texte même.

Chaque objet, quel qu'il soit, est décrit en détail avec ses dimensions, les matériaux qui le composent, et son mode de montage. Chaque outil, instrument, installation, four ou machine est ainsi doté d'une véritable fiche technique. Mais les mots sont généralement insuffisants pour décrire complètement un objet. Pour prendre un cas aussi simple que celui d'un râteau, comment expliquer le positionnement exact de ses dents ? Aussi Agricola recourt-il systématiquement au dessin, 293 gravures[26] sont insérées dans le texte du *De re metallica.* Elles ont fait le bonheur de générations d'historiens des techniques, car souvent elles offrent les premières représentations précises de nombreuses machines et fours. En effet, Agricola, pendant les quelque vingt ans de ses investigations dans les diverses installations minières allemandes, a emmené avec lui des dessinateurs, chargés de dessiner tout ce qu'ils voyaient dans les ateliers ou installations visitées, et d'en relever les dimensions exactes[27]. Chaque objet utilisé au cours d'une des multiples opérations métallurgiques est décrit au moins une fois par une image, fut-ce un objet aussi simple que des seaux pour évacuer le minerai, ou les divers râteaux et rouables des ouvriers. Et le dessin présente, avec un grand soin, toutes les particularités de montage des diverses pièces constitutives de l'outil ou de la machine, souvent grâce à des vues en éclaté.

En outre, chaque atelier est présenté avec tous les outils, instruments et installations nécessaires à la bonne marche des opérations. Pour éviter toute erreur, ces gravures — exécutées par les meilleurs spécialistes du temps — sont insérées dans le texte immédiatement après la description de ce qu'elles explicitent. Une telle disposition, voulue par Agricola, alors qu'elle perd souvent beaucoup de place, ne sera pas reprise dans les éditions postérieures à celle de 1556, après la mort de leur auteur. Chaque image est, en outre, dotée d'une légende indiquant les noms latins de tous les objets, et de leurs diverses parties, ce qui la fait fonctionner comme un véritable glossaire visuel. Un lecteur ignorant des techniques métallurgiques peut ainsi parfaitement comprendre comment fonctionne une machine, en connaître les diverses variantes ; de même peut-il vérifier si l'atelier qu'il visite présente tout l'appareillage nécessaire à son bon fonctionnement.

Mais, au fait, à quel lecteur, est destiné ce livre ? Il est évidemment inutile aux simples manoeuvres. Il n'apprendrait rien aux divers hommes de l'art, chargés de fabriquer fours ou machines : leur métier, eux, ils l'ont appris par la pratique. De surcroît, en lisant avec attention le *De re metallica*, on prend conscience qu'un certain nombre de renseignements techniques fondamentaux n'apparaissent pas : à quel indice concret repérer la qualité de la roche dans

26. Traditionnellement, on considère que cet ouvrage possède 292 gravures. Toutefois, une analyse minutieuse des xylogravures indique bien 293 bois gravés différents. Voir M.-C. Déprez-Masson, *La chose, le mot et l'image : la création d'un langage technique dans le De re metallica de Georg Agricola (1556)*, thèse de doctorat, Montréal, 1996, 363.

27. Voir la Lettre-Dédicace d'Agricola, en tête du *De re metallica.*

une mine, de la gangue dans une opération de grillage, ou de réduction ? Sur quels indices précis mener les feux et les ajouts de flux lors de la réduction ? En fait, tous les tours de main qui ont fait des spécialistes “ allemands ” des experts internationaux sont absents de cet ouvrage. On pourrait penser à un manque d'informations d'Agricola, ou à une méconnaissance quant à l'importance de ces points, alors que tous les autres sont parfaitement explicités. Toutefois, il arrive qu'Agricola détaille des opérations que l'on trouve également dans la *Pirotechnia*, ouvrage qu'il cite comme une de ses sources. On constate alors, dans de tels cas, que ces renseignements techniques précis sont donnés par Biringuccio, un technicien ; mais Agricola ne les reprend pas, alors qu'ils les connaît[28]. Il faut donc chercher une autre explication qu'une éventuelle méconnaissance de l'auteur, et pour cela, examiner de plus près les index — au nombre de trois — qu'il a adjoints à son texte.

Les deux premiers fonctionnent comme des glossaires latin/saxon. Ils présentent les quelque 1.200 mots techniques latins, avec leur traduction en allemand de Meissen (langue d'Agricola, mais aussi allemand *koïné*). Ces mots désignent toutes les opérations et tous les instruments, machines et processus de toutes les opérations minières et métallurgiques. Un troisième index les complète, qui indique, avec quelque 1.500 entrées, où trouver dans le *De re metallica* chaque description d'outils, de machines, *etc*., avec leur “ devis technique ” et l'illustration qui les visualise. De même sont indiqués les passages du livre qui décrivent les divers processus opératoires fondamentaux. A l'aide de ces index, le lecteur peut donc savoir tout ce qui est nécessaire dans un atelier ; quels outils ou machines commander à un charpentier ou à un forgeron, en lui montrant les plans de ce qu'il veut obtenir. En outre, la connaissance du vocabulaire saxon désignant ces *realia*, permet au lecteur de discuter avec les hommes de métier en utilisant leurs vocables techniques. De surcroît, et pour la première fois, grâce aux explications verbales et picturales du *De re metallica*, grâce à la simplicité et à l'univocité du vocabulaire latin mis au point par Agricola, un non-initié peut comprendre les termes techniques de chacun des corps de métier impliqué dans les processus métallurgiques.

A relier ainsi le choix des mots, la précision chiffrée de toutes les descriptions, l'utilisation des gravures et les particularités des index, on se rend compte qu'Agricola a parfaitement mené l'entreprise qu'il s'était fixée, et qu'il signale d'entrée de jeu : réunir sur le papier tous les experts dont les conseils seraient utiles au lecteur[29]. Mais qui a besoin de l'aide d'experts, sinon celui qui, sans être un spécialiste souhaite contrôler les techniciens qu'il emploie ?

28. C'est le cas, par exemple, des détails de la coupellation à grande échelle ; on les trouve chez Agricola, livre X, 376-378 et chez Biringuccio, dans la traduction anglaise (citée n. 14), livre III, ch. 7, 161-169. Agricola omet beaucoup de précisions sur l'aspect des métaux en fusion lorsqu'il devient nécessaire d'y ajouter divers produits, détails fondamentaux qu'il a pourtant lus dans la *Pirotechnia*.

29. *De re metallica*, 2.

On constate donc, par l'organisation même de l'ouvrage, qui va rester pendant deux siècles le livre de chevet des investisseurs et propriétaires de mines, que ce traité est destiné à ceux qui investissent, et non pas aux simples techniciens, ni même aux maîtres-fondeurs. Désormais, ceux qui possèdent le *savoir* grâce à cet ouvrage, vont pouvoir mieux encadrer, et donc mieux contrôler ceux qui ne possèdent que le savoir-faire. Il renforce, en fait, le contrôle du monde des finances sur celui du travail. On explique alors mieux pourquoi Valentin Abel, dans sa préface au *Probier Büchlein* de Schreittman, signalait que les princes et les nobles voulaient que désormais, tous les " arts " (au sens de techniques), soient mis par écrit.

L'imprimé, certes, pourra permettre de faire progresser les techniques puis les sciences qui leur sont sous-jacentes, par la diffusion des procédés et la comparaison de leurs résultats ; mais ce rôle de l'imprimé ne commence vraiment, dans le domaine métallurgique, qu'à la fin du XVIIe siècle et au XVIIIe siècle. Telle ne sera pas la conséquence première de l'imprimerie dans la diffusion des savoirs techniques en métallurgie. Que l'on considère les livres d'essais ou le grand traité d'Agricola, les premiers bénéficiaires de ces ouvrages seront les hommes de pouvoir : la connaissance des savoirs techniques renforcera leur pouvoir sur le simple savoir-faire des techniciens. L'imprimerie n'est pas neutre. Pas plus que ne le sont les nouveaux médias que nous voyons naître avec l'informatique.

LES MÉTHODES D'ANALYSE DES MÉTAUX PRÉCIEUX ET LES VOYAGES DE MARTIN FROBISHER AU CANADA (1576-1578)

Bernard ALLAIRE - Réginald AUGER

Nous avons choisi de vous parler de l'histoire d'une catégorie méconnue de métallurgistes : les essayeurs de métaux précieux à l'époque élisabéthaine. Nous travaillons sur ce sujet depuis 1990 dans un programme de recherche — regroupant des chercheurs de l'Angleterre, des États-Unis et du Canada[1].

L'essayeur du XVIe siècle est un inconnu ; on ne le voit jamais, puisqu'il passe son temps derrière les portes de son laboratoire, afin de répondre aux attentes de ses clients à la recherche de métal précieux. Son rôle est d'évaluer la teneur en métaux précieux du minerai ou la qualité des produits finis[2]. Le savoir-faire de l'essayeur est requis à toutes les étapes du processus, de l'exploitation à la transformation de l'or et de l'argent. Dans les régions minières, il assiste les mineurs dans l'identification des veines les plus prometteuses. À son laboratoire, il analyse les pièces de métal brut achetées et vendues par les affineurs ou teste, au nom du maître de la monnaie, les monnaies étrangères circulant dans le pays.

L'essayeur du XVIe siècle qui nous intéresse particulièrement est celui engagé dans les voyages de Martin Frobisher, cet explorateur anglais dont le sort a largement reposé sur l'intervention d'essayeurs. L'histoire des voyages de Martin Frobisher parti à la découverte d'un passage au nord-ouest de l'Angleterre en 1576 offre un cas remarquable, voire exceptionnel, de complémentarité documentaire à propos des essayeurs. C'est l'existence de sources manuscrites concernant ces voyages d'exploration, de sources imprimées du XVIe siècle sur la métallurgie et de sources matérielles issues de fouilles

1. W.W. Fitzhugh, J.S. Olin (eds), *Archeology of the Frobisher Voyages*, Washington, Smithsonian Institution Press, 1993 ; D.D. Hogarth, P.W. Boreham, J.G. Mitchell, *Martin Frobisher's Northwest Venture 1576-1581 : Mines, Minerals, Metallurgy*, Ottawa, Canadian Museum of Civilization, 1994.

2. B. Allaire, " The gold and silver refiners of Paris in the XVIth and XVIIth centuries ", *Bulletin of the canadian institute of mining and metallurgy*, 89 (1996), 76-82.

archéologiques au laboratoire d'essai de Martin Frobisher, qui nous permettent de porter un regard complet sur ce métier obscur du XVIe siècle.

L'examen des techniques d'analyse des métaux précieux au XVIe siècle permettra de mieux comprendre les origines d'une mégarde historique qui a mené toute l'Angleterre élisabéthaine à croire que le sol de l'Arctique canadien contenait des richesses incroyables.

LES TROIS VOYAGES VERS LE " NORD-OUEST "

Si la recherche de métal précieux est un aspect essentiel des voyages de Frobisher, elle n'est cependant pas le motif direct du premier voyage qui avait pour but de trouver un passage vers la Chine[3]. Entre 1576 et 1578, Martin Frobisher effectue trois voyages dans l'Arctique canadien. Le premier voyage se déroule à l'été 1576 avec deux petits navires et une barque. En somme, une petite escadre qui passe inaperçue aux yeux de la plupart des observateurs étrangers lors de son départ le 7 juin 1576. En route, une tempête fait sombrer la barque et peu de temps après, le deuxième navire abandonne et retourne en Angleterre. Martin Frobisher décide néanmoins de poursuivre le voyage afin d'explorer la partie située entre Terre-Neuve et le Groenland, une zone toujours méconnue des cartographes européens de l'époque. Inspiré par une carte fictive montrant un passage au nord du continent, Frobisher s'imagine qu'une fois passé ce détroit, il débouchera dans le Pacifique et pourra se diriger vers l'Asie. Il arrive un 26 juillet dans une région complètement bloquée par les glaces mais réussit néanmoins à explorer l'une des longues baies du sud de la Terre de Baffin sans prendre le temps de s'engager plus à l'intérieur des terres. Il en conclut (de par la force des marées) qu'il s'agit du passage vers la Chine qu'il recherche. La carte qui en résulte montre effectivement l'Asie à droite et l'Amérique à gauche. Cette conclusion lui est d'autant plus évidente que les personnes rencontrées ont les yeux bridés, les cheveux noirs et sont imberbes tout comme les asiatiques. Frobisher joue cependant de malchance ; il rapporte que cinq de ses marins sont capturés par les autochtones, sans qu'il puisse intervenir. Il attend quelques jours dans l'espoir de revoir ses hommes mais peine perdue. Il capture alors un indigène et rentre en Angleterre avec des spécimens de la flore et un morceau de roche locale ramassé sur la plage. Il arrive le 9 octobre en Angleterre, quatre mois après son départ, à la surprise générale des gens qui le croyaient mort et qui l'accueillent en héros.

Si ce premier voyage s'était déroulé dans la quasi-indifférence des observateurs locaux, au retour en Europe, les analyses du morceau de minerai rapporté attirent l'attention de plusieurs personnes. En effet, les résultats des analyses effectuées par plusieurs métallurgistes et non spécialistes laissent entrevoir la

3. B. Allaire, D.D. Hogarth, " Martin Frobischer, the Spaniards and a Sixteenth Century Northem Spy ", *Terrae Incognitae, Journal for History of Discoveries*, 28 (1996), 46-57.

possibilité d'une forte teneur en métaux précieux. Spécifiquement, tous, à l'exception d'un essayeur, déclaraient que la roche analysée était sans valeur. Ce dernier, Michael Lok, promoteur des voyages de Frobisher et ancien directeur de la compagnie de Moscovie, était très convaincant.

À la suite de l'insistance de l'un des métallurgistes, un deuxième voyage est mis sur pied pour aller chercher une quantité suffisante de minerai et trancher la question à l'aide d'analyses concluantes. Étant donné que la recherche d'or et d'argent devenait l'objectif essentiel de cette deuxième aventure, ce voyage effectué en 1577 suscita l'intérêt de plusieurs notables. L'expédition composée de trois navires mit le cap sur l'endroit que les autorités avaient désormais baptisé *Meta Incognita*. L'aventure est menée rondement avec trois essayeurs et un groupe de mineurs amenés d'Angleterre avec leur équipement pour creuser la roche. Ils prospectent la zone, font des analyses et remplissent les cales des trois navires avec le minerai des environs de l'endroit nommé le détroit de Frobisher.

De retour au pays, les analyses douteuses, le secret donné à cette affaire et les querelles entre les métallurgistes impliqués font que personne n'est encore certain de la valeur réelle du minerai. De plus, la difficulté à trouver les fonds nécessaires à la construction de fourneaux de transformation fait en sorte qu'une troisième expédition est montée sans les garanties nécessaires à un profit assuré. Néanmoins, celle-ci est organisée pour le printemps de 1578, avec l'aide du gouvernement anglais et de la reine Élisabeth qui veulent y adjoindre un projet colonial.

La flotte de Frobisher compte cette fois 15 navires qui ont pour objectif de charger du minerai et de laisser sur place une centaine de colons anglais avec les marchandises nécessaires pour survivre durant une année. Une fois arrivé à la Terre de Baffin, et suite à de multiples déboires dûs aux conditions de navigation périlleuses, l'un des navires transportant des vivres et une maison préfabriquée heurte un iceberg et coule, Frobisher renonce à toute possibilité de fonder une colonie. Il prend cependant la décision de rentrer en Angleterre avec les colons après avoir chargé tout le minerai qu'il était possible de mettre à bord des douze navires restants.

De retour en Angleterre, les analyses sont effectuées selon les règles de l'art à l'aide de fours construits à Dartford au sud-est de Londres. Ces analyses confirment définitivement qu'il n'y a pas, ou peu, de métal précieux dans le minerai. L'association mise sur pied pour ces voyages devient incapable de payer le personnel engagé, les marins, les mineurs. Elle tombe en faillite et les têtes dirigeantes sont emprisonnées.

L'histoire des voyages de Martin Frobisher tombera dans l'oubli général, entre autres, parce que la région visitée par Frobisher sera placée par erreur par les cartographes du XVIIe siècle sur la côte est du Groenland. Frobisher restera inconnu jusqu'à ce que le journaliste/explorateur américain Charles-Francis

Hall, conduit par les Inuit, identifie le site de transformation en 1860. Ceux-ci le mènent à un lieu appelé " Kodlunarn " ou l'île de " l'homme blanc ", nom qu'ils connaissent depuis des générations par tradition orale. Une partie des objets présents en surface seront collectés par Hall, mais le site restera plus ou moins méconnu jusqu'à aujourd'hui.

La question des essais

Deux longues années et trois expéditions maritimes auront donc été nécessaires pour que les personnes concernées s'aperçoivent que le minerai, transporté de *Meta Incognita* vers l'Angleterre à grand frais, était sans valeur. Comment une erreur aussi flagrante a-t-elle pu se produire ? Bien entendu, le travail des essayeurs est le premier désigné pour expliquer l'origine de cette mégarde. Arrêtons-nous un instant sur ce sujet. Pour éviter de s'encombrer inutilement de formules chimiques et de descriptions complexes, voici, brièvement expliqué comment s'effectuaient les essais.

Pour comprendre les conditions dans lesquelles ont été réalisés les essais effectués par les métallurgistes accompagnant Frobisher, nous avons croisé plusieurs types de sources : les sources manuscrites, imprimées et matérielles. L'environnement matériel des voyages de Frobisher comme la description des marchandises ayant servi aux essais, chargées à bord des navires, nous sont connus grâce aux manuscrits conservés au Public Record Office de Londres. Ceux-ci proviennent des nombreuses actions en justice suite à la banqueroute de l'entreprise.

Les ouvrages imprimés sur la métallurgie au XVIe siècle sont aussi très nombreux. En fait, le XVIe siècle voit une floraison d'ouvrages assortis de gravures. C'est entre autres le cas d'une série de petits *Probierbücher* publiés de façon anonyme entre 1500 et 1534, des ouvrages publiés par Vanoccio Biringuccio en 1540 [*Pirotechnia*], par Georg Bauer dit Agricola en 1556 [*De re metallica*], ainsi que des ouvrages de Samuel Zimmerman de 1573, de Lazarus Erker de 1574, de Ciriacus Schreitman de 1578[4]. Tous ces imprimés permettent de connaître les limites de la technologie existant à l'époque des voyages de Frobisher.

Ce sont toutefois les sources d'origine matérielle qui se sont révélées les plus utiles pour connaître les méthodes employées par les essayeurs. Ce sont, entre autres, les vestiges du laboratoire d'essai retrouvés sur le site utilisé par Frobisher qui ont apporté les informations décisives. Il s'agit de plusieurs centaines de fragments, principalement de creusets en pâte réfractaire, de scories d'essai, mais également de scorificateurs, quelques coupelles et d'autres fragments non identifiés, mais liés aux analyses.

4. Voir ici même les textes de M.-C. Deprez-Masson et de V. La Salvia.

Le croisement des sources manuscrites, imprimées et archéologiques a permis de se faire une idée beaucoup plus précise des conditions matérielles dans lesquelles ont été effectués les essais sur le site de Frobisher. Cette étude amène à croire que les essais n'ont peut-être pas été réalisés dans des conditions optimales. La liste des marchandises chargées à bord des navires révèle que les essais effectués ont été réalisés à l'aide de petits fours portatifs en métal démontables qui étaient posés sur une base de brique et alimentés au charbon. Il s'agit vraisemblablement du type de four utilisé par les essayeurs lorsqu'ils prospectaient de nouveaux sites. Ce type de four est facilement transportable, mais ne peut atteindre les degrés de températures requises pour fondre certaines variétés de minerais. Nous croyons que tel était le cas lors des analyses effectuées par les métallurgistes accompagnant Frobisher car, de l'aveu même des essayeurs, le minerai est dur à fondre et les métaux précieux qu'ils ne trouvent pas peuvent être emprisonnés dans les scories. Ce problème technique de chauffage est identifiable par la sous-liquéfaction des scories, ainsi que par les tentatives de refonte des morceaux de scories fusionnés à des fragments de creusets.

Ces mêmes sources semblent indiquer que les essais ont été réalisés avec du plomb métallique plutôt qu'avec un sel de plomb. Il s'agit d'une méthode traditionnelle décrite dans la plupart des ouvrages sur la métallurgie qui n'est cependant pas aussi précise que les méthodes avec des sels de plomb ou d'antimoine. La technique au plomb métallique est cependant plus simple et demande moins de connaissances techniques. On peut en effet travailler rapidement à l'aide de boulettes de plomb d'une taille prédéterminée que l'on emporte avec soi. La méthode avec les sels de plomb est légèrement plus précise, mais nécessite l'utilisation d'une balance. Bien que les sels de plomb ne soient pas toujours purs, ils ont moins de chance d'être contaminés par les métaux précieux, comme c'est souvent le cas du plomb métallique que l'on exploite dans les mines.

Nous avons également la chance d'avoir retrouvé des scorificateurs sur le site. Il s'agit d'un type peu commun de creuset qui pourrait indiquer un désarroi des essayeurs face à des variétés de minerais inconnues. De forme plate, le scorificateur est utilisé pour rôtir un minerai, mais surtout pour observer visuellement les réactions d'un minerai lors de la fonte au four, ce qui est quasiment impossible avec le creuset traditionnel. Enfin, l'utilisation du scorificateur permet de se faire une idée des autres composantes et de préparer la recette d'additifs en conséquence. Une étude des artefacts provenant de la fouille a fait découvrir un aspect inconnu des techniques d'essai du XVIe siècle. En effet, l'analyse détaillée des morceaux de creusets et de scories nous a permis d'émettre l'hypothèse que les essayeurs n'ont pas utilisé de moules coniques pour transvider le minerai fondu, mais qu'ils laissaient plutôt les creusets refroidir puis les brisaient systématiquement au marteau pour récupérer les culots de plomb. Voici à quoi cette technique pouvait ressembler. La destruction des creusets pour récupérer les culots de plomb est moins précise que l'utilisation d'un moule conique, car l'on constate qu'un plus grand nombre d'impuretés demeurent attachées au culot et sont transportées vers la coupelle. Il reste que cette technique n'est pas ou peu décrite dans les imprimés sur la métallurgie du XVIe siècle.

Le *quiproquo* Frobisher

Pourquoi a-t-il fallu trois voyages à Martin Frobisher avant que les promoteurs des voyages ne comprennent que le minerai était sans valeur ? La liste des possibilités au niveau technique que nous avons élaborée pour répondre à cette question est assez longue. Au premier rang des hypothèses se trouve le manque de savoir-faire des métallurgistes. Les Anglais de la Renaissance maîtrisaient encore mal l'analyse et l'extraction des métaux précieux. Leur savoir-faire se limitait aux opérations de transformation des métaux précieux, telles que l'orfèvrerie ou la fabrication de monnaie avec des métaux précieux pour la plupart importés. Le savoir-faire des Anglais dans le domaine de l'extraction repose sur des techniciens étrangers installés en Angleterre. La plupart de ceux-ci viennent des grands sites miniers du continent. Sur trois personnes en charge des analyses du minerai rapporté par Frobisher, deux sont originaires d'Allemagne et plus de la moitié des autres analyses seront effectuées par des techniciens étrangers, dont un Français et un Italien. Les chercheurs qui se sont penchés sur le niveau de compétence des gens impliqués dans les essais de Frobisher ont mis en lumière qu'une majorité des analyses avaient été faites par des non-spécialistes. En fait, la relative simplicité d'un essai le rendait accessible à plusieurs personnes. Tel était le cas des orfèvres, des monnayeurs, voire des pharmaciens, qui utilisaient l'or pour préparer certains médicaments. La principale hypothèse avancée pour expliquer les résultats positifs obtenus sur le minerai d'origine canadienne, est celle de l'utilisation de plomb contenant déjà des métaux précieux, à l'insu des métallurgistes. Il s'agit d'une mégarde possible puisque le plomb exploité dans les mines en contient souvent. Ce phénomène était toutefois connu des spécialistes du XVIe siècle, y compris de plusieurs essayeurs impliqués dans l'aventure de Frobisher.

L'aspect décisionnel

Une autre partie de l'explication de ce déboire doit être recherchée du côté des dirigeants anglais qui sont à cette époque très favorables à ce genre d'expéditions. En temps normal les autorités auraient dû remettre en cause le troisième voyage. Un deuxième voyage était bien sûr nécessaire parce que le morceau de minerai rapporté par Frobisher lors du premier voyage en 1576 n'était pas assez gros pour permettre des analyses concluantes ; la mise sur pied du deuxième voyage dans un but de prospection et d'extraction était donc nécessaire. Cependant, il est logique qu'au retour, l'on aurait dû découvrir la valeur réelle de la roche, mais les choses ne se déroulèrent pas ainsi. L'un des navires revint à Bristol, ce qui retarda le transbordement prévu à Londres. Un secret excessif est maintenu autour de toute cette affaire. Les premières analyses donnent des résultats assez équivoques. De nouveaux métallurgistes font leur entrée et des querelles d'intérêt éclatent entre les essayeurs qui ne s'entendent pas sur les méthodes à utiliser. La lenteur des autorités à mettre sur pied un troisième voyage incite même l'un des essayeurs (croyant naïvement en la valeur du minerai) à fausser son analyse, ce qui ajoute à la confusion.

Le seul aspect sur lequel les essayeurs s'entendent est qu'il faut des fours plus puissants pour atteindre les températures nécessaires à des essais définitifs. Le seul problème est que la construction de ces fours allait durer jusqu'au mois de septembre suivant, date trop tardive pour envoyer une expédition dans les régions nordiques. En mars 1578, les autorités anglaises se retrouvent donc face à un dilemme. Personne n'est certain de la valeur du minerai, mais il est impossible de faire de bonnes analyses avant juin, date prévue pour le départ des navires. Dans de telles conditions d'incertitude, les autorités auraient dû reporter la mise sur pied du troisième voyage d'une année, d'autant plus qu'ils avaient assez de minerai à Londres pour faire leurs futures analyses. En mars 1578, la reine Élisabeth tranche en faveur de l'envoi d'une troisième expédition à laquelle elle adjoint le projet de fonder la première colonie anglaise en Amérique. La tentative coloniale meurt dans l'oeuf, mais Frobisher ramènera néanmoins en Angleterre douze navires remplis de minerai qu'il croit chargé des métaux précieux.

ÉPILOGUE : L'ÉTUDE D'UN ÉCHEC

La dimension métallurgique des voyages de Frobisher reste encore à creuser. L'étude de la teneur naturelle du plomb européen en métal précieux, la comparaison avec des sites métallurgiques du même type en Europe, l'analyse détaillée des artefacts et des scories sont des voies de recherches envisagées pour les années à venir. La recherche concernant l'histoire des voyages de Martin Frobisher ne se limite cependant pas à des dimensions métallurgiques. Au contraire, les marchandises enterrées sur place en vue d'un quatrième voyage sont demeurées congelées durant plus de 400 ans dans le sol de la Terre de Baffin qui ne dégèle qu'en surface durant quelques semaines par an. Ces objets tels des tonneaux, des paniers, ou des tuiles ; des denrées alimentaires telles des pois, biscuits, viande ou poisson issus du site de Martin Frobisher deviendront bientôt des références internationales pour la culture matérielle européenne du XVIe siècle.

Remerciements

La réalisation de ce programme de recherche a été rendue possible grâce à la participation financière du Conseil de recherche en Sciences Humaines du Canada et du Fonds pour la Formation des Chercheurs et l'Aide à la Recherche.

Bibliographie

B. Allaire, " The gold and silver refiners of Paris in the XVIth and XVIIth centuries ", *Bulletin of the canadian institute of mining and metallurgy*, 89 (1996), 76-82.

B. Allaire, D.D. Hogarth, " Martin Frobisher, the Spaniards and a Sixteenth Century Northem Spy ", *Terrae Incognitae, Journal for History of Discoveries*, 28 (1996), 46-57.

R. Auger, " Le premier établissement anglais au Nouveau-Monde : les expéditions de Martin Frobisher (15761578) chez les Inuit de la Terre de Baffin ", *L'Archéologie et la rencontre de deux mondes*, Québec, Musée de la civilisation, 1992.

M.B. Donald, " Burchard Kranich (1515-1578) : miner and Queen's physician... ", *Annals of Sciences*, 6 (1950), 308-322.

W.W. Fitzhugh, J.S. Olin (eds), *Archeology of the Frobisher Voyages*, Washington, Smithsonian Institution Press, 1993.

D.D. Hogarth, P.W. Boreham, J.G. Mitchell, *Martin Frobisher's Northwest Venture 1576-1581 : Mines, Minerals, Metallurgy*, Ottawa, Canadian Museum of Civilization, 1994.

B. Neumann, " Die Anfänge der Probierkunst und die ältesten deutschen Probiervorschriften ", *Metall und Erz : Zeitschrift für Metallhüttenwesen und Erzbergbau*, vol. 17, 7 (1920), 168-173.

THE TRANSFER OF INDIAN TEXTILE PRINTING TECHNIQUES TO EUROPE

Walter ENDREI

Lynn White called the medieval Western culture a " molten society ", Mumford wrote about its well ploughed and harrowed soil ready to conceive foreign seed and Needham made the point that : " Technical inventions show slow but massive infiltration from east to west throughout the first fourteen centuries... Until the 15th century, Western European technology may be said to have been less advanced than that in other Regions of the Old World "[1]. Major innovations such as stirrups, paper, the wind-mill, gun-powder, letter-printing, horse-collar, Indian figures, distillation or in the field of textile-industry : the introduction of silk and cotton, the spinning-wheel, treadle-loom and knitting-devices, were all taken over and adapted to European conditions during the Middle Ages although some late-coming inventions still had serious difficulties in this respect.

The most important among these was textile printing, products of which were well known in Western markets. In medieval times, the oriental fustian, damask, and knitted hosiery were used long before the Occident became familiar with their " know-how ". While due to lack of evidence the way these early innovations were transferred, is rather vague, 17th and 18th century endeavours to acquire cotton-printing techniques are more clear. Let us look at the main acknowledged facts :

a. From time immemorial, oriental textile makers were familiar with methods of printing with fast colours. I mention only Herodotos' amazement when he writes about painted designs on textiles — a technology used by Caucasian peoples — which " do not fade by washing " and look " as if they were woven into the fabric " (I. 203).

b. These methods were manifold, varying from Coptic Egypt to modern Indonesia and differ from each other — in part in the mechanical equipment

1. J. Needham, *Science and Civilisation in China*, Cambridge, 1954, I, 222.

used (i.e. printing with blocks, stencil, painting with a brush, fountain-pen like instruments *etc.*) and in part in the chemical processes put into practice.

c. Nevertheless all had one thing in common : not one could fix the dyestuff while patterning. This was done during subsequent dying.

d. European textile-printers however used only blocks and fixed pigments with adhesives on the fabric. Rohich became mostly dull and would have disappeared by washing.

I wish to emphasise point c. To the European craftsman, the staining of cotton or linen fabrics by an almost invisible mordant or wax was inconceivable, the appearance of the pattern after dying stupefied him ; early on, Pliny the Elder (23-79) is amazed by the " marvellous " method used by Egyptians to obtain different colours on the fabric from the same vat which hints at the use of mordants (*Historia naturalis*, XXXV).

However, despite the fact that dyers were secretive, innovations spread like wildfire : an English recipe brought by a Flemish embroiderer to Italy was recorded by the Frenchman Jean Alcherius in 1410[2]. Likewise, the new American dyestuffs (cochineal, logwood) were accepted over all of Europe within a few decades (*ca.* 1540). It is not the purpose of this lecture to clear up the problem of why, where and how interplay of certain forces caused the boom in cotton printing in the last quarter of the 17th century. Dozens of scholars from Mantoux to Aiolfi have devoted attention to it and here I prefer to allude to some matters of record rarely taken notice of : a) Chintz was well known in Western Europe long before it came into fashion. It had to be imported from India because of b) a shortage of cheap, light fabrics especially in England. Here no significant flax-culture existed so that linen goods had to be imported from the continent[3].

Referring to a. quotes come mostly from Pepy's Diary (5th September 1663), Molière (*Le bourgeois gentilhomme*, 1670) and Defoe (referring to Queen Mary's calico-dress, 1709). However, pintadoes are first recorded in 1602, chintz in 1614 and we know that in 1644 over twenty-five thousand yards of chintz were sold in London.

Sometimes W. Sherwin's patent of 1676 has been considered a starting point of the transfer of know-how as he emphasises to know " the only true way of East India printing…never till now performed in our Kingdom ". This is erroneous. As early as 1646, the first Indian printers were documented in Marseilles ; it is uncertain whether it was the same two masters who engaged two Armenians in 1672 with the object of *peindre d'indianes de la façon du Levant et Perse*. In addition the inventory of a woodcarver and producer of

2. M.P. Merrifield, *Original treatises dating from the XIIth to the XVIIIth centuries on the art of painting... dying..., etc.*, London, 1849 ; H. Cassebaum, " Der Ursprung der Indigofarberei ", *Melliand Textilbericht*, 6 (1965), 625.

3. S. Aiolfi, *Calicos und gedrucktes Zeug*, Stuttgart, 1987, 19-22.

chintz is mentioned in Avignon (1677)[4]. Cotton printers emerged all over Europe in the 1670s and 1680s. Therefore, it seems likely that there was an even earlier adaptation of oriental methods.

There are those two mysterious merchants with their Turkish expert from Ammersfort (1678), the French Huguenots who were awarded a four-year privilege in Geneva (1687) and who must have had considerable experience already. In Hungary, the first documented printer of indigo-reserves died in 1692, so that at least a decade must have gone by since he had learned his trade[5].

Turning to the circumstances of the transfer it would seem useful to list the main features acknowledged by most writers :

a. Although different kinds of fast colour printing and similar patterning processes were known throughout the Orient it was India that gave those techniques to Europe.

b. The first recipients Dutchmen and Englishmen, placed orders in the beginning. This was followed by pattern or sample incitation (after 1662) which often involved elements of European (or even Chinese) taste, while

c. the first (poor) imitations, based upon the spot experience were produced using the " trial and error " method.

Together with the first statement we should also consider the fact that although Indian printed cottons had been exported since early medieval times from all directions and reached e.g. Egypt (Fostat, Yuseir el Quadim), in contrast to Chinese silk fabrics they did not become common in Europe till well into the 16th century. The question intrudes of whether the gorgeously colourful Mughal style of the golden age of the imperial workshops established under Akbar (1556-1605) triggered the " Indian craze " in the following decades ?

It is remarkable that though Turks and Armenians were sporadically engaged in the industry, not a single Indian name appears in the documents. By the 17th century, the English and Dutch printing-techniques were considered the best. It may be assumed that civil servants belonging to the two East India Companies were the main observers and transmitters. The French company *Compagnie des Indes* was founded much ratter and therefore perhaps their efforts to transplant Indian methods were restricted to the acquisition of written reports. It is indicative that while at least half a dozen such descriptions composed by French merchants and travellers have survived just an no English one is known. In contrary to the so far mentioned cases, we possess a series of (so called " learned ") detailed description, mostly French preserved in the archives to one of them even samples of the subsequent phases of the printing-

4. R.M. Schwarz, " Les toiles peintes indiennes ", *Bulletin de la Société Industrielle de Mulhouse*, 709 IV (1962), 10 (BSIM).

5. O. Domonkos, W. Endrei, " Európai textilnyomás és hazai kékfestés ", *Technikatörténeti Szemle* 1-2 (1962) 39.

process is attached[6]. A typical example of transfer seems to be the case of the factory of the Duke of Bourbon-Condé in the castle of Chantilly[7].

As prime minister the duke was well acquainted already with the chemist Cisternay du Fay, at that time inspector of dye-works and mines, who had carried out various experiments with dyestuffs. Du Fay gave an order to the naval officer Beaulieu, employee of the *Compagnie des Indes* to prepare a circumstantial survey concerning the printing methods used in Pondichery (1734-35). The coincidence of time and personal contacts make it plausible that the short-lived chintz production of 1733-40 was mainly based upon the recipes of Beaulieu. The next point in question must be : which were the main components of oriental know-how transferred to Europe ? We may exclude from these the knowledge of carved wooden printing-blocks, of the main dyestuffs indigo and madder, even that of the black pigment ferro-sulphate or acetate and some important mordents, though, their use was limited to the wool and silk craft.

Also uncertain is whether patterning using a brush had been in use or not. Let us sum up the main acquisitions concisely :

a. the preparation of grey cotton cloth not only by bleaching but also to make it susceptible to mordants ;

b. printing and painting as well as the mixing of a whole gamut of mordants and the subsequent dying process ;

c. the application of the wax-resist technique for indigo blue printing and its combination with the previous procedure.

Although these are the main achievements the transfer was much more complex, for different reasons. Amongst others was the fact, expressed by Ryhiner (1776) as follows : *La façon des Indiens ne peut convenir en Europe, parce que l'on ne peut se servir des mêmes drogues, et que la main d'oeuvre y est trop chère*. This is only one of numerous causes, which induced Ryhiner to grumble over the inferiority of European chintz. In order to give an idea of what know-how was transferred, let me describe the main steps in the sophisticated procedure :

1. First bleaching
2. Treatment with buffalo-milk + tanning agent
3. Drying, rinsing
4. Pounding with wooden club
5. Transferring pattern by paper stencil
6. Outlining of contours with ferroacetate by brush

6. V. Berinstein gives a summary concerning *French Travellers, Bulletin du CIETA Lyon*, 66 (1988), 53-56. We also know about an early (1727) series of printing-recipes of Dutch origin L.A. Driessen, " Altholländische Kattundruckrrezepte ", *Melliand*, 21 (1948), 191.

7. P.R. Schwarz, " La fabrique d'indiennes du Duc de Bourbon (1692-1740) au château de Chantilly ", *BSIM*, 722, 1 (1966).

7. Treatment with sheep-dung
8. Treatment with rice-water
9. Resist-way application
10. Indigo-dying, drying
11. Removing wax by boiling
12. Second treatment with buffalo-milk
13. Printing and painting of mordants. Ferri-, chrome- zinc *etc.* alum-cupri salts and/or oxides
14. Red-dying
15. Drip-drying, rinsing
16. Adding of yellow (green) shades
17. Second bleaching

We may assume that initially only a distorted, comparatively simple method was adapted to the European situation. That means, certain elements were neglected or substituted for. However further developments took place as well.

As far as we know neither the kalam, a sort of fountain-pen with which fine lines of wax are applied to the cloth, nor the punched paper-stencil was adapted. Wax as a resist-material was however abandoned as well. Instead a starch-colloid became common. A major problem was the thickening agent for mordents and adjacent dyestuffs. The gum employed was not identified neither in the manuscript of Ahmedabad (1678) nor in the patent of Sherwin (1767). Floud attaches importance to this unsettled problem, though Arabic gum is mentioned in 1737.

In my opinion therefore, the transfer though it aimed to imitate chintz perfectly, succeeded earlier and spread faster by fabricating simple indigo reserve and madder mordant prints. Obvious reasons exist for this including :

a. European printers were accustomed to work with one or at most two colours and

b. in the latter case the wooden blocks often did not match.

c. the use of different mordants and dye-baths combined with resist-printing represented a level of sophistication which could only be achieved by considerable investment and/or teamwork, moreover their improper use caused often bad colour-fastness

d. artists and wood-carvers had to readjust themselves to the new method.

e. the supplementary painting required the training of a new type of skilful female auxiliaries (*pinçoteuses*).

On the other hand European printers already had some advantages, techniques quite unknown in India and these were quickly incorporated into the process :

- since the 16th century they used dyestuffs (logwood, cocheneal) of American origin ;
- they knew copper plate printing ;

- they used a series of sophisticated machines, like horse driven mangels, pumps, raspmills, hoisting whinches.

But the main advantage of the European industry was its flexibility. The complicated indigo-reserve method was after some decades replaced by the direct, so called English-blue (ab. 1745), new and more suitable thickening agents were invented by which the process could be speeded up and last not least roller-printing swept away all other printing techniques by its high productivity.

In Needham's view, quoted in the introduction, the east-west advance of technological modernisation may be characterised as a sort of " infiltration ". Inevitably, the parallel with a semipermeable membrane arises caused by an absence of movement in the opposite direction or any kind of feedback.

One wonders why the Islamic block, wedged between Europe and the Far East since the 7th century, and serving as a perfect one-way transmitter, has never adopted or sent on the results achieved in the West ? It is not only the clockwork, Gothic skeletal architectural structures and camshaft mills which were missing, but upgraded forms of many of the Western innovations never made their way back to the east either.

Innovations in the textile industry fall exactly within this category.

The sophisticated technology of silk Processing was taken up relatively rapidly in Europe. Thus, the silk throwing mill invented in Italy never spread back to the east. Although the spinning wheel is Indian in origin, its improved form, the flyer-wheel never re-crossed the eastern border of the Balkans. Hosiery is another oriental invention as well but William Lee's knitting frame reached its ancestral region only by the last century.

Textile printing too is a typical example. At the beginning, the Industrial Revolution hit the Indian textile industry hard which could not come up with the necessary innovations on its own. In the end, it was in the financial interests of English machine builders as well as the colonial powers, to export roller printing machines, other related modern equipment and chemical procedures to India during the second half of the 19th century.

Inventivité technique et naissance d'industrie innovante en Belgique, 1860-1910[1]

Michel Oris

Le titre de ce papier est quelque peu menteur, car notre contribution porte presque exclusivement sur la ville de Huy, située le long de la Meuse à quelque 30 km en amont de Liège. Cette étude de cas permet cependant de dégager des leçons générales. Sur un mode qui peut sembler *a priori* paradoxal, nous essaierons en effet de montrer que Huy est un cas de marginalité exemplaire et significative, un cas où la marginalité et ses nuances sont constitutives du système innovateur. En outre, étudier l'innovation technologique à l'échelle d'une localité d'une dizaine de milliers d'habitants est une approche originale. Ce thème fait l'objet d'une littérature abondante, mais l'essentiel est constitué soit d'analyses d'indicateurs quantitatifs au niveau national (brevets, production scientifique, *etc.*), soit de discussions théoriques sur les systèmes innovateurs et les opinions discordantes des grands économistes sur la question. Bien que l'ensemble soit d'une qualité intellectuelle évidente, on peut lui reprocher de négliger la réalité humaine de l'innovation, qui ressort mieux dans un cadre plus restreint.

Une marginalisation technique et économique toute en nuances, 1793-1860

Une localité comme Huy était un des fleurons de la vieille sidérurgie wallonne d'ancien régime. Sous le régime français, à l'extrême fin du XVIIIe, début du XIXe siècle, son potentiel humain et technique sort d'une longue léthargie[2]. Entre 1803 et 1810, les forges de la région expérimentent toute une série de

1. Le dépouillement des volumineux registres de brevets d'invention n'a pu être réalisé que grâce au concours amical de Françoise Brandt, licenciée en histoire, qui dans le contexte quelque peu difficile d'une fin de thèse, a apporté une aide précieuse. Tous mes remerciements !

2. D. Morsa, " Une croissance manquée : Huy (1600-1800). Quelques hypothèses sur un effondrement économique ", *Revue belge de Philologie et d'Histoire,* 59-2 (1981), 325-381.

progrès partiels qui vont mener à la technique de l'affinage de la fonte par puddlage. Dès 1793, donc moins d'une dizaine d'années après leur mise au point dans l'Est de la France, des laminoirs à fer ont été installés le long du Hoyoux, rivière torrentueuse qui rejoint la Meuse à hauteur de Huy. Deux de leurs propriétaires, Bastin et Dautrebande, sont remarqués lors de l'exposition impériale de Paris en 1806, pour la qualité de leurs tôles. A cette occasion, ils restent cependant dans l'ombre de leur allié, Nicolas Delloye, rejeton d'une vieille famille de notables hutois qui a développé une remarquable fabrique de fer blanc destinée à remplacer les produits anglais. Il reçoit même le monopole de la fabrication et de la commercialisation à travers tout l'empire de Napoléon Ier. Dès l'an XII, ses usines sidérurgiques ont un produit brut annuel de 1,5 millions de francs qui en font un géant industriel. Dans tout l'Est de la Belgique, il ne cède le pas qu'à une seule personne : rien moins que William Cockerill[3].

William Cockerill est arrivé à Verviers en 1799 à l'invitation de deux puissantes familles de drapiers, les Simonis et les Biolley. Il était chargé de mettre au point des “ mécaniques à l'anglaise ”, puis d'en équiper leurs établissements. Après divers aléas, ses efforts sont couronnés de succès et il est rapidement concurrencé par des fabricants locaux. Les ateliers de construction mécanique vont constituer, selon l'expression de Pierre Lebrun, un “ point de convergence technologique ”. Il va réagir sur la sidérurgie par le biais d'une demande accrue de fer. C'est dans ce créneau que s'installe à partir de 1817 le fils de William, John Cockerill. Il érige en 1820 un premier haut fourneau au charbon de terre. Une solidarité étroite, qui durera plus d'un siècle, se dessine alors entre l'industrie des métaux et l'exploitation houillère. L'ensemble de ce processus constitue une chaîne de fondation technico-économique, fil d'Ariane de la première révolution industrielle en Wallonie[4].

Le fer blanc de Nicolas Delloye est en dehors de cette chaîne, tout comme le zinc mis au point par Jean-Daniel Dony, qui a quand même connu une destinée industrielle remarquable. Cependant, alors que la mécanisation réalisée sous le régime français permet aux drapiers verviétois de faire face à la concurrence anglaise après 1815, il n'en a pas été de même dans le secteur du fer blanc qui s'étiole aussitôt. De 1820 à 1850, un écart décisif se creuse alors entre Liège et Huy. Il faudra attendre 1837 pour qu'un des fils de Nicolas Delloye, Clément (1782-1845), tente d'inscrire la région hutoise dans le processus de révolution industrielle. A cette date, il n'y a encore que quatre hauts fourneaux au coke dans la province de Liège, mais onze sont en construction. Clément Delloye met sur pied la *Fabrique de Fer du Hoyoux* dans le but explicite de moderniser les outils existants autour d'un de ces hauts fourneaux modernes. Les principaux industriels locaux s'associent dans cette société anonyme

3. G. Hansotte, “ Contribution à l'histoire de la métallurgie dans le bassin du Hoyoux aux temps modernes ”, *Bulletin de l'Institut Archéologique Liégeois,* 80 (1967), 65-66.

4. P. Lebrun, *Essai sur la révolution industrielle en Belgique (1770-1847),* Bruxelles, Académie royale de Belgique, 1979, 585.

qui bénéficie des conseils et du soutien de John Cockerill en personne. Pourtant, dès 1842, c'est la faillite. La pauvreté du sous-sol local, l'éloignement des gisements de houille, rendaient le projet irréaliste.

Il faudra attendre une dizaine d'années et un mariage heureux avant que Charles Delloye-Matthieu (1816-1896), fils de Clément, puisse récupérer une partie des sites et développer des laminoirs. Parallèlement, depuis 1822, les Godin investissent dans la fabrication du papier. Ces deux familles n'ont plus essayé de bâtir de grandes usines modernes. Elles se sont contentées d'associer des sièges de taille réduite, éparpillés le long du Hoyoux dont ils continuent à tirer leur énergie hydraulique, bien suffisante pour une production modeste assurée par de petites équipes traditionnelles d'ouvriers hautement qualifiés. Visant le créneau limité des produits de qualité supérieure dans lequel elles excellent — comme le prouvent les récompenses qu'elles conquièrent dans les expositions universelles tout au long du siècle — ces entreprises utilisent au mieux des structures technologiques archaïques comme atout économique. Il s'agit de tactiques et non d'archaïsmes des mentalités, car profitant de la mort du fondateur, dès les années 1840, les Delloye-Dodémont-Dautrebande et les Godin-Preud'homme ont pris des parts dans le capital de Cockerill. Les représentants de ces deux familles seront régulièrement en majorité au sein du conseil d'administration de cette entreprise phare de l'économie belge, devenue en quelque sorte une firme hutoise... Il est donc évident que les industriels de Huy avaient accès à des ressources techniques et assez de moyens financiers pour se moderniser s'ils l'avaient jugé utile. Instruits par l'échec de la *Fabrique de Fer*, ils ont préféré accepter la polarisation industrielle sur le bassin houiller liégeois et une marginalisation relative, somme toute assez confortable, pour Huy[5].

Confortable, car Huy est ainsi passée à côté des affres d'une expansion urbaine effrénée commandée par l'afflux massif d'immigrants. Les conditions sanitaires désastreuses des banlieues industrielles se résument simplement : au milieu du XIXe siècle, il y a 5 années d'écart entre les espérances de vie à la naissance à Seraing, cité de Cockerill, et dans une petite ville tranquille comme Huy. La distorsion entre la population d'une part, les structures d'encadrement, les services publics et privés d'autre part, exprime une dualisation extrême du monde urbain. Plus fondamental que l'hygiène pour notre propos, l'état de l'enseignement en témoigne. Dès 1846, les écoles hutoises fournissent une formation scolaire de base à 95 % des enfants de la localité, alors que deux ans plus tôt, le taux de scolarisation primaire sérésien n'était que de l'ordre de 27 %. Cet indicateur atteint 122 % à Huy, qui importe des élèves des villages voisins dès 1866, alors qu'en 1900, il n'est encore que de 81 % à Seraing. De

5. M. Oris, *Economie et démographie de Huy au XIXe siècle,* Thèse inédite de l'Université de Liège, 1991, vol. 1 et 2 ; et *Id.* " Une culture économique originale ", dans J.L. Doucet (dir.), *Huy. Hommes de fer et de fonte,* Huy, 31-32.

même, Huy a fondé vers 1827-1829 une des premières écoles industrielles du Royaume, alors qu'un grand centre de production comme Seraing ne bénéficiera d'un établissement semblable qu'en 1858, et que les puissants bassins charbonnier du Hainaut attendront 1868[6]. Globalement, dans l'ensemble de la Belgique, les villes traditionnelles conservent et même renforcent leurs " privilèges culturels " tout au long du XIXe siècle[7].

On nous pardonnera d'avoir consommé le début de ce papier à fixer ainsi le cadre hutois avant 1860, mais le processus de la deuxième révolution industrielle est indissociable de ce milieu et ne peut se comprendre en dehors de lui.

Le développement de la métallurgie légère de 1860 à 1914

A Huy, la deuxième révolution industrielle s'identifie largement à la montée de la métallurgie légère, productrice de biens finis. En 1846, le recensement industriel ne repère dans la ville que 231 métallurgistes, soit moins de 20 % de la main-d'œuvre travaillant en atelier. En 1910, ils forment le premier secteur d'emploi qui réunit 40 % des ouvriers hutois, soit 1.639 personnes réparties entre pas moins de 57 établissements. Cette ascension s'est réalisée à partir des années 1860, mais ne s'est vraiment accélérée que durant les 20 ou 25 années qui ont précédé la première guerre mondiale[8].

Dès les environs de 1850, le cubilot et la technique de la seconde fusion ont permis à cette branche d'activités de s'affranchir des hauts fourneaux et d'éclater en plusieurs industries, de la mécanique à la fabrique des outils ménagers. Pour servir le modeste marché industriel local, des mécaniciens comme Berbuto s'installent à partir de 1845 au plus tard. Les Godin supportent la création de la firme Thiry-Dautrebande en 1855. Elle était spécialisée dans la mise au point de " continues ", chef-d'œuvres technologiques de l'époque qui assuraient la transformation de la pâte en longues bandes de papier[9]. Plus tard, en 1868, une branche des Preud'homme, descendants et héritiers, par leur mère ou grand-mère, des fondateurs des papeteries Godin, renforce ce secteur avec une fabrique de fournitures ferroviaires et industrielles. Une autre branche créera une usine de confection de machines-outils à la fin du siècle.

Parallèlement, des firmes de quincaillerie visant un marché privé s'efforcent de dépasser le stade de la production strictement artisanale. C'est le cas de

6. N. Caulier-Mathy, " Le patronat et le progrès technique dans les charbonnages liégeois, 1800-1914 ", dans G. Kurgan-Van Hentenryk et J. Stengers (éd.), *L'innovation technologique, facteur de changement (19e-20e siècles)*, Bruxelles, 1986, 52-53 ; J. Puissant, " A propos de l'innovation technologique dans les mines du Hainaut au 19e siècle, ou la guerre des échelles n'a pas eu lieu ", dans *Ibidem,* 86-87.

7. M. Crubellier et M. Agulhon, " Les citadins et leur culture ", dans G. Duby (éd.), *Histoire de la France urbaine,* t. 4 : *La ville de l'âge industriel* (1983).

8. M. Oris, " Une culture économique originale ", *Op. cit.,* 33, 35.

9. M. Daumas, *Histoire générale des techniques,* Paris, 1968, t. 3, 639-641.

Vandenkieboom dont Natalis Briavoinne[10], dans son célèbre portrait des industries belges au lendemain de l'indépendance, note qu'il était le seul industriel hutois équipé d'une machine à vapeur. Son entreprise a pourtant végété jusque dans les années 1860. Un troisième sous-groupe se forme un peu plus tard, centré sur la fabrication de ce que les économistes appellent des " biens de consommation durables ", qui sont principalement des pièces de l'ameublement domestique, poêles, cuisinières, *etc.* Dans cette branche, le nom le plus célèbre est celui de Nestor Martin, qui abandonne Saint-Hubert pour fonder un atelier à Huy vers 1849. Une quatrième branche réunit les ferronniers, qui ont bénéficié des nouveaux courants architecturaux, en particulier de l'*Art Nouveau.* Les magnifiques escaliers en fer forgé de Nicolas Porta en sont un exemple[11].

Séparer la métallurgie légère en quatre groupes d'activités, comme le propose ce descriptif, est une simplification outrancière. Outre les articles de bâtiments et les spécialités d'ornement pour la grande construction, Porta concurrençait Nestor Martin avec ses poêles et cuisinières, et fabriquait encore des pièces pour serruriers, de la quincaillerie et des articles funéraires (principalement des croix en fer forgé)[12]. Il s'attaqua même à la mécanique industrielle puisqu'en 1873, il fit breveter une machine à plier et coller les enveloppes. Cet éparpillement apparent en fait un personnage tout à fait représentatif de ce milieu dynamique. Le véritable trait distinctif de toutes ces firmes, que révèle très bien cette source méconnue que sont les brochures et encarts publicitaires, c'est leur polyvalence. Comme les laminoirs de la vallée du Hoyoux, elles gardent des sièges de production de taille moyenne et fondent leur réputation sur la qualité plus que sur la quantité, mais elles ajoutent encore une corde à leur arc : la diversité de leurs offres, la gamme très large de produits qu'elles proposent. C'est ce caractère souple, polyvalent, cette flexibilité technique, cette capacité à s'adapter à des marchés en mutation, qui vont définir de nouveaux rapports avec l'innovation.

Innovation et brevets d'invention

La législation sur les brevets remonte à la loi du 25 janvier 1817 qui offre des garanties financières et juridiques inexistantes auparavant. Le recours à ces protections légales, plus modernes mais limitées dans le temps, se généralise à partir des années 1840. Il en résulte diverses réorganisations d'ordre administratif (commission d'experts, tarifs, durée, sanctions, *etc.*), sans modification des principes inscrits dans la norme de 1817, dont le principal est l'exploitation

10. N. Briavoinne, " Sur les inventions et perfectionnements dans l'industrie depuis la fin du 18e siècle jusqu'à nos jours ", *Mémoires couronnés par l'Académie royale des Sciences et Belles-Lettres de Bruxelles* (1838), t. 13, 135.

11. Voir J.M. Doucet, *Huy. Hommes de fer et de fonte, op. cit.,* 36 et *passim.*

12. J. Gougnard, *Huy-pittoresque. Guide de l'excursionniste à Huy,* Huy, 1891.

économique de l'invention sous peine d'annulation du brevet. La loi du 24 mai 1854 renforce cette option tout en imposant deux modifications de fond : la publication du " secret " et l'absence d'un examen préalable portant sur le caractère innovateur. Autrement dit, pour contester comme pour faire respecter un brevet, il faut recourir au débat contradictoire devant les tribunaux[13]. Cela n'empêche pas les octrois de passer de moins de 2.000 à près de 12.000 entre 1855 et 1913, les graphiques établis par Paul Servais[14] montrant que l'essentiel de cette progression se situe entre 1870 et 1910.

L'imposition de publier une description du procédé ou de l'outil a produit à partir de 1854 une source imprimée extraordinairement volumineuse et détaillée. Bien que les renseignements soient moins complets auparavant, il est possible de remonter jusqu'en 1830 à partir du recueil de Dujeux[15] et des archives du Service de la Propriété Intellectuelle du Ministère des Affaires Economiques. Ce type de document a déjà été étudié dans plusieurs pays pour y rechercher le reflet de l'inventivité, les dynamiques de l'innovation[16]. Il va cependant de soi que les brevets ne dessinent jamais un portrait exhaustif de la recherche industrielle. L'attitude traditionnelle du " secret de fabrication " jalousement caché a certainement subsisté[17], mais la prise de brevet s'est manifestement de plus en plus popularisée. C'est en soi un phénomène culturel qui prouve que, surtout après la première révolution industrielle, l'innovation fait l'objet d'une reconnaissance sociale et le statut d'inventeur reconnu est recherché comme composante d'une image de marque[18].

13. Un aperçu plus complet de la législation sur les brevets a été brossé récemment par P. Servais " Les brevets d'invention en Belgique de 1854 à 1914 ", dans *LIe Congrès de la Fédération des Cercles d'Archéologie et d'Histoire de Belgique et 4e Congrès de l'Association des Cercles francophones d'Histoire et d'Archéologie de Belgique. Congrès de Liège 20-23.VIII.1992. Actes* (Liège, 1994), t. 2, 362-365.

14. *Ibidem,* 368.

15. J.B.C. Dujeux, *Catalogue des brevets d'invention délivrés en Belgique du 01/11/1830 au 31/12/1841,* Bruxelles, Demanet, 1842. Voir aussi *Recueil spécial des brevets d'invention,* Bruxelles, 1854-1896, t. 1-44, suivi du *Recueil des brevets d'invention,* Bruxelles, 1897-1900, t. 45-48.

16. A. Fleischer, *Patentgesetzgehung und Chemisch-pharmazeutische Industrie in deutschen Kaiserreich (1871-1918),* Stuttgart, 1984 ; J.P. Hirsch, " A propos des brevets d'invention dans les entreprises du Nord au XIXe siècle ", *Revue du Nord,* 67-265 (1985), 447-459 ; R.J. Sullivan, " The revolution of ideas : widespread patenting and inventions during the industrial revolutions ", *The Journal of Economic History,* 50-2 (1990), 349-362 ; K.L. Solokoff et B.Z. Khan, " The democratization of invention during early industrialization : evidence from the United States, 1790-1846 ", *The Journal of Economic History,* 50-2 (1990), 363-378 ; *etc.*

17. P. Lebrun, *Essai sur la révolution industrielle..., op. cit.,* 606-608.

18. Paul Servais (" Les brevets d'invention en Belgique de 1854 à 1914 ", *op. cit.,* 363) note d'ailleurs que dans la loi néerlandaise de 1817, des primes et récompenses étaient prévues pour " l'encouragement de l'art et de l'industrie nationale ". Dans le Nord de la France, Hirsch (" A propos des brevets d'invention dans les entreprises du Nord..., *op. cit.,* 458) raconte l'histoire très significative de cet inventeur qui donne littéralement son secret à ses concurrents locaux, et propose même de le communiquer à ses collègues d'un département voisin parmi lesquels se trouve le fils d'un ministre, parce que la seule chose qu'il veuille absolument comme récompense pour son invention, c'est la croix de la Légion d'Honneur...

Pour dresser un portrait global de l'inventivité hutoise telle qu'elle ressort du dépouillement des brevets, une double typologie a été utilisée. La première clé est le secteur servi par l'objet du brevet (usage indéterminé, domestique, agricole et industriel, ce dernier décomposé par branche d'activités autant que possible). La seconde clé est le " type de création ". Cette expression mérite quelque explication. Ont été distingués :

- les machines, " destinées à transformer l'énergie et à utiliser cette transformation " ;
- les appareils, industriels ou ménagers, qui ne font qu'utiliser l'énergie ;
- les outils, distincts des précédents par l'absence d'un mécanisme (pelles, pioches, vis, limes, *etc.*) ;
- les ustensiles de ménage, qui sont repris séparément pour servir l'analyse, mais sont en fait des outils ;
- les biens d'équipement (briques, charpentes, lits, meubles divers, *etc.*) ;
- et une rubrique Divers qui accueille par exemple les produits chimiques (tableau ci-après).

Les effectifs ne permettent pas de décomposer l'analyse outre mesure ; aussi seules deux périodes sont-elles distinguées, avant et après 1880. Le choix de cette date se comprend au vu du graphique qui reprend le nombre de brevets hutois par période quinquennale. Il n'y a à peu près rien avant 1850, puis une poussée presque continue, à peine un peu freinée entre 1866 et 1870, jusqu'à un maximum de 61 en 1876-80. Le recul n'en est que plus spectaculaire et brutal : 16 brevets seulement de 1886 à 1890. Les événements sociaux de 1886 sont bien connus en Belgique pour avoir brutalement replacé la question sociale au centre de la vie politique belge. Pour plagier le titre d'un ancien article de J. Ruwet, au vu d'un indicateur de l'inventivité, la crise fut aussi morale à Huy. Il y a reprise après 1890, mais à la fin du XIXe siècle, les niveaux des années 1860 sont à peine dépassés. Cette chronologie est particulière puisque, rappelons-le, dans l'ensemble du Royaume, hormis un pic vers 1895 suivi d'un léger affaissement, la tendance à une progression continue est dominante depuis 1870[19].

19. P. Servais, " Les brevets d'invention en Belgique de 1854 à 1914 ", *op. cit.*, 368.

RÉPARTITION DES BREVETS PRIS PAR DES HUTOIS SELON LE TYPE DE CRÉATION ET LA DESTINATION, 1831-1900

Type de création :	*Machines*	*Appareil indéter.*	*Appareil domestique*	*Appareil industriel*	*Outils*	*Ustensiles*	*Biens d'équip.*	*Divers*	*Total n*	*Total %*
DESTINATION	1831-1880									
Indéterminée	1	3	0	6	0	0	1	5	16	7,77
Usage domestique	0	0	23	0	0	8	0	7	38	18,45
Agriculture	1	0	0	6	3	1	0	1	12	5,83
Usage indust. indét.	7	0	0	12	1	0	0	0	20	9,71
Mines-carrières	0	0	0	6	0	0	0	0	6	2,91
Métallurgie	0	0	0	22	1	1	0	6	30	14,56
Bois	0	0	0	2	0	0	0	0	2	0,97
Papeterie	0	0	0	16	2	0	0	0	18	8,74
Imprimerie	0	0	0	2	0	0	0	0	2	0,97
Chimie	0	0	0	6	0	0	0	2	8	3,88
Alimentation	0	0	0	29	3	0	0	1	33	16,02
Cuir	0	0	0	1	0	0	0	0	1	0,49
Textile	0	0	0	1	0	0	0	0	1	0,49
Bâtiment	0	0	3	0	0	0	13	0	16	7,77
Transport	1	0	0	2	0	0	0	0	3	1,46
Total n	10	3	26	111	10	10	14	22	206	100,00
Total %	4,85	1,46	12,62	53,88	4,85	4,85	6,80	10,68	100,00	

DESTINATION	1881-1900									
Indéterminée	1	5	0	11	0	0	0	5	22	16,67
Usage domestique	0	0	23	1	4	0	1	8	37	28,03
Agriculture	1	0	0	1	2	0	0	0	4	3,03
Usage indust. indét.	3	0	0	5	3	0	0	0	11	8,33
Mines-carrières	0	0	0	4	0	0	0	0	4	3,03
Métallurgie	0	0	0	5	0	0	0	4	9	6,82
Bois	0	0	0	0	0	0	0	0	0	0,00
Papeterie	0	0	0	10	0	0	0	0	10	7,58
Imprimerie	0	0	0	0	0	0	0	0	0	0,00
Chimie	0	0	0	2	0	0	0	0	2	1,52
Alimentation	0	0	0	10	0	0	0	0	10	7,58
Cuir	0	0	0	0	1	0	0	0	1	0,76
Textile	0	0	0	0	0	0	0	0	0	0,00
Bâtiment	0	0	3	2	0	0	11	1	17	12,88
Transport	0	0	2	1	1	0	0	1	5	3,79
Total n	5	5	28	52	11	0	12	19	132	100,00
Total %	3,79	3,79	21,21	39,39	8,33	0,00	9,09	14,39	100,00	

MOUVEMENT QUINQUENNAL DE LA PRISE DE BREVETS PAR DES HUTOIS, 1831-1900

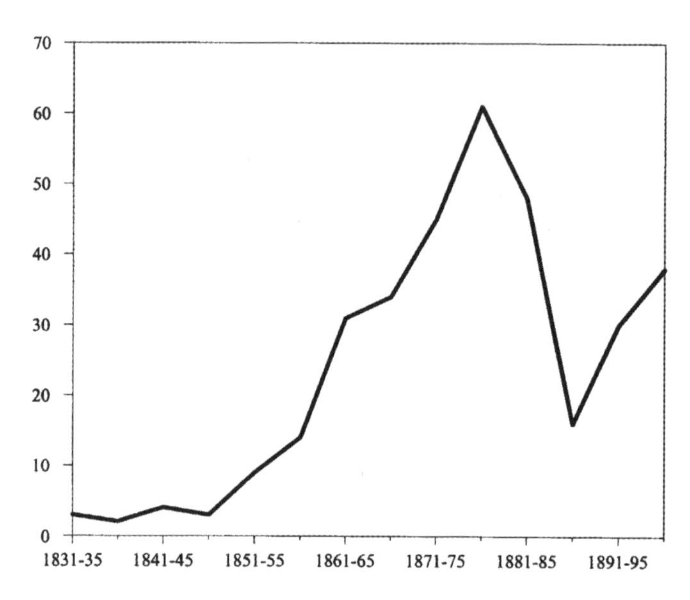

Durant la période d'expansion, jusqu'en 1880, l'usage industriel domine avec un peu moins de 75 %. Les branches les plus importantes sont la métallurgie au sens large (14,6 %) et le papier (8,7 %), précédés de manière un peu surprenante par l'alimentation (16 %). Ceci montre bien l'impact somme toute limité des secteurs de la première expansion hutoise. Parmi les lamineurs, Hyacinthe Delloye a pris un brevet en 1849 pour un nouveau système d'affinage du fer, en 1854 pour un mode de serrage des laminoirs, en 1855 pour une technique d'introduction d'air comprimé dans les fours à puddler et à chauffer le fer. Les Smal-Delloye reviennent aussi à plusieurs reprises dans les registres de brevets, mais ces noms disparaissent avant 1860. Dans la papeterie, les Godin se sont toujours montrés très attentifs aux nouveautés. Ils mènent régulièrement de savantes études préalables, n'hésitant pas à envoyer des missions à l'étranger, notamment en Angleterre, puis ils importent[20]. C'est d'ailleurs à ce titre seul qu'ils figurent sur la liste des brevets belges. C'est exactement le même comportement prudent, retenu, qu'observe J.P. Hirsch[21] dans le secteur textile du Nord de la France, et qui se traduit par un fort recours aux brevets d'importation.

Bien qu'elle ait moins marqué les mémoires que la papeterie ou les laminoirs, la distillerie-brasserie a sa place parmi les branches dynamiques dès la

20. M. Oris, *Economie et démographie de Huy au 19e siècle, op. cit.*, 297-300.

21. " A propos des brevets d'invention dans les entreprises du Nord... ", *op. cit.*, 449.

deuxième moitié du XIXe siècle. Peu importante sur le plan de l'emploi, davantage sur le plan de la valeur ajoutée (un sixième de celle de la ville de Huy à en croire le verdict des patentes en 1847), elle s'est mécanisée avec vigueur dans les années 1850-1860 d'après le témoin que représente l'adoption des machines à vapeur[22]. Cette modernisation des outils est confirmée par 32 brevets sur les 40 qui se rapportent au groupe *Alimentation*. Ce sont avant tout les industriels intéressés qui sont engagés dans cette entreprise d'adaptation des techniques brassicoles ou de distillation : la famille Bernard est à l'origine de 18 brevets et les Springuel de 7. Il est d'ailleurs manifeste qu'un effet de concurrence joue. A deux ou trois reprises, les Springuel demandent un brevet quelques mois, voire quelques jours après les Bernard pour un procédé similaire...[23]

En matière de brasserie, les apports hutois se situent bien dans la multitude des essais empiriques qui avaient pour objet de mieux contrôler le processus de fermentation en le protégeant à la fois des variations de température et des infections. Bernard-Martin met au point un appareil pour refroidir et oxygéner la bière (brevet du 31 janvier 1860) qu'il perfectionne à plusieurs reprises avec l'aide de B.J. Springuel, avant apparemment que les deux compères se brouillent. A partir du milieu des années 1870, Louis Pasteur met en évidence le rôle des bactéries dans les infections, et il montre qu'elles peuvent être annihilées par un échauffement. C'est la technique de *pasteurisation* qui se répand rapidement[24]. Les Hutois l'adoptent et l'adaptent très vite en proposant des systèmes à la fois réchauffant et refroidissant (Springuel le 31 décembre 1880 ou le 31 janvier 1885).

Dans la distillerie, outre l'introduction de la vapeur et l'adaptation des méthodes appliquées dans les brasseries, les efforts portent principalement sur la macération des matières premières, en particulier du riz, qui permettait de rencontrer le développement de la demande pour les alcools blancs[25]. Il ne semble pas qu'il y ait de relâchement des efforts avant le début des années 1880, mais la faiblesse des effectifs ne permet guère de certitude en la matière.

22. Cl.M. Christophe, " De l'introduction de la machine à vapeur dans la région de Huy ", *La Vie wallonne,* 58 (1984), 69-97.

23. L'exemple le plus flagrant : F.J. Bernard obtient le 15 mai 1868 un brevet pour un *stuikmande danaïde à l'usage des cuves matières des brasseries*, et le 30 mai 1868, A. Springuel-Bollinne pour un *panier dit stuikmande à l'usage des cuves matières des brasseries*. On voit là les effets pervers de l'acceptation des brevets sans examen préalable portant sur la réalité de l'innovation.

24. G. Dejongh, " Une boisson nationale et une industrie nationale : le développement de l'industrie de la bière en Belgique ", dans B. Van der Herten, M. Oris, J. Roegiers (dir.), *La Belgique industrielle en 1850. Deux cents images d'un monde nouveau,* Deurne, Editions MIM-Crédit Communal, 1995, 277.

25. M. Goosens et K. Dries, " L'agriculture belge et deux exemples précoces d'agrobusiness : l'industrie sucrière et les distilleries ", dans B. Van der Herten, M. Oris, J. Roegiers (dir.), *La Belgique industrielle en 1850..., op. cit.,* 269.

En tout cas, par la suite, la perte d'inventivité va de pair avec le déclin du nombre de firmes et du personnel employé[26].

En termes relatifs, la part de l'industrie dans les destinations des brevets recule globalement après 1880, et ceci pour chacune des branches qui composent ce secteur, à l'exception du *Bâtiment*. Il est vrai qu'il n'y a pratiquement rien qui soit rangé sous cet intitulé et que nous n'aurions pu plutôt considérer comme destiné à l'usage domestique. En additionnant ces deux rubriques, se dégage un groupe en pleine expansion relative, d'environ 25 à quelque 40 % de 1831-80 à 1881-1900.

La répartition selon le type de création permet de dégager les mêmes tendances. Entre 1831 et 1880, plus de la moitié des brevets ont pour objet des appareils industriels (près de 54 %). Entre 1881 et 1900, ils ne représentent plus que 39 %. Ce recul profite en ordre principal aux appareils domestiques (surtout des poêles et des cuisinières), qui réalisent une progression de 12,6 à 21,2 %. Même si la rubrique *Divers* croît elle aussi de 11 à 14,4 %, ce classement, ainsi qu'un simple parcours des données de base, montrent à quel point l'emploi des métaux se popularise au cours du XIXe siècle, pour finir par être partout présent à tous les stades de la vie quotidienne. Des machines aux biens d'équipement, si le secteur destinataire varie dans les proportions susdites, le secteur producteur reste pratiquement toujours la métallurgie légère, celle des ateliers de construction mécanique et des fonderies[27]. En ce domaine, l'expérience hutoise est parfaitement représentative d'une lame de fond qui traverse tout le XIXe siècle[28].

UNE INNOVATION DÉBORDANTE AU SERVICE DU MARCHÉ

Entre 1830 et 1900, les habitants de Huy ont été à l'origine de quelque 338 brevets. C'est minuscule par rapport à Bruxelles, notamment en raison de la présence de bureaux et sociétés spécialisés dans la capitale, mais c'est beaucoup pour une petite ville d'une dizaine de milliers d'habitants. Ce débordement d'inventivité, à la fois apparent et réel, doit être considéré comme un indicateur culturel de modernisation, mais aussi comme une composante au sein d'une stratégie industrielle et commerciale. De ce point de vue, ce qui est clair, c'est la préparation de l'expansion par une dynamique qui la précède et dont l'inventivité est un signe.

Le pionnier est sans conteste Jacques Vandenkieboom. Briavoinne[29] écrit de lui : “ M. Vandenkieboom de Huy s'est fait connaître par ses ustensiles de cui-

26. M. Oris, *Economie et démographie de Huy au 19e siècle, op. cit.*, 162-173, 218-225.

27. La liste de tous les brevets pris par des Hutois de 1830 à 1900 est publiée dans *Idem*, vol. d'annexes, 72-80.

28. F. Caron, *Le résistible déclin des sociétés industrielles,* Paris, Perrin, 1985, 57.

29. N. Briavoinne, “ Sur les inventions et perfectionnements dans l'industrie… ”, *op. cit.,* 135.

sine en tôle, estampés d'une seule pièce, qui présentent une grande économie sur les mêmes ustensiles en fer coulé ; il est breveté depuis 1833 pour ce procédé ; il est parvenu à estamper des chaudrons de tôles d'une seule pièce d'une grande dimension "[30]. Inutile de dire que ce procédé ne lui vaut pas une place d'honneur dans les encyclopédies d'histoire des sciences, mais il participe d'une logique globale. La première révolution industrielle a énormément développé la production de fonte brute. Vers 1850, un pays comme la Belgique manque cruellement d'outils de transformation en biens semi-finis ou finis. C'est cette situation qui a fondé la croissance des tôleries du Hoyoux, mais aussi les multiples essais des ferronniers traditionnels pour conquérir de nouveaux marchés, voire pour les inventer. Parmi les noms qui renvoient à une firme connue, un Nestor Martin a déposé pas moins de 32 brevets, ses concurrents directs de la famille des Laurent une douzaine, les Bovy une quinzaine, Nicolas Porta plus modestement 5, *etc.*

Il s'agit d'apports certes nombreux mais toujours d'importance mineure. Dans la continuité des recherches de Vandenkieboom, L.J. Laurent-Tihange prend le 28 février 1861 un brevet pour un procédé de moulage des coquemars en fonte de fer au sable vert ; Lelong-Herla met au point une " marmite pour toute espèce de poêle cuisinière " (15-10-1862) ; Nestor Martin un " système de construction de marmites, casseroles et d'ustensiles de cuisine en fonte brute émaillée " (15-11-1869) ; *etc.* Comme certains intitulés l'indiquent, ces produits ménagers sont pour une bonne part associés à de nouvelles techniques de cuisson sur poêle ou cuisinière. En ce domaine, ce sont les progrès de la tôlerie et de la fonderie qui permettent, à partir du milieu du XIXe siècle, de réaliser une production commerciale de types de plus en plus perfectionnés[31]. Une demi-douzaine de modèles de poêles sont " inventés " à Huy dès les années 1860 : un " sphérique fumivore " (Bovy-Delcourt, 1-12-1864), un autre " à buse plate " (Petit-Jamagne, même date), un " à air chaud ", *etc.* De multiples améliorations ponctuelles leur sont apportées par la suite, comme le " système de registre en fonte destiné à régler le courant d'air dans les tuyaux de poêles ", breveté par Nestor Martin le 16 décembre 1872.

Dans pratiquement tous les cas, la prise d'un brevet vise surtout à protéger le produit mis au point, à s'en assurer durant dix ou quinze ans une relative exclusivité sur le marché national. L'expérience hutoise supporte l'interprétation proposée par J.P. Hirsch pour le Nord de la France[32] : " Une part au moins de ces brevets représentait une sorte de droit d'entrée dans le métier, destiné à prévenir contestations et procès en contrefaçon, sans produire aucune amélioration bouleversante ". Il est d'ailleurs significatif qu'à Huy, la période 1886-

30. En fait, le brevet dont il est question dans ce témoignage est du 28 juin 1834.

31. M. Daumas, *Histoire générale des techniques, op. cit.*, t. 3, 517.

32. J.P. Hirsch, " A propos des brevets d'invention dans les entreprises du Nord... ", *op. cit.*, 452.

1900, caractérisée par un très net développement de l'emploi et de la production dans la métallurgie légère, le soit aussi par un recul du rythme de dépôt des brevets. Les marchés se sont créés et se développent d'eux-mêmes ; le succès étant au rendez-vous, il est désormais moins nécessaire de chercher fébrilement et en tous sens.

Cependant, il ne faut pas exagérer ce recul ; des brevets continuent à être pris. En 1893, Nestor Martin dépose coup sur coup des dossiers sur un système de réchaud à gaz, une cuisinière mixte au coke et au gaz, le poêle " L'indispensable ", le poêle " Eureka "... Ces recherches font partie d'une tactique de singularisation du produit ; ils participent à la construction de son image de marque. A partir d'une technologie de base empruntée, les fabricants hutois améliorent et surtout particularisent. Par là même, ils adaptent leurs produits aux attentes de clients qui sont encore aussi loin de la consommation de masse qu'eux-mêmes le sont de la production de masse. En somme, ils réalisent l'adéquation souple et exigeante entre l'offre et la demande à un stade transitionnel. Le caractère mixte de la production, mi-artisanal, mi-industriel, renvoie à la nouvelle composition de la clientèle, où la bourgeoisie traditionnelle, exigeante et individuelle, commence à voisiner les " petits bourgeois " et les ouvriers d'élite.

A moyen terme, la révolution industrielle a modifié la structure sociale de la population, moins sans doute en raison d'une hausse des salaires réels parmi les prolétaires, qu'en supportant la croissance de " cette classe moyenne et de cette petite bourgeoisie devenues typiques de la société belge "[33]. C'est à ces groupes sociaux intermédiaires qu'il faut attribuer l'ouverture de marchés comme ceux des biens de consommation durables, franchement ou relativement coûteux, tels que les automobiles ou les poêles, cuisinières, *etc.* Ils forment une clientèle moins riche que la bourgeoisie traditionnelle, plus nombreuse cependant, et malgré tout très exigeante. Les " grands magasins " qui apparaissent à la fin du XIXe siècle se posent en modèles et symboles de modernité, mais ils ressuscitent le vieux *truck-system* en s'attachant les services de dizaines de cousettes et cordonniers[34]. Ce sont ces artisans qui fabriquent sur commande les vêtements de clients qui n'imaginent même pas autre chose que du " sur mesure ". Partout, la publicité met l'accent sur la confection pièce par pièce, sur l'originalité et la diversité des produits, sur la présence de l'atelier en arrière-cour qui permet de venir surveiller la réalisation du bien commandé, *etc.*[35]

33. J. Gadisseur, " Le triomphe industriel ", dans *L'industrie en Belgique. Deux siècles d'évolution 1780-1980* , Gand, 1981, 83.

34. G. Kurgan-Van Hentenryk, " Les patentables à Bruxelles au 19e siècle ", *Le Mouvement social,* 108 (1979), 80-81.

35. La Veuve Mossoux et ses fils, petits métallurgistes hutois, possèdent aussi une fabrique de pianos et d'harmoniums. Ils croient bon de préciser dans une publicité de 1891 que " les ateliers étant à la maison, les clients pourront suivre et contrôler l'exécution de leurs commandes ". Le 15 février 1885, ils ont breveté un " appareil de précision servant à accorder les pianos ".

Ce sont exactement les mêmes désirs que s'efforce de rencontrer la métallurgie des biens d'équipement domestique qui, à Huy, est sans nul doute l'activité de production de biens finis qui évolue le plus vers une structure industrielle[36]. L'analyse des brevets éclaire ce secteur d'un jour nouveau. En effet, il utilise l'innovation comme outil de singularisation du produit, comme une méthode pour réaliser l'adéquation entre la rentabilité de la fabrication et la satisfaction du client exigeant.

Les inventeurs, un produit social et culturel

Au XIXe siècle, le personnage de l'inventeur est inséparable des idéaux de mérite et de promotion sociale. Victorieuses de révolutions sanglantes qui ont abattu les privilèges héréditaires de la noblesse, les bourgeoisies glorifient le mérite personnel. Dans le domaine industriel, innombrables sont les biographies de fondateurs partis de rien, qui ont gravi tous les échelons à la force du poignet, grâce à leur prévoyance, leur ténacité, leur intelligence, leur travail. Ces récits à vocation " hagiographique " et pédagogique enjolivent la réalité en noircissant l'origine sociale de leur modèle. Jean-Marie Doucet[37] l'a bien montré dans le cas de Nestor Martin notamment.

En fait, le père de Nestor Martin était un joaillier de Saint-Hubert, puis petit fondeur à Huy, et avant de voler de ses propres ailes, Nestor avait d'abord créé une entreprise avec ses deux frères aînés. Les Vandenkieboom étaient actifs dans la métallurgie hutoise depuis au moins le XVIIIe siècle[38]. En 1847, Jacques était un des dix principaux contribuables de la ville, tout comme Jean-Baptiste Berbuto ou le distillateur Springuel. Dans la fabrication de machines pour papeteries, Dautrebande était issu d'une vieille famille patricienne de maîtres de forge, et son associé Thiry était l'ingénieur principal des Godin. Eux et leurs enfants produiront quelque 17 brevets d'invention ou de perfectionnement.

Cette alliance de la tradition bourgeoise et du savoir moderne se répétera régulièrement à l'intérieur même des familles. Le fils et homonyme du constructeur de machines à vapeur Jean-Baptiste Berbuto, a été en 1852 le premier Hutois diplômé ingénieur de l'Université de Liège. Il avait choisi l'option mécanique, ce qui l'a mis en excellente position pour reprendre l'atelier paternel et le maintenir jusqu'à sa mort en 1892. Joseph Bernard a conquis le même

36. Il faut d'ailleurs noter que, des vêtements des grands magasins aux poêles et cuisinières des petits métallurgistes, il est une autre logique, celle de la formation ou du développement d'idéaux de la féminité bourgeoise, avec la femme en son coquet foyer, symbole d'aisance. Voir le récent essai de Erika Rappaport " The halls of temptation : gender politics and the construction of the department store in late Victorian London ", *Journal of British Studies* (1996), 58-83.

37. J.M. Doucet, *Huy. Hommes de fer et de fonte, op. cit.,* 74-75.

38. A. Budo, " L'émaillerie hutoise ", dans J.L. Doucet (dir.), *Huy. Hommes de fer et de fonte, op. cit.,* 80.

grade académique en 1864 avant de diriger les ateliers Bernard-Martin. Trois ans plus tard, il a d'ailleurs été suivi par Bernard Martin qui ouvre sa fonderie en 1869. Au fil du *Mémorial du Cinquantenaire 1847-1897* de l'*Association des ingénieurs sortis de Liège*, bien d'autres noms connus dans le milieu des métallurgistes hutois ressortent encore : Dautrebande, Preud'homme, Smal, Thiry, Springuel, Wilmotte, Beco, *etc.* Ils forment un socle résistant de Hutois formés à Liège pour dominer les techniques les plus modernes, et qui une fois leurs études terminées, reviennent dans leur ville natale mettre leurs compétences au service de l'entreprise familiale. A Lyon, à une époque à peine plus tardive (1880-1890), P. Cayez[39] a lui aussi observé le remplacement des artisans mécaniciens par des ingénieurs constructeurs qui, dans ce cas, " étaient plus souvent les gendres que les fils ".

Le choix de rester à Huy n'allait pas de soi, car plusieurs ingénieurs originaires de la ville ont réalisé, à Liège, dans le Hainaut, jusqu'en Russie, des carrières brillantes comme gestionnaires de puissantes sociétés anonymes. La dissociation entre les détenteurs du capital et les techniciens qui assurent la gestion courante s'est étendue à travers toute la grande industrie belge au cours de la seconde moitié du XIXe siècle, et a offert à cet égard beaucoup d'opportunités[40]. Ce divorce implique la " prédominance d'une perspective négociante qui tend à placer technique et techniciens à des places subordonnées "[41]. Les ingénieurs issus des familles de métallurgistes hutois ont préféré " s'enterrer " dans une petite ville pour rester leur propre maître et suivre l'exemple des pionniers de la révolution industrielle en cumulant direction technique et économique.

Evidemment, leur présence permet de mieux comprendre la floraison de nombreux brevets à Huy. Il ne faut pour autant négliger le rôle de la popularisation, de la relative démocratisation des savoirs et donc de l'inventivité industrielle. Le fondeur Michel Thonar, un des fondateurs de la fédération hutoise du *Parti Ouvrier Belge*, a déposé pas moins de quatre brevets entre 1874 et 1888, dont un " système de foyer économique chauffant les appartements en utilisant la chaleur perdue des cheminées " (16-4-1879), qui démontre son sens social[42]. Ce géant truculent n'était ni un ingénieur, ni un pur autodidacte. Il avait été formé durant trois ans dans cette Ecole Industrielle de Huy que nous avons évoquée plus haut. Ce fut le cas également des mécaniciens Jules et Arthur Moussiaux, et on a même quelque raison de croire que le célèbre inven-

39. " Quelques aspects du patronat lyonnais pendant la deuxième étape de l'industrialisation ", dans M. Levy-Leboyer (éd.), *Le patronat de la seconde industrialisation*, Paris, 1979, 195.

40. G. Kurgan-Van Hentenryk, " Banques et entreprises ", dans H. Hasquin (éd.), *La Wallonie. La Pays et les Hommes,* Bruxelles, La Renaissance du Livre, 1976, 48-51.

41. J.P. Hirsch, " A propos des brevets d'invention dans les entreprises du Nord… ", *op. cit.,* 448.

42. Ses autres " inventions " sont un poêle calorifère à air chaud à l'usage des écoles (16-2-1874), une amélioration du " système économique " (16-4-1883), et un poêle calorifère en fonte avec ventilateur.

teur de la dynamo, Zénobe Gramme en personne, a bénéficié de cet enseignement technique[43].

Si Huy, dans sa position économique marginale, a pu se transformer au cours du XIXe siècle en une véritable pépinière d'inventeurs, elle le doit avant tout à ses privilèges culturels en matière de système éducatif. Cet atout a été doublement consistant car il a aussi offert à des entrepreneurs imaginatifs et changeants une main-d'œuvre hautement qualifiée.

L'intégration de l'inventivité dans le tissu social

L'extension de la diffusion de connaissances élémentaires dans de plus vastes couches de la population est indéniable (*cf. supra*). L'interprétation sociale classique de ce mouvement a été remise en cause par l'historien anglais Childs[44]. Pour lui, les progrès de la scolarisation ont été un facteur de prolétarisation, un des éléments constitutifs du déclin des élites ouvrières dont la formation — dans les deux sens du terme — était basée sur un long et précoce apprentissage. Ce jugement est erroné parce qu'il se veut universel. Il apparaît tout à fait pertinent appliqué à des élites traditionnelles d'origine préindustrielle, comme les verriers du Val-Saint-Lambert par exemple, ou les lamineurs du Hoyoux. Mais de nouveaux groupes se constituent à partir du milieu du XIXe siècle.

C'est qu'après 1845-1850, quand la première révolution industrielle peut être considérée comme accomplie dans l'espace belge, la technologie ne s'est pas figée. Au contraire, elle se complexifie sans cesse au point que sa mise en œuvre finit par exiger une formation spécifique qui, elle-même, est impensable sans le préalable d'une formation générale. Mis en évidence d'une manière globale par plusieurs auteurs[45], ce point a déjà fait l'objet de quelques démonstrations à l'échelle régionale. Dans le cas alsacien, Michel Hau[46] note que : " L'industrie manufacturière suppose la détention, par un nombre croissant de

43. Dans l'état actuel de l'inventaire des archives de la ville de Huy, il n'existe pas de listes d'élèves de l'Ecole Industrielle. Les personnes évoquées ci-dessus n'ont été retrouvées que parce qu'elles figuraient parmi les étudiants brillants récompensés par la commune. Cf. *Rapport sur l'Administration et la Situation des Affaires de la Ville de Huy 1888-89*, 53 ; *Ibidem 1893-94*, 52 ; *Ibidem 1899-1900*, 47. Sur Zénobe Gramme, voir Cl.M. Christophe, " L'industrialisation à Huy jusqu'à la seconde guerre mondiale ", dans *Huy : 985-1985, le livre du millénaire,* Liège, Vaillant-Carmanne, 1985, 108.

44. M.J. Childs, " Boy labour in late Victorian and Edwardian England and the remaking of the working class ", *Journal of Social History,* 24-4 (1990), 783-802.

45. P. Bairoch, " Histoire des techniques et problématique du démarrage économique ", dans *L'acquisition des techniques par les pays non-initiateurs. Pont-à-Mousson 28 juin-5 juillet 1970,* Paris, CNRS, 1971, 174-176 ; F. Greenaway, " Individual action and group action : their roles in promoting adjustement to technological change in Great Britain, dans *Ibidem,* 239-250 ; M. Daumas, " Le mythe de la révolution technique ", *Revue d'Histoire des Sciences et de leurs Applications* (1963), 291-302.

46. *L'industrialisation de l'Alsace (1803-1939),* Strasbourg, Publication des Universités de Strasbourg, 1987, 434.

collaborateurs de l'entreprise, d'un savoir considéré jusque là comme sans intérêt pour le monde de la production ". Dans le Nord de la France, Jean-Pierre Hirsch met en épingle l'opinion exprimée dès 1828 par Théodore Barrois, chargé de mission de la Chambre de Commerce de Lille, pour lequel : " Ce n'est pas à sa législation [sur les brevets d'invention] que l'Angleterre doit la supériorité de ses manufactures mais à la manière dont l'instruction est répandue dans la classe ouvrière ; toutes les conceptions nouvelles trouvent de suite des mains propres à les mettre à exécution en y ajoutant les petits détails qui pourraient manquer, au lieu d'une force d'inertie souvent invincible que rencontre le manufacturier français "[47].

Certes, au cours de la seconde moitié du XIXe siècle, la variance spatiale des taux d'alphabétisation s'est réduite ; les centres de l'industrie lourde ont comblé une partie de leur retard[48]. En outre, dans ces pôles industriels, l'application du taylorisme ou de ses prémices tend à contrebalancer les effets négatifs de formations insuffisantes par la spécialisation des tâches assignées aux travailleurs. Tout cela limite la portée des privilèges culturels des villes traditionnelles comme Huy, mais ceux-ci restent importants pour comprendre l'épanouissement de petites et moyennes entreprises performantes, comme celles de la mécanique et de la quincaillerie à la fin du XIXe siècle. Compte tenu de leur large gamme de produits, ces entreprises requièrent une main-d'œuvre plus qualifiée que spécialisée. Bien sûr, le sens du mot " qualification " n'est plus celui dont se flattent les tôliers ou les papetiers du Hoyoux ; son contenu s'est " modernisé ".

Une main-d'œuvre qualifiée doit être fixée, " attachée " à l'entreprise qui ne peut se permettre de la licencier au moindre soubresaut conjoncturel, faute de ne plus la retrouver disponible au moment de la reprise. Lors de la crise qui affecte la production entre 1881 et 1885, la productivité diminue, signe que l'emploi est préservé. En 1878, l'atelier Dautrebande-Thiry en est réduit à ne plus travailler que quatre jours par semaine mais ne licencie pas[49]. Le dépouillement exhaustif des registres de population de Huy en 1890, réalisé pour une autre étude, montre que 67 % des métallurgistes hutois et 71 % des papetiers étaient nés dans la ville, alors que 55 % des travailleurs du bâtiment, 58 % de ceux de l'alimentation, 62 % de ceux du commerce et des services, étaient des immigrés.

Ce serait cependant gravement dénaturer la politique sociale des ateliers de mécanique, fonderies, *etc.*, que de la réduire à des motivations économiques rationnelles. Elle repose fondamentalement sur une morale sociale qui s'articule autour du mythe unanimiste. Le patron fondateur, ancien ouvrier ou petit

47. J.P. Hirsch, " A propos des brevets d'invention dans les entreprises du Nord… ", *op. cit.,* 453.

48. M. Oris, *Economie et démographie de Huy au 19e siècle, op. cit.,* 364-371.

49. R. Dion, *Histoire du socialisme dans la région hutoise,* Huy, s.d., 19.

artisan, est magnifié pour sa réussite personnelle. D'emblée, une large part est faite au fantasme, car souvent, comme nous l'avons vu plus haut, l'origine sociale n'est pas si basse et l'ascension est le produit du travail de deux ou trois générations. Mais l'image de l'homme qui s'est fait tout seul, à partir de rien, par ses seuls mérites, est une des plus puissantes qui traversent les mentalités du XIXe siècle[50]. La montée d'un individu bénéficie aussi d'une reconnaissance sociale car elle profite à ses ouvriers, auxquels il s'impose naturellement comme modèle et figure paternelle. Il sait ce qui est bon pour l'entreprise, comme le démontrent ses succès, et ce qui est bon pour la firme est bon pour ceux qui y travaillent, qui doivent accepter ses décisions et même les supporter[51].

De nombreux témoignages décrivent effectivement des ouvriers fiers d'un patron qu'ils considèrent comme issu de leurs rangs, et des rapports sociaux relativement harmonieux. Le plus bel ensemble est constitué par les réponses à l'enquête organisée en 1891 sur l'érection d'un Conseil de l'Industrie et du Travail à Huy. Pour Nicolas Porta et Cie : " Notre établissement existe depuis bientôt 24 ans et jamais nous n'avons eu la moindre grève. Nos rapports avec nos ouvriers sont très bons et nous faisons tout ce qui est en notre pouvoir pour améliorer leur condition. Nous croyons qu'il en est de même chez nos confrères de la ville "[52]. De fait, Thiry et Cie, Vandenkieboom et fils, Nestor Martin, Jean-Baptiste Berbuto, mais aussi divers groupements de travailleurs, confirment. Il est aussi significatif qu'entre 1894 et 1914, un contraste net existe entre les succès électoraux honorables du *Parti Ouvrier Belge* et la médiocrité du mouvement syndical hutois, dont les seules actions réellement couronnées de succès sont les grandes grèves politiques de 1902 et 1913 pour l'obtention du suffrage universel[53]. En 1902 d'ailleurs, dans la plupart des entreprises de la métallurgie légère, les patrons supportent plus ou moins franchement le mouvement[54].

Il est clair que ce climat social fut en partie le produit d'une économie flexible basée sur l'introduction d'innovations modestes mais continues, qui impliquait une participation active des travailleurs. En retour, l'harmonie des

50. P. Lebrun, *Essai sur la révolution industrielle..., op. cit.,* 637 ; M. Oris, *Economie et démographie de Huy au 19e siècle, op. cit.,* 1030-1032.

51. Jean-Marie Doucet (*Huy. Hommes de fer et de fonte, op. cit.,* 96) évoque ainsi ce discours d'un ouvrier de Nestor Martin qui, en 1877 lors d'une manifestation en l'honneur du patron, prouve sa culture et sa docilité en déclarant : " Nous mettrons en action le vieil apologue d'Agrippa : que les membres doivent servir la tête et se laisser diriger par elle ".

52. Archives de l'Etat à Huy, *Ville de Huy*, n° provisoire 20130.

53. M. Satinet-Demet, " Naissance et développement du mouvement socialiste dans la région hutoise ", dans *Fédération des Cercles s'Archéologie et d'Histoire de Belgique. XLIVe session. Congrès de Huy. Annales* (Tielt, 1978), t. 1, 274-279 ; J.J. Messiaen et A. Musick, *Mémoire ouvrière 1885/1985. Histoire des fédérations. Huy-Waremme,* Bruxelles, PAC, 1985, 25.

54. M. Oris, " La gestion de la grève générale de 1902 à l'échelle locale. Edition du dossier de police de la ville de Huy ", dans *Annales du Cercle hutois des sciences et Beaux-Arts* (1994), 211-212.

rapports sociaux a elle aussi favorisé la souplesse et l'adaptation des fonctions, l'intégration des nouveautés techniques. Ceci posé, il faut tout aussi clairement souligner les limites de cette structure solidaire en apparence parfaite. En 1891, plusieurs grèves ont déjà eu lieu dans des entreprises hutoises, y compris certaines de celles qui affirment explicitement le contraire. Les premières grèves connues dans la région ont d'ailleurs eu lieu en 1871 dans des firmes métallurgiques, chez Nestor Martin et aux ateliers Thiry. Il s'agissait de protester contre une durée de travail portée à 13h30, dont 12h30 effectives, pour répondre à la formidable croissance de la demande entre 1870 et 1873[55]. La souplesse, la flexibilité de ces firmes, avaient aussi des côtés négatifs ! Leur omission dans des portraits idéalisés montre bien que, hors de toute considération politique, l'unanimisme est un mythe : considérer que les intérêts des patrons et des ouvriers ne sont jamais en contradiction est un leurre porteur de conflits d'autant plus exacerbés qu'ils se placent sous le sceau de l'incompréhension réciproque.

Conclusions

A côté des grandes inventions qui tracent le portrait d'une science triomphante, l'examen des quelques centaines de brevets hutois montre que la réalité humaine de l'innovation au XIXe siècle fut aussi et sans guère de doute surtout faite de beaucoup de tâtonnements, que “ le changement technique est un processus cumulatif par essence ”. Les habitants de Huy ont sans conteste participé à cette “ densification ” du réseau d'inventions qui caractérise le XIXe siècle[56]. Le fouillis des initiatives individuelles ne doit cependant pas cacher qu'elles s'inscrivent dans un système innovateur local dynamique, extrêmement cohérent, mais aussi très fragile.

Au départ se trouve la rencontre d'une tradition industrielle qui met l'accent sur le beau, sur la qualité, et de privilèges culturels que la ville doit à son ancienneté, à son poids politique persistant et à son expansion modeste, grâce à laquelle ses structures éducatives ne sont pas soumises à une énorme pression démographique, comme cela s'observe dans les grands centres industriels. La géographie des institutions culturelles et d'enseignement produit un paradoxe fondamental : les mentalités qui ont permis la première révolution industrielle et qui ont triomphé avec elle ont été mieux diffusées sur les marges que dans les pôles mêmes de cette révolution industrielle ; l'idéologie du mérite, du progrès et de l'innovation opposés au conservatisme et à la tradition, s'est mieux diffusée dans un centre urbain tranquille comme Huy, que dans un foyer bouillonnant comme Seraing[57].

55. J. Messiaen et A. Musick, *Mémoire ouvrière..., op. cit.,* 10, 15.
56. F. Caron, *Le résistible déclin des sociétés industrielles, op. cit.,* 53.
57. M. Oris, *Economie et démographie de Huy au 19e siècle, op. cit.*

A la tête d'un réseau dense de petites et moyennes entreprises, des autodidactes comme Nestor Martin, des techniciens qualifiés comme Michel Thonard, des ingénieurs au fait des techniques les plus modernes, comme Berbuto fils, se mettent à l'affût du marché. François Caron, dans une large synthèse des travaux sur l'innovation, a montré que l'inventivité industrielle était beaucoup plus déterminée par le souci de rencontrer une demande, de satisfaire les aspirations des consommateurs, plutôt que par des opportunités nouvelles issues de la recherche scientifique " pure ". Une proportion 3/4 vs 1/4 est couramment observée tant au XIXe qu'au XXe siècle[58].

Les métallurgistes hutois appartiennent manifestement au courant dominant. C'est ce qui explique la multiplicité de leurs productions et de leurs recherches, la polyvalence de leur entreprise. Solokoff et Khan ont également observé la multiplicité des intérêts des inventeurs nord-américains. *Even the great inventors, who might be expected to possess highly specialized knowledge, were only slightly more specialized than the average*[59]. Il en résulte une grande capacité d'adaptation, une souplesse dont l'exemple le plus flagrant est fourni par les Springuel. Nous les avons vu actifs dans la brasserie-distillerie hutoise. Quand ce secteur décline à la fin du XIXe siècle, ils se reconvertissent tout simplement dans une branche d'activités complètement différente : la construction automobile. Jules Springuel-Wilmotte commence à fabriquer des automobiles en 1902 ou 1903, fonde sa firme en 1907, s'associe à la maison *Impéria* de Nessonvaux en 1912, fait triompher ses produits dans plusieurs compétitions avant la première guerre mondiale[60].

Pour réaliser des adaptations continuelles, les entrepreneurs ne s'appuient pas seulement sur leur capital intellectuel et financier, mais aussi sur le capital humain d'ouvriers d'élite, eux aussi produits des privilèges culturels de la ville. Une forme évidente de complicité entre patrons et travailleurs s'inscrit sous le sceau du paternalisme. C'est une notion rabâchée qu'il ne faut pas interpréter de manière simpliste. D'abord, le " maître " assume un rôle paternel parce qu'il s'y sent moralement engagé. Rappelons qu'une des clés du succès de la métallurgie légère hutoise, c'est la décision, *a priori* saugrenue, d'une génération de jeunes ingénieurs d'investir leurs talents et leurs savoirs sur place. Plus tard, le lamineur Delloye-Matthieu ou le fondeur Nestor Martin resteront profondément attachés à Huy et à leurs ouvriers du cru, quand bien même ils auront déplacé ailleurs l'essentiel de leurs intérêts.

Enfin, le système innovateur est cohérent avec le cadre spatial dans lequel il s'inscrit. Il recrute des ouvriers spécialisés et sert une clientèle bourgeoise et

58. F. Caron, *Le résistible déclin des sociétés industrielles, op. cit.,* 153-157.

59. K.L. Solokoff et B.Z. Khan, " The democratization of invention during early industrialization... ", *op. cit.,* 371.

60. Y. Kupelian et J. Sirtaine, *Histoire de l'automobile belge,* Bruxelles-Paris, 1979, 46-47, 184-186.

petite bourgeoise principalement urbaine, dont la proximité lui est précieuse. En outre, la taille relativement modeste des ateliers leur permet une insertion à peu près aisée dans le tissu urbain. Ils participent donc à ce que l'historienne française J. Gaillard appelle " le nouveau pacte entre la ville et l'atelier "[61]. Ce mouvement méconnu permettra à Bruxelles de devenir, à la fin du XIXe siècle, la première région industrielle de Belgique, plus puissante que l'agglomération liégeoise, les bassins hennuyers ou la zone portuaire d'Anvers ![62] Nestor Martin avait d'ailleurs dès 1868 bâti une succursale dans un faubourg de la capitale, Molembeek-Saint-Jean, le " Manchester belge ". De ce point de vue, le cas hutois est tout à fait exemplaire ; il présente de nombreux points communs avec l'expérience de la ville de Liège proprement dite, de Bruxelles ou de Paris[63].

Ce système innovateur dynamique est cependant aussi très fragile. Face à une demande en croissance mais toujours très exigeante, les producteurs sont dans une situation délicate. Leur inventivité consacrée par des brevets participe autant d'une stratégie commerciale de singularisation de leurs produits que d'une amélioration qualitative réelle. Les manufacturiers hutois sont capables de répondre à des besoins à la fois nouveaux et traditionnels, mais cela même les rend peu aptes à assurer une production massive, uniforme et à bon marché, une fois achevée la transition entre la consommation d'élite et la consommation de masse. Par exemple, " l'histoire du cycle et celle de l'automobile montrent qu'au fur et à mesure que le produit se diffuse dans une clientèle de plus en plus large, sa nature se transforme : dans les deux cas que nous venons de citer, un objet 'sportif' au départ, dont l'usage correspond largement à un désir de faire preuve d'originalité, devient progressivement un produit utilitaire "[64]. C'est un phénomène de mutation très général qui s'étend de plus en plus au XXe siècle, parallèlement à la hausse des revenus qui modifie la hiérarchie des accès à la consommation.

Quand les petites et moyennes entreprises de la métallurgie, à Huy ou ailleurs, doivent cesser d'être un secteur de pointe pour se transformer en une industrie traditionnelle de production massive de biens peu différenciés au plus bas prix, elles payent sévèrement la rançon des qualités qui firent leur succès. Jadis considérés comme des novateurs, les patrons les plus " traditionnels " succombent avec leur firme. Les plus prévoyants imposent des mutations qu'ils jugent inévitables, au grand dam de leurs ouvriers d'élite qui ressentent très

61. J. Gaillard, *Paris, la ville (1825-1870)*, Lille-Paris, Atelier de Reproduction des Thèses-H. Camprion, 1976.

62. J. Puissant, *Bruxelles, une ville industrielle méconnue*, Bruxelles, Dossier de la Fonderie, 1994, 72 p.

63. J. Gaillard, *Paris, la ville (1825-1870), op. cit.* ; J. Puissant, *Bruxelles, une ville industrielle méconnue, op. cit.* ; C. Bauwens, *Architecture industrielle : les ateliers Jaspar à Liège*, Mémoire inédit de l'Université de Liège (Histoire de l'Art et Archéologie), 1993.

64. F. Caron, *Le résistible déclin des sociétés industrielles, op. cit.*, 194-195.

mal leur prolétarisation. Le mythe unanimiste renforce l'incompréhension réciproque, et le paternalisme pimente les conflits d'une sauce émotionnelle qui n'arrange rien. Dans *L'Europe technicienne*, David Landes dressait un parallèle éclairant : " Le fabricant paternaliste du continent croyait qu'il était un père pour ses hommes. Et c'était la sincérité même de cette croyance qui le rendit souvent inflexible dans ses rapports avec la main-d'œuvre organisée. Pour l'employeur britannique, un syndicat était sans doute un adversaire, une grève était contrariante et coûteuse, l'effort des travailleurs pour relever les salaires, chimérique. Pour l'employeur du continent, au contraire, un syndicat était une conspiration contre l'ordre public et la moralité publique ; une grève, un acte d'ingratitude ; l'effort des travailleurs pour relever les salaires, l'indiscipline d'un fils impatient. Tout cela était le mal. Et l'on ne négocie pas avec le mal "[65].

Ce cocktail explosif explique des conflits sociaux aigus qui ont parfois pris une tournure proprement suicidaire, comme la grande grève qui bloqua les papeteries Godin de la mi-décembre 1922 à la fin août 1924, ou celle qui frappa durant six mois les ateliers de l'ami liégeois de Zénobe Gramme, l'électricien et fabricant d'ascenseurs Jules Jaspar, en 1957.

En outre, l'expansion d'une activité industrielle plus classique remettait en cause le pacte avec la ville. Dans le cas précisément des ateliers Jaspar, une remarquable étude d'archéologie industrielle a récemment montré quelle imagination la firme a dû déployer pour parvenir à étendre ses installations dans un cadre urbain, s'imposant une dénivelée de plusieurs mètres entre ses bâtiments, allant même jusqu'à construire sur le tunnel du chemin de fer...[66]

Tout cela ne signifie pas nécessairement l'échec inévitable à moyen terme, mais augure certainement d'une adaptation difficile. A Huy, les nombreuses faillites de l'entre-deux-guerres confirment...[67] Une seule firme, celle de Nestor Martin, a survécu à tous les aléas pour devenir, vers 1950, une multinationale implantée en Belgique bien sûr, mais aussi en France, en Allemagne, en Argentine et au Japon, employant dans 15 usines 6.500 personnes auxquelles s'ajoutaient 22.500 sous-traitants. Un des prix fut cependant la fermeture des vieilles usines belges, dont celle de Huy en 1938, au profit de bâtiments modernes érigés à Berchem-Bruxelles en 1929. En outre, en 1979, le groupe a été absorbé par une autre multinationale plus puissante encore, *Electrolux*[68].

Ces expériences sont porteuses de leçons pour le présent. A beaucoup d'égards, il existe un parallélisme fascinant entre les mécaniciens et fondeurs hutois de la deuxième moitié du XIXe siècle, et les secteurs de pointe de notre

65. D.S. Landes, *L'Europe technicienne. Révolution technique et libre essor industriel en Europe occidentale de 1750 à nos jours,* Paris, Gallimard, 1975, 266.

66. C. Bauwens, *Architecture industrielle : les ateliers Jaspar à Liège, op. cit.*

67. Cl.M. Christophe, " L'industrialisation à Huy... ", *op. cit.,* 109-110.

68. J.M. Doucet, *Huy. Hommes de fer et de fonte, op. cit.,* 59, 70.

fin du XXe, par exemple dans le domaine des nouvelles technologies, de l'informatique et du multimédia. A l'heure où les P.M.E. sont considérées à juste titre comme les moteurs principaux des dynamiques de l'innovation et de l'emploi, il serait dangereux de tout miser sur elles sans les préparer aussi à sortir de leur propre transition, et à affronter les soubresauts destructeurs qui accompagnent la croissance.

Calling into Question the Current Knowledge about Zénobe Gramme and his Inventions

Philippe Tomsin

Introduction

Since the end of the 19th century, Zénobe Gramme has been the topic of a lot of biographical summaries or obituaries ; until now, I have listed about fifty.

Fifteen years ago, Maurice Daumas questioned himself about Gramme's personality. First, he noticed many biographical uncertainties about Zénobe Gramme, and showed that his life is still unknown[1]. How is our exact knowledge about Gramme and his work ? What is its impact in the field of the history of technology ? Must we demystify Gramme and his inventions ? Was Gramme a brilliant self-taught, as his historiographers said, or was rather the intellectual context in Paris decisive ?

In this text, I shall not try to answer these questions, because I have not had enough time in order to take advantage of all the documentation I have been able to find. However, I shall analyse his main biographers' works, I shall try to find a reason for the birth of the Gramme myth, and I shall give some pieces of information for future researches.

Oscar Colson's biography

Until today, articles about Gramme have been copied from one another. The original information source would be Oscar Colson's book *Zénobe Gramme, sa vie et ses oeuvres*, published in Liège in 1903. Until 1913, five reissues were published. A stern critical analysis of this book is absolutely necessary. In his book of ninety-two pages, the author gives a lot of stereotypes about his figure. For example : " The origin of Gramme is modest ; he came from a large fam-

1. M. Daumas, " Zénobe Gramme : incertitudes biographiques ", *Technologia*, VI (1) (1983), 3-35.

ily. A dunce but intelligent child, he was clever and ingenious. All his life, he has been a self-taught. His genius was born in precarious moral and material conditions. Although his works were acknowledged by the most important scientific institutions, Gramme stayed particularly modest. Generous towards the working class, which he stemmed from, he had a funny spirit, he was facetious and just a bit irreverent. As a teenager, he was both a dandy and virtuous ; he exposed his republican opinions, and could be irreverent towards authorities, but he has never been an outlaw ".

Colson gave a caricature of Gramme's character, which recalls that of Tchantchès, a typical popular figure of Liège.

Oscar Colson was born in 1866 in the Liège suburb. When he was a young teacher, he adopted radical political opinions. In 1893, he launched out the review *Wallonia*, which was devoted to the study of local popular and folk traditions, and to the cult of regional artists[2].

Colson did not give any indication in order to prove his assertions. Nowhere in his book are archives documents or oral investigation mentioned, and the few bibliographical references in the footnotes are rarely on the subject.

Colson builds the archetype of the lonely and misunderstood hero who can overcome the general incomprehension and who is carry shoulder high, but who stays fair, upright, unbending and humble. After the First World War, a Gramme worship grew in the Walloon region. Statues appeared, crystals were engraved with Gramme's face, and even barrels, made in Damas steel, were engraved with his name and sold. Legends about him and his work began to spread.

Of course, Colson was responsible for this phenomenon, which is a real myth ; until today it has always kept a relative power. In order to understand the origin of this myth, I must explain briefly the political situation of Belgium in the end of the 19th century. From the beginning of the 1880s to the First World War, Belgium was disturbed by political quarrels. The people of Wallonia became aware of their own reality. A wish for autonomy increased, which knew its climax in 1912, with the famous *Lettre au Roi sur la séparation de la Wallonie et de la Flandre* (Letter to the King about the separation between the Walloon and Flemish regions), written by Jules Destrée, a politician of Charleroi[3].

In order to demonstrate that a union between the Flemish and the Walloons in the same country was impossible, some separatists tried to elaborate a political, economic and scientific history of the Walloon region. Thus, Victor Dwelshauvers-Dery, professor of the University of Liège, published a paper

2. C. Godefroid, " Fréres d'armes en cette campagne. La correspondance échangée par Oscar Colson et Arille Carlier entre 1919 et 1925 ", *La Vie wallonne*, LXVIII (1994), 7-9.

3. Fr. Joris, " Les étapes du combat wallon ", dans Fr. Joris, N. Archambeau, *Wallonie. Atouts et références d'une région* (1995), 37-40.

about the Marly hydraulic machine, which was also a panegyric of his builder, Rennequin Sualem, a mechanic of Huy.

Colson's goal appears clearly when one reads the last sentence of his biography on Gramme : *C'est glorifier notre beau pays et c'est honorer notre Race que de revendiquer hautement Zénobe Gramme comme une fleur spontanée du fécond terroir de la Wallonie* ![4] (it is glorifying our beautiful country and honouring our race to claim strongly Zénobe Gramme as a spontaneous flower of the fertile Walloon region).

Colson gave a lot of details without interest and funny, or grotesque, anecdotes, but he evaded completely the most important subjects for the history of technology, or the most embarrassing ones for his nationalist thesis. For example, the circumstances of the dynamo invention are very woolly. And the influences of the other scientists, who were contemporary with Gramme, are never examined.

Jean Pelseneer's biography

The second important biography on Gramme is Jean Pelseneer's, which was published in 1941, and re-issued in 1944. In this booklet of seventy-nine pages, the author's personal contribution is short (p. 7-35). He added three appendices to his work. The first one (p. 36-40) is a bibliography about Gramme and dynamoelectric machines that he has probably not completely consulted, or at least not exploited. The second one (p. 41-74) is the reprint of four of Gramme's articles, published in the *Comptes Rendus hebdomadaires des séances de l'Académie des Sciences*. The third one (p. 57-74) is the reprint of rare articles contemporary with Gramme.

Jean Pelseneer was born in 1903 in Brussels. He was a doctor in Physics and Mathematics of the University of his native city. From the end of the 1920s, he took an interest in the history of science and worked the main part of his career on this subject[5].

Of course, Pelseneer took his inspiration from Colson's book. Indeed, he was sterner and less subjective, but he did not give any new information, and, above all, he did not analyse the impact of Gramme's inventions in the field of the history of electricity. In fact, the ideology of his work was not Colson's one. Pelseneer published his book during the Occupation, in the *Collection Nationale* (national collection) of the Brussels editor *Office de Publicité*. He saw in Gramme a man who was *épris de justice et imprégné d'un sens réaliste*

4. O. Colson, *Zénobe Gramme, sa vie et ses oeuvres*, Liège, 1913, 92.

5. H. Elkhadem, “ In Memoriam. Jean Pelseneer 1903-1985 ”, *Archives Internationales d'Histoire des Sciences*, 36 (116) (1986), 162-163.

de la liberté et de la démocratie[6] (who liked justice and was in favour of liberty and democracy).

At this time, showing off like this one's patriotism was an important source of risk. In 1944, the Gestapo put Pelseneer through the severe strain of six months' imprisonment.

OTHER LATER BIOGRAPHIES

In order to be complete, I must also mention Chauvois' and Brien's biographies. The first one is an astonishingly naïve booklet of hundreds of pages[7]. As Louis De Broglie wrote in his preface, Chauvois was Arsène d'Arsonval's collaborator. Now, d'Arsonval would have known Gramme. Thus, it is a book principally written by a very old author, with the memory of an older scientist as basic source…

The absence of a final bibliography, and even the booklet's title *Histoire merveilleuse de Zénobe Gramme. Inventeur de la dynamo* (Marvellous story of Zénobe Gramme, inventor of the dynamo) speaks for itself. Chauvois did not give any new information ; on the contrary, as I shall explain, he gave new anecdotes which are deceptions for many authors, including Daumas.

As to the biographical summary published by Brien, it is a study solely founded on Pelseneer's[8]. One more time, no new information came from it.

SOME DOUBTS ABOUT GRAMME

His stay in the L'Alliance company

According to his historiographers — and firstly Colson — Gramme came to Paris in 1856 in order to work as a carpenter. He took an interest in electricity after his hiring by *L'Alliance*, in 1860[9]. This company sold magnetoelectric machines as power supplies for arc lamps. Gramme worked in this company for the construction of wood patterns, which were used for the iron casting of machines' frames. His passion for electricity was born on contact with these machines.

At this stage, Gramme's real function in this factory is already woolly. Of course, in the 19th century, foundries had their own carpentries for making wood patterns, but *L'Alliance* was not really a foundry. Every day, it did not

6. J. Pelseneer, *Zénobe Gramme*, Brussels, 1941, 10.

7. L. Chauvois, *Histoire merveilleuse de Zénobe Gramme. Inventeur de la dynamo*, Paris, 1963, 60-92.

8. P. Brien, " Zénobe Gramme 1826-1901 ", dans *Florilège des sciences en Belgique pendant le XIXe siécle et le début du XXe. Académie Royale de Belgique. Classe des Sciences* (1967), 227-241.

9. O. Colson, *op. cit.*, 46-47 ; J. Pelseneer, *op. cit.*, 10-11 ; L. Chauvois, *op. cit.*, 67-70.

conceive new models of machine, and new wood patterns was not a permanent need.

As for most of the workers of this time, the vocational training of Gramme was strong and varied enough to make him a polyvalent technician. According to Hyppolyte Fontaine, who knew Gramme very well, he was a very skilful worker, as much with metals as with woods[10]. Then, during his stay in the *L'Alliance* company, he was maybe integrated in the magnetoelectric machines assembly crew.

His so-called stay in the Christofle goldsmith's trade

According to Daumas, Gramme worked as a carpenter in the Christofle goldsmith's trade. This society used the Ruolz patent of 1840 in order to silver metallic pieces. Gramme was obliged to build wood patterns, which were used after for the metal casting of pieces. In order to silver metallic pieces, Christofle did a galvanic coating with power supplied by Bunsen's batteries. For Gramme, this represented an important contact with an industrial application of electricity. An anecdote says that he was frightened by the dirtiness of the batteries[11].

It is difficult to believe this story because I do not understand what a carpenter could do in a goldsmith's trade (wood patterns for cutlery and jewellery ?). Moreover, Daumas did not understand an anecdote from Chauvois. According to this author, in 1862, Gramme built up banisters in the Christofle factory, and he was horrified at the batteries coated with verdigris. From this observation would come his goal : making DC " cleanly ", like Chauvois wrote. In fact, by an electromagnetic way rather than by a chemical one[12].

I am sure that this anecdote is completely imaginary ; Gramme has never seen this scene, because it does not exist. Of course, in the middle of the 19th century, the electrical power necessary for gild or silver metals was supplied by Bunsen's batteries[13].

The positive electrode of this battery is a little coal stick, introduced into a jug full of nitric acid. The negative electrode is a zinc tube introduced into an another jug full of sulphuric acid diluted with water. The contact between these elements gives off nitrous vapours, but never verdigris, which is carbonate of copper[14].

10. H. Fontaine, " Zénobe Gramme ", *L'Industrie électrique*, X (1901), 53.

11. M. Daumas, *op. cit.*, 7.

12. L. Chauvois, *op. cit.*, 69.

13. Edm. Becquerel, *Traité d'Electricité et de Magnétisme*, II, Paris, 1855, 220, fig. 136 and 262, fig. 140.

14. F. Lucas, *Traité pratique d'électricité à l'usage des ingénieurs et des constructeurs*, Paris, 1892, 235.

This anecdote, which is mentioned the first time by Chauvois, led not only Daumas but also some other authors astray[15]. Since then, they have seen Christofle's company as the second factory where Gramme met electrical technology.

His stay in the Ruhmkorff factory

One more time, according to his historiographers, Gramme was hired by Ruhmkorff[16]. This event is presented as a banal anecdote said Daumas. However, Ruhmkorff was a famous scientific devices maker, in communication with many physicists, among them Louis Breguet and Marcel Deprez[17]. Maybe Gramme was already in touch with this club of technicians.

It was after his stay in the Ruhmkorff factory that Gramme devoted an essential part of his time to the tuning of his machine, therefore since 1865 or 1866. However, in 1861, Gramme took a patent relative to improvements (# 51.023) for an arc lamps mechanism. He was already concerned with supplying these with AC rather than DC. At this time, only DC of batteries was used. As electromagnetic machines supplied wave currents, it was necessary to rectify these with commutators, somehow or other. Nevertheless, DC gives an unequal wear of the arc lamps coal sticks, and AC does not.

Then Gramme would have worked on electricity for his stay in *L'Alliance*, therefore for maximum one year, and he would have been able to understand soon afterwards that using AC gives an equal wear of the arc lamps coal sticks. As Daumas wrote, this is a proof of an understanding of electrical phenomenon and machines running, which amazed us after what his biographers said about his so-called simple-minded's ignorance[18].

CONCLUSION

We must absolutely question ourselves about Gramme's interest for electrical machines. I am sure that he had a large know-how about its building, but where did this know-how come from ? It is one of the most fascinating questions.

Maybe, as Pelseneer wrote, he attended the free Becquerel electricity lessons, at the *Conservatoire Impérial des Arts et Métiers*, in Paris[19]. Until 1890, this high school gave free public lessons on electricity. But, were these elaborate enough in order to give Gramme a sufficient knowledge ? Indeed, these

15. P. Aigrain, " Zénobe Gramme, un manager de l'innovation ", *Science et Avenir*, 309 (1972), 950.

16. O. Colson, *op. cit.,* 49 ; J. Pelseneer, *op. cit.,* 11 ; L. Chauvois, *op. cit.,* 70.

17. M. Daumas, *op. cit.,* 11-12.

18. *Ibidem*, 11.

19. J. Pelseneer, *op. cit.,* 11.

consisted in public demonstrations without examination and practices in laboratories ; according to Christine Blondel, these lessons were more cultural manifestations of a time than a real vocational training[20].

Maybe Gramme was also a regular listener of lectures and scientific evenings about electricity ? Maybe he met mechanics who, as he did, tried to conceive electrical devices ?

Around 1849, Gramme was also a student in the industrial high school of Liège. If they always exist, consulting the programs of this school would probably give some information about Gramme's original background on electricity.

20. Chr. Blondel, " L'électricité au Conservatoire des arts et métiers : des physiciens aux électrotechniciens (1850-1940) ", dans L. Badel, *La naissance de l'ingénieur-électricien. Origines et développement des formations nationales électrotechniques*, 3rd International Congress of the Association pour l'Histoire de l'Electricité en France, Paris, 1997, 26.

Modern Influences on Rural Building in Northwest Germany (1880-1930). The Effect of Vocational Training[1]

Michael SCHIMEK

Introduction : industrialization and rural building

The sight of Heinemann's farm in Dalsper near Oldenburg in Northwest Germany represents a very striking example how deeply rural building could be affected by modern influences during the last decades of the 19th and the first decades of the 20th century (Fig. 1). In 1904 when Johann Heinemann built his new massive constructed dwelling house the traditional half-timbered farm house of the 18th century, in which the farmer's family and domestics used to live together with the cattle under one roof, was reduced to a stable without a dwelling function. Which influences do we have to consider in order to explain why rural population changed their living habits in such a way between the 1880s and 1920s ? In fact there are many different developments that led to modern building in the rural areas[2]. For example, agriculture made remarkable progress in intensifying its production. Soil improvement, the use of artificial manure or concentrated breeding created higher crop yield and life stock of cattle or pigs. Consequently farmers needed new stables and barns. With the increase of their profits many farmers had the money to build quite representative farming buildings especially before World War I. On the other hand urban building standards were brought closer to the countryside as communication between the cities and the countryside improved because of new built streets and railways and due to the newspapers. The rural population aimed at better, more comfortable and if possible at more representative dwell-

1. Slightly revised version of the lecture held during the XXth International Congress of History of Science in section 8.6.2, 1997, 24th July at Liège. I have to thank Elke Preul for her proof-reading.

2. *Cf.* O. Fok, " Tradition und Wandel am Bauernhaus ", in R. Wiese (ed.), *Im Märzen der Bauer. Landwirtschaft im Wandel*, Hamburg, 1993, 117-136, here 131-136 (= Schriften des Freilichtmuseums am Kiekeberg, vol. 13).

ing conditions, too. Considering the improved traffic system one has also to take into account that only by railways the transport of modern building materials like iron girders, tarboard or Portland cement over long distances from the industrialised cities to the countryside was made possible and economically reasonable. A further influence is the change in building legislation[3]. Between the 1890s and the 1920s most states enacted special laws and regulations in order to control and channel building. New installed building authorities exercised influence especially on the appearance of buildings and on security standards. Another development affected the workmen in building trades. More and more theoretical knowledge was added to their professional training as cities and communities founded vocational training schools. After their apprenticeship many journeymen attended special academies, schools of civil engineering, in German called Bauschulen or Baugewerkschulen.

The effects all these developments listed above exercised on rural building are not yet fully explored. But since 1995 the Deutsche Forschungsgemeinschaft (DFG ; German Research Community) finances a special research project on this subject, which is run by the Lowersaxonian open-air-museum Museumsdorf Cloppenburg in co-operation with the open-air-museum Freilichtmuseum am Kiekeberg of the county Harburg near Hamburg[4]. As the project is not finished yet it is only possible to present preliminary results here. Moreover, the following will focus exclusively on the influence of the workmen's professional training on rural building. Particularly the changed training contents in the building trades were often and are often declared as the most important influence that led rural constructing into a modern direction. The loss of regional traditional building patterns in the countryside is mostly traced back to this development[5]. If and in which degree the training of the building professions changed rural building, shall be demonstrated on the basis of the rural areas of Northwest Germany. This essay will concentrate on the conditions of the Northern parts of the former independent state of Oldenburg.

3. *Cf.* A. Buff, *Die bestimmenden Faktoren der deutschen Bauordnungen im Wandel der Zeit*, dissertation Uni Hannover, Wuppertal, 1971.

4. *Cf.* to this project : M. Schimek, " Ländliches Bauen im nördlichen Oldenburg zwischen 1890 und 1930 — ein Zwischenbericht ", in *Gebaute Welten. Beiträge der Herbsttagung 1996 der Gesellschaft für Volkskunde in Schleswig-Holstein e.V.*, Großbarkau 1997, 69-106 (= Schriftenreihe der Gesellschaft für Volkskunde in Schleswig-Holstein, vol. 3).

5. E.g. : J. Kleinmanns, " Hausbau im kurkölnischen Sauerland. Die Entwicklung der ländlichen Architektur von 1600 bis 1900 ", in S. Baumeier and Ch. Köck (eds), *Sauerland. Facetten einer Kulturregion*, Detmold, 1994, 34- 47, here 34 (= Schriften des Westfälischen Freilichtmuseums Detmold - Landesmuseum für Volkskunde, vol. 12). The influence of vocational education in Baugewerkschulen on rural building was already noticed in the beginning of the 20th century. Buildings planned and constructed by absolvents of these schools were mainly criticised as loss of traditional building patterns which represented in the eyes of the critics like Paul Schultze-Naumburg the " good old " pre-industrialized epoch. *Cf.* G. Grüner, *Die Entwicklung der höheren technischen Fachschulen im deutschen Sprachgebiet. Ein Beitrag zur historischen und zur angewandten Berufspädagogik*, Braunschweig, 1967, 87.

THE INFLUENCE OF VOCATIONAL TRAINING ON RURAL BUILDING

Vocational training carried out at special schools means modernity in different ways. On the one hand it is a modern phenomenon itself as it will be shown in a short survey later on. On the other hand it is considered as a modern effect on rural building. " Modern effect " understood as using new building materials, using engineer's constructions, using urban exterior and interior arrangements.

In order to the federal structure of the second German Empire, school education was carried out by each state according to its own ideas and prescriptions. This caused a confusion not only concerning the terms which were used for these schools but also referring to the standards, contents and the aims they conveyed and the titles they gave to their graduates[6]. However generally spoken we have to consider three branches of schools that dealt with the training of apprentices and workmen in Germany at the end of the last century. They stood in a hierarchical relation to one another. Basic knowledge was imparted by industrial continuation or vocational training schools (*Gewerbliche Fortbildungsschulen*). Schools of civil engineering or technical schools (*Baugewerkschulen, Technika*) taught their students on a higher level of theory. On the highest level colleges of advanced technology (*Technische Hochschulen*) educated architects almost exclusively for the higher civil service or for large building enterprises in the cities[7]. As their graduates hardly worked in rural areas, we can neglect the colleges of advanced technology in this reflection.

The influence of industrial continuation schools

Basic professional knowledge was educated by so called industrial continuation schools, in German Gewerbliche Fortbildungsschulen. Literature mentions two roots for them : The older one is the continuation school that was mostly established during the era of Enlightenment in order to refresh elementary school knowledge. Subjects were basic reading, writing and calculation. Lessons were held on Sundays or in the evenings and the intensity of attendance was quite low[8].

6. K. Kümmel, " Einleitung ", in K. Kümmel (ed.), *Quellen und Dokumente zur schulischen Berufsbildung 1918-1945*, Köln, Wien, 1980, 1-43, here 6, 10 and 13 (= Quellen und Dokumente zur Geschichte der Berufsbildung in Deutschland, Reihe A, vol. 2). H. von Seefeld, " Die gesetzliche Reglung und Verwaltung des Berufsschulwesens ", in A. Kühne (ed.), *Handbuch für das Berufs- und Fachschulwesen*, Leipzig, 1922, 91-107, here 99 and 102. R. Meyer-Braun, " Gründung und Anfänge des Bremer Technikums ", *Bremisches Jahrbuch*, 69 (1990), 133-157, here 151.

7. *Cf.* K.-H. Mangold, " Geschichte der Technischen Hochschulen ", in L. Boehm and Ch. Schönbeck (eds), *Technik und Bildung*, Düsseldorf, 1989, 204-234 (= Technik und Kultur, vol. 5).

8. A. Kühne, " Entwicklungsstufen der Berufserziehung ", in A. Kühne (ed.), *Handbuch für das Berufs- und Fachschulwesen*, Leipzig, 1922, 1-23, here 10. A. Barth, " Gewerbliche Berufsschulen (Gewerbeschulen, gewerbliche Fortbildungsschulen) ", in A. Kühne (ed.), *Handbuch..., op. cit.*, Leipzig, 1922, 142-151, here 143-144.

During the 19th century the states turned these schools more and more into industrial continuation schools. Improved education of craftsmen was recognized as a proper means to lift standards in the trades and to further economy. Lifting trades' standards was also the reason for the foundation of craftsmen schools the second root of industrial continuation schools. Especially since the 1820s single craftsmen educated apprentices after-work primarily in structural drawing, but also in technical matters, in calculating and writing[9]. The more traditional craft was led into a crisis by cheap industrial mass production, the more communities and states pushed the expansion of professional training ahead[10]. In 1910, 3.600 industrial continuation schools worked in the German Empire and about 540.000 pupils attended these schools[11]. But public financial support turned out quite differently : for instance Prussia spent 28 Reichsmark per head, Baden, a rather small state in the South-west of Germany, spent 73 Reichsmark[12]. And in the end the industrial continuation schools — or vocational training schools, in German : Berufsschulen as Prussia renamed them in 1920[13] — were almost totally limited to towns and cities. Such schools were hardly established in the countryside[14].

Oldenburg — the state we are concentrating on — was and is mainly rural. Located in the Northwest of Germany it belonged to the small states of the German Empire. In 1900, twenty industrial continuation schools existed in Oldenburg's towns and communities[15], in 1912 already 74[16]. As in many other German states the duty to attend these schools after elementary schools depended on a corresponding decree declared by the local authorities[17]. If a continuation school was founded, communities were entitled to oblige each male workman under eighteen years to attend the school and employers had to allow their workers and apprentices to attend the school during working time. However a time-table of the continuation school of the Oldenburgian town of Brake shows that lessons were held mainly in the late afternoons, for instance

9. H. Von Seefeld, *op. cit.*, 91. A. Barth, *op. cit.*, 143 -144. G. Grüner, *op. cit.*, 80.

10. A. Kühne, *op. cit.*, 20. F. Lenger, *Sozialgeschichte der deutschen Handwerker seit 1800*, Frankfurt/Main, 1988, 112 (= Neue Historische Bibliothek ; = Edition Suhrkamp, Neue Folge, vol. 532).

11. A. Barth, *op. cit.*, 145.

12. A. Barth, *op. cit.*, 145. Figures refer to 1912.

13. A. Barth, *op. cit.*, 142.

14. A. Kühne, *op. cit.*, 20.

15. Stadtarchiv (city archives) Brake Nr. 785, " Akte betr. : Statut der gewerblichen Fortbildungsschule ".

16. G. Lüschen, " Das Schulwesen ", *Heimatkunde des Herzogtums Oldenburg*, ed. by Oldenburgischer Landeslehrerverein, Redaktion W. Schwecke, W. v. Busch and H. Schütte, Bremen, 1913, vol. II, 387-444, here 432. H. Rasche, *Die Entwicklung des Berufs-, Berufsfach- und Fachschulwesens im Lande Oldenburg von den Anfängen bis zur Gegenwart*, Inaugural-dissertation, Uni Münster, Münster, 1951, 92.

17. § 120, § 142, § 150 Reichsgewerbeordnung. *Cf.* K. Kümmel, *op. cit.*, 9-10.

the apprentices of the building trades were educated between half past six and half past eight pm (Fig. 2)[18].

A curriculum which was worked out by a special commission in 1907 suggested a maximum of six and a minimum of four lessons a week[19]. Subjects were vocational lore (Berufskunde or Gewerbekunde) including legal relationships in trade, raw materials, tools and machines : one hour, German including business correspondence : one and a half hour, calculation and geometry including estimates : one and a half hour, structural drawing including unwinding of geometric bodies and timber-works : two hours. In the framework of the so-called " dual-system "[20] vocational training in continuation schools took three years and was accompanied by vocational training or work in workshops. Due to the small number of lessons, due to the fact that the classes often consisted of apprentices of different trades[21], due to the differing foreknowledge of pupils and due to the differing training of the teachers the influence of continuation schools on possible improvement of training standards is difficult to consider. One can assume that especially in the countryside, where vocational school attendance was not compulsory, apprenticeship in the workshops still played the leading role[22]. This means that — referring to rural building — continuation schools probably did not have the power to influence an average apprentice in the direction of modern building ideas. It depended very much on the single apprentice, on his interests and on his aims in life if he desired to spent more time on additional training. In this case training in vocational schools formed only the basis for further training. This leads to the second branch of vocational training we have to think of in connection with the changing of rural building.

Schools of civil engineering

Schools of civil engineering — in German Baugewerkschulen or Technika — formed the second branch of vocational training for workmen in the building trades. Foundation of these schools started in the 1820s[23], but

18. Stadtarchiv Brake Nr. 785, " Akte betr. : Statut der gewerblichen Fortbildungsschule ". Time-table of about 1910.

19. *Grundzüge eines Lehrplanes für die gewerblichen Fortbildungsschulen des Herzogtums Oldenburg*, aufgestellt von der vom Verein für das Fortbildungswesen im Herzogtum Oldenburg damit beauftragten Kommission, Oldenburg i. Gr., 1907, 5-6.

20. K. Kümmel, *op. cit.*, 7.

21. In Oldenburg 46 of the 74 in 1912 existing continuation schools consisted only of one class. H. Rasche, *op. cit.*, 93.

22. *Cf.* H. Rasche, *op. cit.*, 106.

23. O. Peters, " Die deutschen Baugewerkschulen ", in A. Kühne (ed.), *Handbuch für das Berufs- und Fachschulwesen*, Leipzig, 1922, 279-290, here 279. Schools of civil engineering were founded 1820 in Munich, 1828 in Weimar and 1831 in Holzminden. G. Grüner, " Entwicklung der technischen Fachschulen ", in L. Boehm und Ch. Schönbeck (ed.), *Technik und Bildung*, Düsseldorf, 1989, 175-203, here 187 (= Technik und Kultur, vol. V).

most were established during the last three decades of the 19th century. In 1913 67 public schools of civil engineering which were run by the states or communities existed in Germany[24]. In addition to them existed a remarkable number of private schools. Many public schools were founded by private persons[25], and because of the enormous demand for them by workmen the running of a school of civil engineering was a profitable business[26]. The rising number of this kind of school shows mainly one thing : the traditional way of teaching technical knowledge from master to journeyman in the workshop and on the building site did not suffice anymore in the course of the 19th century[27]. New building materials, new forms of buildings with new ways of construction and new instructions made by the states required additional vocational training of a workman if he wanted to work in a leading position or if he aimed at founding or taking over a firm[28]. Many states prescribed special examination for the master craftsmen's diploma particularly in theoretical matters. And public building authorities required a certificate of a school of civil engineering for the middle civil career[29]. Since the 1890s states reinforced their efforts to improve the vocational training by standardization of the schools of civil engineering. As Prussia was the largest and most important state in the German Empire, its ideas became also relevant for those schools that were located in other states[30].

Basic requirement for the attendance of such a school was the final examination of an elementary school and twelve months of working practice in a building trade. Until 1908 students had to attend these schools for four, thereafter for five semesters, but often it was possible to leave school after one year with a foreman's degree (Polier)[31]. Classes were 44 hours per week and sixteen subjects were taught. About 45% of the lessons were drawing, about 25% technical knowledge and another 20% were set for the mathematical realm. Not even 4% were spent on matters of language[32].

In Oldenburg the first school of civil engineering was founded in 1881 in

24. O. Peters, *op. cit.*, 279. *Cf.* there also the table on p. 288-289.

25. E.g. Holzminden 1831. Nienburg 1831. *Cf.* W. Müller, " Die Entwicklung von der Bauschule Quaet Faslems zur Fachhochschule ", in *Festschrift. 125 Jahre Ausbildung von Architekten und Bauingenieuren in Nienburg/Weser. 1853-1978*, ed. by Fachhochschule Hannover, Fachbereiche Architektur und Bauingenieurwesen in Nienburg, Nienburg, 1978, 13-29, here 13. O. Peters, *op. cit.*, 279. Grüner, *op. cit.*, 188.

26. *Cf.* G. Grüner, *op. cit.*, 196. R. Meyer-Braun, *op. cit.*, 149.

27. G. Grüner, *op. cit.*, 179.

28. *Cf.* W. Müller, " Die Entwicklung von der Bauschule Quaet Faslems zur Fachhochschule ", in *Fachhochschule Hannover, Fachbereiche Architektur und Bauingenieurwesen in Nienburg/Weser. 125 Jahre Bauausbildung,* 1853-1978, Nienburg, 1978, 13-29, here 13.

29. G. Grüner, *op. cit.*, 189. O. Peters, *op. cit.*, 282.

30. G. Grüner, *op. cit.*, 190. O. Peters, *op. cit.*, 279-280. *Cf.* R. Meyer-Braun, *op. cit.*, 146.

31. O. Peters, *op. cit.*, 281. G. Grüner, *op. cit.*, 188-190.

32. *Cf.* O. Peters, *op. cit.*, 287.

Brake. But due to missing demand it had to close already after three years[33]. A permanent school was established in 1882 in the capital Oldenburg. Three years later a branch for mechanical engineers was added which guaranteed a sufficient number of students especially during the summers. Many building students were not able to attend school in the summer as they had to earn money in order to pay their school fees[34]. During its most prosperous time before World War I Oldenburg's " Grand-ducal school of building and mechanical engineering " had about seventy students. In the main lines it developed according to the described Prussian pattern[35].

The Bauschule Rastede of Carl Rohde

A private school of civil engineering with a very special character was founded in 1905 by Carl Rohde in the small town of Zetel. He started with six students, but during the following years and decades his school quickly turned into an enormous successful and popular institute (Fig. 3). After the initial period and excluding the time of World War I the school was attended by about one hundred students each year. During the late 1920s even 180 students were taught yearly and until 1930 2127 students had received vocational education by Carl Rohde[36]. About half of them came from the state of Oldenburg or the neighbouring province of Hannover that belonged to the Prussian state[37]. However the remaining half came from all other parts of Germany. What made the *Bauschule Rastede*, how Rohde called his school after moving to the town of Rastede in 1907, so popular ? Above all Rohde was a man of practice. He was born in 1880 in the rural structured parts of Northern Oldenburg. During his apprenticeship in the carpenter's trade between 1895 and 1899 he got an insight into rural as well as into urban construction. After he had worked during the following three years as a journeyman for different masters, Rohde

33. A. Eckhardt, " Gründung und Aufstieg der Stadt Brake ", in A. Eckhardt e.a. (ed.), *Brake. Geschichte der Seehafenstadt an der Unterweser,* (= Oldenburgische Monographien), Oldenburg, 1981, 119-246, here 229-232. Stadtarchiv Brake Nr. 782, " Akte betr. : Technikum Brake ". *Niedersächsisches Staatsachiv Oldenburg Best.*, 230-6, Nr. 26, " Großherzoglich Oldenburgisches Amt Brake. Acta betreffend : Die Baugewerk- und Maschinen-Bau-Schule zu Brake ".

34. G. Grüner, *op. cit.*, 80.

35. K.-H. Jung, " Von der Winterbauschule zur Fachhochschule ", in *1877-1977. 100 Jahre. Von der Winterbauschule für Bauhandwerker zur Fachhochschule Oldenburg. Fachbereiche Architektur, Bauing.-Wesen, Vermessungswesen, Seefahrt,* ed. by Fachhochschule Oldenburg, Oldenburg, 1977, 11-23, here 11-15. F. Weiß, " Handwerkerschule - Bauschule - Ingenieurschule - Ingenieurakademie ", in *Die staatliche Ingenieurakademie Oldenburg und ihre Ingenieure. Jubiläumsschrift anläßlich der Wiederkehr der Wiedereröffnung der Ingenieurakademie Oldenburg am 15.Oktober 1970*, 11-46, here 11-13. J. Pühl, " Geschichte des Technikums ", in *Technik und Verkehr. Beilage des Gemeinnützigen*, Varel i.O., 29.08.1925/202. *Programm der Großherzoglichen Baugewerk- und Maschinenbauschule in Varel a.d. Jade Großherzogtum Oldenburg* (about 1907).

36. *25 Jahre Bauschule 1905/30. 6.-7. Dezember,* 3-4.

37. " 25 Jahre Bauschule Rastede ", in *2. Beilage zu Nr. 334 der " Nachrichten für Stadt und Land " vom Montag, dem 8. Dezember 1930*.

attended the above-mentioned " Grand-ducal school of building and mechanical engineering " in the winter-semesters. During the summers he worked as a journeyman or foreman. In 1905 he finished the school of civil engineering with the degree of a constructional engineer and founded in the same year his own school. Many documents and the books of his library prove that Rohde kept himself informed about new developments in the building trades. For instance he attended the school of civil engineering in Friedberg in 1910 and the Oldenburgian engineer's academy in the summer-semesters of 1927 and 1928[38].

Another proof for the school's stress on practical matters is : Rohde offered in the first place a training that imparted knowledge that was needed to pass the exams for obtaining a master craftsmen's diploma. This diploma formed the basis for founding or taking over a firm. And this training lasted only five months, a fact which made the *Bauschule Rastede* particularly attractive ![39] Rohde could only keep the promise of successful and quick training by concentrating on the very necessary practical and theoretical knowledge (Fig. 4). In addition to this three years of working practice were required to be accepted at Rohde's school of civil engineering[40]. Just for comparison : public schools of civil engineering required only twelve months of practice. A long list of names of successful masters and foremen given in the school's program proves that Rohde's education system worked[41]. His success is based on the fact that he filled a gap in the market. Many journeyman wanted to rise in their professions, wanted to become a foreman or aimed at setting up on themselves without having the financial means to attend a school of civil engineering for four or even five semesters. After all the cost of attending Rohde's school for one semester amounted similar to others to 410 Marks in 1911 including school-fees, drawing materials and board and lodging[42]. This explains that most of the students must have been highly motivated to shorten school attendance by good performance in order to keep the expenses low. Nevertheless Rohde offered a second course that was based on the first one for those who intended to become a structural draughtsman or building supervisor. As in public schools of civil engineering Rohde's students had to attend 44 lessons a week. Subjects were German, calculation, geometry, structural drawing, freehand drawing, calligraphy, modelling and statics. Obviously Rohde's offer oriented

38. Archiv Museumsdorf Cloppenburg Inv.-Nr. 21639.

39. *Programm der Bauschule Rastede i. Oldbg.*, Oldenburg i. Gr. 1911, 3-6.

40. *Programm der Bauschule Rastede i. Oldbg.*, Oldenburg i. Gr. 1911, 6.

41. *Programm der Bauschule Rastede i. O. von C. Rohde,* (about 1929), 14-29.

42. *Programm der Bauschule Rastede i. Oldbg.*, Oldenburg i. Gr. 1911, 8. This amount is quite typical : the attendance of the Grand-ducal school of building and mechanical engineering in Varel cost 400 Marks in 1907 per semester. *Cf. Programm der Großherzoglichen Baugewerk- und Maschinenbauschule in Varel a.d. Jade Großherzogtum Oldenburg*, (about 1907), 4-5.

itself by the subjects taught in industrial continuation schools what he deepened by adding contents of public schools of civil engineering.

THE INFLUENCE OF VOCATIONAL TRAINING ON RURAL BUILDING

It is quite difficult to find out which contents the students learned in detail and — as a second problem — it is even more difficult to disclose which of the learned contents the students used during their later practice. Fortunately some documents and drawings of Rohde's first student Karl Gerdes (1880-1959) were handed down to us[43]. His school drawings show that particularly the training of the three-dimensional imagination was strongly emphasized. This observation is supported by lot of corresponding drawings that are handed down from other students of other schools. The graphic determination of the real size of structural components was important especially to the work of carpenters. But also the appropriate construction and design of buildings are reflected in these drawings — in details like different forms of brickwork as well as by means of the example of one house in its entirety (Fig. 5). Another important means to convey knowledge was the making of models to which Rohde attached great importance (Fig. 6).

However the school drawings as the examples given in an architectural pattern book that Karl Gerdes bought in 1906 during his time at Rohde's school of civil engineering hardly correspond with the mass of those houses that were actually built in the rural realm. The buildings on school drawings and in the pattern books are up-to-date in their stylistic design, most of them are totally or at least mainly plastered. But Karl Gerdes and his colleagues had to solve totally different architectural problems in their rural surrounding (Fig. 7). This impression is confirmed by those 300 houses we have documented during our research work so far[44].

It is true that buildings like Heinemann's farm mentioned in the beginning were realized, buildings that represent modern and — at least relating to their outward appearance — urban style. All of them were planned by a graduate of a school of civil engineering. But these houses — almost exclusively dwelling houses — are rather rare exceptions and were only realized by members of the rural upper class, for example like Johann Heinemann who was the mayor of his village. Although houses of lower social classes are also ornate with stylistic means of the Historicism, we have to state generally that for the most part buildings were still constructed according to more traditional and regional patterns. Most rural buildings still connected dwelling and stable beneath one roof

43. Archiv Museumsdorf Cloppenburg Inv.-Nr. 21890-033000. We have to thank Hans Gerdes in Wiefelstede-Heidkamperfeld for leaving us his grandfather's documents and drawings for the purpose of documentation and research.

44. *Cf.* to our method of documentation : M. Schimek, *op. cit.*

and the toilet was situated in the pigsty. The traditional house types of the so called Niederdeutsches Hallenhaus and especially the Gulfhaus dominated until World War II. A slight modernization can be observed in the used materials and constructions. The cellars' floors are massive built in form of the so called Prussian vaults with iron double-T-girders. Most buildings are provided with a hollow cavity brickwork in their dwelling parts. Yet ordinary red bricks without plaster formed the basic material to build walls, wooden pillars carried not only the roofs but also the whole buildings in their farming parts[45].

Summing up it seems that vocational training schools and schools of civil engineering did not influence rural buildings as strongly as it is assumed so far for the period of the late decades of the last and the first decades of this century. Probably their influence was more decisive for the planning and calculating of rural buildings and the management of the rural building firms. Bookkeeping and business dealings became more professional, the standards of structural drawing were improved. Workmen in the building trades obtained more knowledge and became more versatile. But most building owners still preferred rather traditional solutions probably as they did not want to change their customs in living and working or just in order to save money.

FIGURES

1. Heinemann's Farm in Elsfleth-Dalsper in 1997. Dwelling house of 1904 (left) and farmhouse of the so called *Niederdeutsches Hallenhaus* of the 18th century (Foto : Museumsdorf Cloppenburg).

45. M. Schimek, *op. cit.*, 104-105.

2. Timetable of the vocational continuation school of Brake, about 1910 (Stadtarchiv Brake Nr. 785).

Stundenplan

für die

Gewerbliche Fortbildungsschule der Stadt Brake i. O.

Sonntag	7½—9½ Uhr morgens	Gemischte Klasse	Fachzeichnen	Lehrer Potthast
Montag	6½—7½ Uhr abends	Bauhandwerker	Geschäftsaufsatz	Lehrer Witthold
	7½—8½ „ „	Bauhandwerker	Sommer: Geometrie Winter: Buchführung	„ Witthold
	6½—7½ „ „	Metallarbeiter	Geschäftsaufsatz	„ Kunst
	7½—8½ „ „	Metallarbeiter	Sommer: Geometrie Winter: Buchführung	„ Kunst
	6½—8½ „ „	Vorklasse	Freihandzeichnen	„ Potthast
Dienstag	6½—7½ Uhr abends	Bauhandwerker	Berufskunde	Lehrer Witthold
	7½—8½ „ „	Bauhandwerker	Rechnen	„ Witthold
	6½—7½ „ „	Metallarbeiter	Berufskunde	„ Kunst
	7½—8½ „ „	Metallarbeiter	Rechnen	„ Kunst
Mittwoch	2—2¾ Uhr nachmitt.	Gemischte Klasse	Berufskunde	Lehrer Blohm
	2¾—3½ „ „	Gemischte Klasse	Buchführung	„ Blohm
	3½—4¼ „ „	Gemischte Klasse	Geschäftsaufsatz	„ Blohm
	4¼—5 „ „	Gemischte Klasse	Rechnen	„ Blohm
	5½—6½ „ abends	Vorklasse	Rechnen	„ Speckmann
	6½—7½ „ „	Vorklasse	Geschäftsaufsatz	„ Speckmann
	7½—8½ „ „	Vorklasse	Deutsch	„ Speckmann
	5½—6½ „ „	Unterstufe	Rechnen	„ Gerdes
	6½—7½ „ „	Unterstufe	Deutsch	„ Gerdes
	7½—8½ „ „	Unterstufe	Geschäftsaufsatz	„ Gerdes
Donnerstag	6½—8½ Uhr abends	Metallarbeiter	Fachzeichnen	Lehrer Wessels
	6½—8½ „ „	Bauhandwerker Metallarbeiter	Projektionszeichnen	Lehrer Auffarth
Freitag	6½—8½ Uhr abends	Bauhandwerker	Fachzeichnen	Lehrer Auffarth

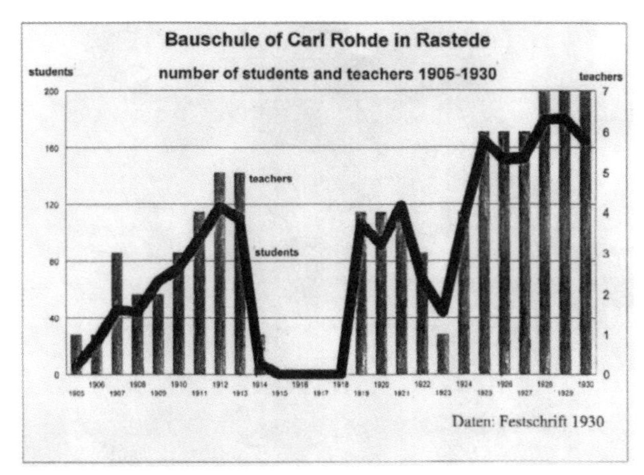

3. School of civil engineering *Bauschule Rastede* of Carl Rohde : number of students and Teachers, 1905-1930.

4. Class-room of the Bauschule Rastede with students learning (Foto : Museumsdorf Cloppenburg).

5. Construction drawing of Karl Gerdes made during his studies at the Bauschule Rastede in 1906 (Archiv Museumsdorf Cloppenburg Inv.-Nr. 21890-033025)

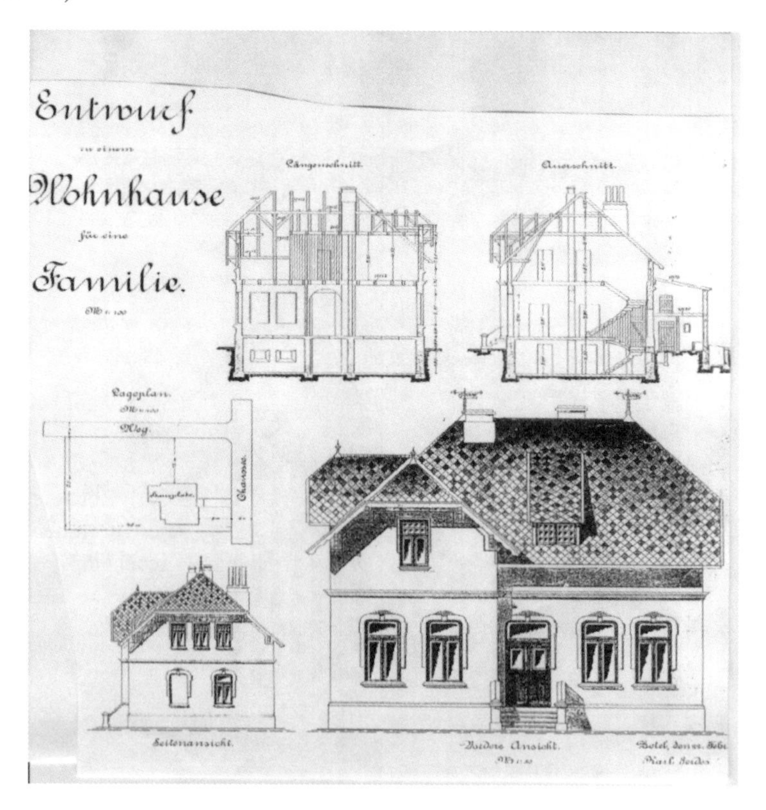

6. Director of the *Bauschule Rastede* Carl Rohde (5th from left) with his first students in 1906. Among them Karl Gerdes (3rd from left) (Foto : Museumsdorf Cloppenburg).

7. Construction drawing of Karl Gerdes. Farmhouse in form of the so called Niederdeutsches Hallenhaus for Heinrich Neumann in Heidkamp in 1924 (Archiv Museumsdorf Cloppenburg Inv.-Nr 21890-033035).

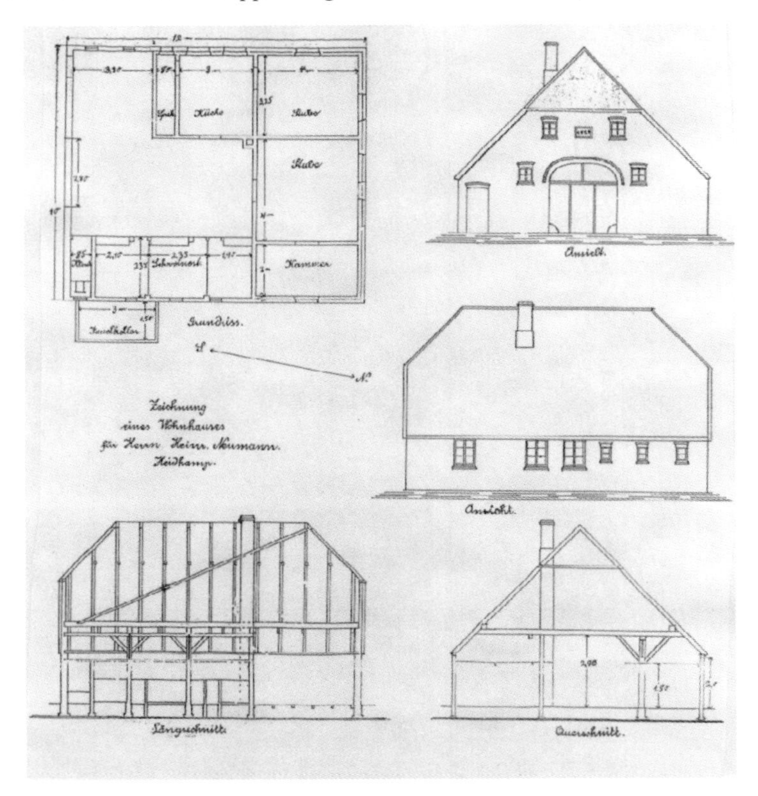

Le rôle de l'entreprise et de l'entrepreneur dans l'introduction du béton précontraint : Eugène Freyssinet et les Entreprises Campenon Bernard ou l'histoire d'une rencontre (1920-1939)

Dominique Barjot

L'histoire du béton précontraint offre un exemple particulièrement intéressant pour qui s'interroge sur les rapports entre la science et l'entreprise[1]. Les progrès de ce matériau nouveau (le béton précontraint), de ce procédé nouveau (la précontrainte), n'auraient pu être aussi rapides sans la prise de risques assumée par quelques entrepreneurs. Deux entreprises ont joué un rôle prépondérant en ce domaine : en France, les Entreprises Campenon Bernard (ECB) ; en Allemagne, Dyckerhoff und Widmann AG (Dywidag). A la fin des années 1950, l'essentiel des ouvrages d'art réalisés dans le monde l'étaient à partir de brevets et de licences déposés par ces deux sociétés (deux tiers pour la firme française, un tiers pour son homologue allemande). Nées en 1920, les ECB jouèrent un rôle prépondérant dans l'introduction du béton précontraint comme matériau de construction. Le point de départ en fut la rencontre d'un entrepreneur ayant le goût du risque, mais efficace gestionnaire, Edmé Campenon (1872-1962), et de l'un des ingénieurs les plus doués de sa génération, Eugène Freyssinet (1879-1962). En optant plus massivement qu'aucun de leurs concurrents en faveur du béton précontraint, les ECB s'imposèrent, dès la seconde moitié des années 1930, comme une des plus grandes entreprises françaises de travaux publics. Elles connurent aussi, dès la même époque, un succès certain hors de leur frontières : ainsi, en Allemagne, grâce à Wayss und Freytag.

À l'origine, les conceptions de Freyssinet

Entre les deux guerres, le béton armé apparaissait comme l'un des matériaux les plus aptes à répondre aux besoins du temps, grâce à des performances

1. Cette étude se fonde sur le dépouillement des archives des Entreprises Campenon Bernard aujourd'hui conservées par Campenon Bernard SGE.

accrues et à la gamme toujours plus large des solutions qu'il permettait. Toutefois, la logique à laquelle obéissaient ces progrès favorisa l'apparition d'un matériau nouveau, plus économique et plus souple d'emploi : le béton précontraint. Le premier en France, E. Freyssinet décida de sauter le pas[2].

Une première association fructueuse : Limousin et Freyssinet

Principal élève de Charles Rabut, avec Albert Caquot (1881-1976), Eugène Freyssinet naquit à Objat, en Corrèze, le 13 juillet 1879, puis intégra l'" X " et l'Ecole des Ponts et Chaussées. Nommé en 1905 ingénieur ordinaire à Vichy-La Palisse, il y obtient rapidement d'importants succès, qui attirèrent sur lui l'attention de l'entrepreneur François Mercier (1858-1920), ami personnel de Clemenceau. En 1910, Mercier offrit au conseil général de l'Allier l'argent pour la réfection des trois ponts de Boutiron, de Châtel-de-Neuvre et du Veudre. E. Freyssinet y mit au point un nouveau procédé de décintrement des arcs et en conclut que les articulations de voûte constituaient la cause première de la faiblesse des arcs. Peu de temps après, en janvier 1914, il opta pour le privé en rejoignant son camarade de promotion Claude Limousin (1880-1953) au sein de la société Limousin, Mercier et Cie, en tant que directeur technique. En 1916, Limousin rompit avec Mercier, pour fonder sa propre entreprise, avec le concours de Châtillon-Commentry et de Neuves-Maison. E. Freyssinet apporta à la nouvelle société toute une gamme de procédés nouveaux, grâce auxquels l'entreprise s'imposa comme le numéro un français du béton armé : les grandes couvertures voûtées en béton armé à nervures extérieures, permettant l'utilisation d'échafaudages roulants ; les voûtes en shed conoïdes ; le décintrement par vérins ; la mise en compression des voûtes par tension préalable.

Le système Freyssinet donna à C. Limousin les moyens de réaliser une série de ponts exceptionnels : Villeneuve-sur-Lot (1919), longtemps record mondial des voûtes en béton non armé ; Cauderlier (1921), premier grand pont français en béton armé construit sous voie ferrée ; Saint-Pierre-de-Vauvray (1923), record mondial de portée des ouvrages en béton armé jusqu'en 1928 ; Plougastel, sur l'Elorn. Pourvu de trois arches de 172 mètres de portée chacune, ce dernier pont détint, de 1928 à 1934, le record mondial des ponts en arc en béton armé. Il donna lieu à une intéressante innovation : on coula, successivement, ses trois arcs sur un même cintre de 180 m de portée, transporté sur deux chalands en béton armé. E. Freyssinet fut tout autant un concepteur de structures à grande portée. En 1914, il réalisa à Bourges et à Istres ses premiers hangars d'aviation en béton armé, reproduits de nombreuses fois par la suite. 1921 vit la construction des célèbres hangars à dirigeables d'Orly (voûtes de 88 m de portée), à quoi s'ajoutèrent, en 1930, les voûtes translucides de la gare de Reims. Il s'agissait de son dernier ouvrage avec Limousin. Dès le 1er janvier 1929, E. Freyssinet quittait l'entreprise. Le 2 octobre 1928, il venait de déposer

2. J.A. Fernandez Ordonez, *Eugène Freyssinet,* Paris, Editions 2 C, 1979.

le brevet dans lequel il définissait pour la première fois avec précision le concept de précontrainte. Mais Limousin n'y croyait pas, ce qui entraîna le départ de Freyssinet et le déclin de l'entreprise.

La précontrainte : une remise en cause des conceptions du temps

La précontrainte recherche une maîtrise plus complète de la distribution des contraintes dans le béton armé. Alors que dans le béton armé la résistance aux efforts de compression est exigée du béton, l'acier supportant les efforts de traction, dans le béton précontraint la résistance à tous les efforts (traction, compression, cisaillement) est demandée à une matière unique, le béton, préalablement comprimé de façon à y supprimer les tractions dangereuses[3] et, par conséquent, la fissuration. Il existe deux grandes méthodes de précontrainte : par câbles et par fils adhérents. Dans la première, les câbles, logés dans des gaînes souples à l'intérieur desquelles ils peuvent se déplacer, ne sont tendus que lorsque le béton de la poutre, préalablement coulé, présente une résistance suffisante pour supporter les efforts de compression qui résultent de la précontrainte. On injecte dans la gaine, après que les câbles soient ancrés, un coulis de ciment qui fait prise et bloque définitivement les câbles sur toute leur longueur à l'intérieur des gaines. Cette première méthode est généralement adoptée pour la construction des ouvrages d'art et de poutres préfabriquées à grande portée. Quant à la seconde, qui évite l'usage d'un système d'ancrage, elle ne peut s'utiliser que pour la préfabrication de poutrelles ou de poutres destinées à la réalisation de planchers d'immeubles.

Freyssinet dut mener de durs combats avant d'imposer son idée. En 1929, il proposa l'exploitation de son brevet à la société Forclum. Celle-ci n'avait pas le choix, car il lui fallait améliorer de toute urgence la qualité lamentable des poteaux en béton armé. Si les premiers poteaux en béton précontraint s'avérèrent bien meilleurs, le procédé se révéla vite excessivement coûteux, au grand dam de Freyssinet lui-même, qui y engagea une grande partie de sa fortune personnelle. L'occasion d'imposer le nouveau procédé intervint plus tard, en 1934, à l'occasion des travaux de sauvetage de la gare maritime du Havre. Il s'agissait de résoudre un problème technique d'une extrême urgence. La gare s'enfonçait en moyenne de 25 mm par mois dans une couche de limon de 30 m de profondeur. Le sol de fondation n'étant pas homogène, on risquait l'affaissement et, à terme, l'effondrement. On avait bien essayé de fonder des pieux, mais le remède avait aggravé le mal. Fin 1933, l'effondrement total apparaissait probable. L'architecte Urbain Cassan fit alors appel à Freyssinet. Ce dernier réussit à former un ensemble monolithique avec les vieux massifs de formation, en les reliant les uns aux autres par de longues poutres en béton

3. Voir, sur ce point, E. Freyssinet, " Exposé d'ensemble de l'idée française de précontrainte (1949) ", *Un demi-siècle de technique française de la précontrainte*, Travaux, 327-358.

précontraint. Ce fut un succès, car, fin 1934, les tassements étaient déjà presque arrêtés.

Cette réussite suscita rapidement un mouvement en faveur de la précontrainte. Il s'agissait certes d'une idée ancienne, mais qui n'avait pas abouti, malgré les recherches, avant la Première Guerre mondiale, d'ingénieurs comme Jackson (Etats-Unis), Doering (Allemagne), Rabut ou Considère (France), puis, dans les années 1920, Wettstein et Emperger (Allemagne) ou Dill (Etats-Unis). Au contraire, Freyssinet exploita aussitôt sa découverte : étude de la construction du bâti de la machine d'essai du Laboratoire du Bâtiment et des Travaux publics (1935) ; mise au point d'un procédé de construction de tuyaux précontraints par fils adhérents (1936) ; réalisation de caissons de fondations de quais précontraints pour le port de Brest (1937-1939) ; étude du marché couvert de Francfort-sur-le-Main, construit par Wayss und Freytag, peu avant la Seconde Guerre mondiale. L'efficacité du nouveau procédé suscita de nombreux imitateurs : en France, l'ingénieur André Coyne (1891-1960) ; à l'étranger, les Allemands Dischinger, Finsterwalder et Rüsch, constructeurs de ponts précontraints par câbles extérieurs, ainsi que les Américains Hervett et Crown, auteurs, à partir de 1934, d'une série de réservoirs cylindriques précontraints. Mais, grâce à Edmé Campenon, les entreprises françaises préservèrent leur avance.

Les ECB, une entreprise dynamique portée par la précontrainte

Fondées par E. Campenon, les ECB effectuèrent une spectaculaire percée durant les années 1920[4]. Ayant fait le choix de la précontrainte, elles s'imposèrent, durant les années 1930, parmi les leaders de la profession, grâce à leur avance technique.

La percée

Durant les années 1920, les ECB se développèrent à un rythme très élevé : entre 1920 et 1929, leur chiffre d'affaires TTC augmenta de + 15 % par an en moyenne et en francs constants. Elles s'identifiaient à la personne de leur fondateur[5]. Edmé Campenon associait non-conformisme et goût des affaires. Né à Tonnerre le 27 décembre 1872, dans une famille de vieille noblesse, il fut un élève doué. Mélomane et excellent pianiste, il fit ses études à Louis-le-Grand, où il présenta le Concours général avant d'entrer en Math. sup., puis Math. spé. Il refusa cependant de se présenter au concours de Polytechnique, suivit quel-

4. D. Voldman, *Le béton, la technique et la précontrainte. Histoire de l'entreprise de BTP Campenon Bernard 1920-1975*, Paris, 1987, 89 p., dactyl. ; *Entreprises Campenon Bernard. Références*, Paris, Sapho, 1951 ; *Entreprises Campenon Bernard : Références*, Paris, Presses des Imprimeries de Bobigny, 1963 ; *Campenon Bernard. Références 1971,* Paris, Imprimerie Desgrandchamps, 1971.

5. *Edmé Campenon* (1872-1962), Paris, Ed. Vendel, 1962.

ques cours à la faculté de Droit et à Langues-O. Après avoir effectué son service militaire, il roula sa bosse à travers le monde : il visita successivement la Corse, l'Espagne, l'Autriche, l'Amérique, l'Indochine, la Chine et le Japon, s'y occupant d'affaires variées (topographie, commerce, concessions de mines, de chemins de fer et de ports). De retour en France, il fonda une famille, à Lyon, où il s'occupa d'aviation, puis s'établit en Gascogne. En 1910, il entra chez Thouvenot, principal entrepreneur de la Compagnie du Midi, alors engagé dans de très gros chantiers. Bien que Campenon se fût imposé assez vite comme le second de son patron, il songeait à voler de ses propres ailes. En 1920, il fonda sa propre entreprise : la société en nom collectif Entreprise Campenon et Bernard, où il se trouvait associé à André Bernard, un jeune ingénieur des Arts et Métiers.

Ce fut une réussite immédiate, car Campenon savait s'entourer des meilleurs ingénieurs, faire des choix techniques heureux, se reconvertir très vite. De 1920 à 1925, l'entreprise fonda sa croissance sur le bâtiment (reconstruction du Nord-Est, HBM, maisons ouvrières), mais le génie civil représentait déjà une part importante (ouvrages d'art, terrassements pour les compagnies ferroviaires, aménagement de chutes). Adoptant, dès 1921, le statut de commandite par actions, la société transféra son siège d'Albi à Paris. Mais elle connut une crise grave en 1924-1925. Elle s'en arracha en se réorientant, de 1926 à 1929, vers l'hydraulique et l'hydroélectricité. Elle réalisa alors de grands chantiers : extension du port de Strasbourg (1926-1928) et construction des centrales de Sabart, Point-de-Rivières et la Gentille-Saint-Sernin (1926-1931). En 1927, elle enleva, face à quinze concurrents, l'adjudication des travaux du barrage du Chambon, alors le plus haut d'Europe. Un an plus tard, elle se transforma en SA par actions, sous la raison sociale de Société Campenon Bernard.

Accès au rang de " major "

Les ECB connurent, dans les années 1930, une croissance plus forte encore : de 1930 à 1939, son chiffre d'affaires TTC augmenta de + 16,4 % en moyenne et en francs constants. Cette expansion — la plus forte de tout le secteur des travaux publics — se manifesta pour l'essentiel à partir de 1934-1935, c'est-à-dire à partir de l'application effective de la précontrainte aux travaux de génie civil. En effet, de 1929 à 1933, la société se heurta à de sérieuses difficultés. La première fut la mort inopinée d'A. Bernard, suite à une embolie qui privait l'entreprise de son principal organisateur et désorganisa le chantier du Chambon. Campenon crut trouver la solution dans un rapprochement avec la Société Dufour-Constructions générales, dont la Banque de l'Union parisienne (BUP) était le principal actionnaire. La fusion s'opéra à l'instigation de la BUP en janvier 1930, Campenon Bernard absorbant la Société Dufour. Il s'agissait d'une opération risquée, car elle s'accompagnait de la prise en charge de deux très gros chantiers : la construction du barrage algérien de l'Oued Fodda

(1926-1932) ; celle de la tranche de Rochonvillers (1930-1934), l'un des lots les plus importants de la ligne Maginot. La société surmonta les difficultés techniques. Dès 1932, les ECB atteignaient des rythmes de bétonnage record au Chambon, à Rochonvillers et à l'Oued Fodda surtout. Elles s'imposèrent ainsi comme l'une des plus importantes entreprises françaises dans le domaine des aménagements hydrauliques et hydroélectriques ainsi que du génie militaire, y surclassant les spécialistes.

A partir de 1934-1935, la croissance reprit à un rythme très vigoureux. Deux facteurs y contribuèrent. Les ECB bénéficiaient, en premier lieu, du soutien très appréciable de la Banque de l'Union parisienne, qui leur ouvrit de larges découverts et favorisa, en février 1937, la fusion de la Société Campenon Bernard et des Entreprises Jean Hesbert au sein d'une puissante SA de 20 millions de francs de capital. En second lieu, les ECB firent d'emblée le choix de la précontrainte : en 1934, ayant visité le chantier du Havre, E. Campenon fit à E. Freyssinet l'offre de le rejoindre dans son entreprise. Ce fut le début d'une collaboration poursuivie jusqu'à la mort des deux hommes. Campenon décida d'appliquer d'emblée la précontrainte à de très grands ouvrages. Deux réussites le fortifièrent dans ses convictions : les 44 km de conduites forcées à gros diamètre de l'Oued Fodda et les caissons du port de Brest. De plus, l'arrivée de Freyssinet au sein des ECB incita les grandes administrations, notamment militaires, à faire confiance aux Entreprises Campenon Bernard. Elles acceptèrent ainsi de prendre des risques, mais à bon escient : les conduite forcées de l'Oued Fodda, pourtant situées en plein épicentre du séisme d'El-Asnam, en 1954, lui résistèrent sans dommage.

Les ECB renforcèrent ainsi leurs positions dans le domaine des travaux hydrauliques. En Algérie, elles édifièrent les barrages du Hamiz (1934-1935), des Portes-de-Fer (1937-1939) et des Beni-Bahdel (1939-1941). La surélévation de ce barrage à voûtes multiples donna lieu à la première application de la précontrainte à ce type de travaux. Les ECB renforcèrent aussi leur avance en matière de génie militaire, avec l'achèvement de la tranche de Latiremont de la ligne Maginot (1935), puis la réalisation de deux nouveaux chantiers, l'un sur la ligne Maginot, l'autre près d'Orléans (un aérodrome et un camp militaires). Surtout, elles entrèrent en force sur le marché des grands équipements portuaires. Leur maîtrise de la précontrainte rendait les ECB capables de réaliser d'énormes ouvrages à des coûts largement inférieurs à ceux que l'on aurait pu atteindre avec le béton armé. L'entreprise obtint ainsi une série de gros marchés. A Brest, elle édifia, de 1935 à 1939, 1,25 km de quais à grande profondeur et une bonne partie du troisième bassin de radoub de Laninon. A Mers-el-Kébir, à partir de 1937, elle construisit 2 km de quais ainsi qu'une jetée de 1,75 km de longueur pour 32 m à 36 m de profondeur. Signe évident de leur réussite, les ECB se situaient, en 1939, au quatrième rang de l'industrie française des travaux publics par l'importance de leur chiffre d'affaires.

Durant la Seconde Guerre mondiale, les ECB parvinrent à maintenir leur avance technique. Bien que leur chiffre d'affaires, exprimé en francs constants, eût alors connu une forte contraction, elles poursuivirent d'importants travaux de génie civil en Algérie (Beni-Bahdel, Oued Fodda) et en métropole (déblaiement et reconstruction d'Orléans, construction de l'émissaire Sèvres-Achères). Elles exécutèrent un volume substantiel de travaux allemands, exécutés seuls ou en participation avec l'entreprise Julius Berger de Berlin (consortium Bercamp). Mais, en définitive, leur chiffre d'affaires demeura très inférieur à celui de l'avant-guerre. Ses dirigeants opposèrent un refus poli et ferme à toutes les demandes allemandes de prêt de matériel ou de coopération technique, tandis que se développait, au sein de la société, une activité de renseignement au profit des Alliés. Elles maintinrent, grâce à l'importance des redevances perçues sur les brevets Freyssinet, une position financière enviable. Elle leur permit de poursuivre leurs investissements et de se diversifier, vers le bâtiment et les travaux routiers en particulier. Surtout, elles se dotèrent, au travers de la Société technique pour l'utilisation de la précontrainte (STUP), d'une filiale d'ingénierie chargée d'exploiter et de développer les brevets Freyssinet. Grâce à elle et au soutien que lui apportèrent les Ciments Lafarge, la STUP constitua désormais, avec les ECB, le fer de lance de l'expansion internationale du groupe Campenon Bernard, expansion demeurée vigoureuse jusqu'au milieu des années 1970. Ensuite, les ECB perdirent progressivement leur identité dans les multiples restructurations qui semblent enfin atteindre leur terme aujourd'hui. En revanche, la STUP, devenue Freyssinet International, demeure aujourd'hui l'un des plus beaux fleurons de l'ingénierie française à l'étranger : pour preuve, son implication dans la réalisation du plus grand chantier en cours dans le monde aujourd'hui, celui de l'aéroport de Chek Lap Kok à Hong Kong.

Le poteau et le voile : essor et apogée du béton armé dans le logement collectif en France

Dominique Theile

Malgré quelques avancées significatives, l'histoire des techniques de la construction reste largement à écrire, tout au moins en France. Plus particulièrement, la diffusion de l'emploi du béton armé dans la construction de bâtiments résidentiels y reste encore mal connue, alors même que la France se distingue actuellement de ses voisins européens par l'emploi massif et monolithique de béton armé dans le logement collectif. Comment est-on parvenu à cette domination du voile de béton armé dans le logement en France ?

Je commencerai par quelques précisions sur l'étude dont est issue la présente communication. Ensuite, je retracerai le processus qui conduit à une émergence du béton armé, dans le résidentiel collectif, sous la forme poteau dans les années 1930. Enfin, je tenterai d'expliquer comment on passe du poteau porteur au voile porteur entre 1930 et 1960.

Le point de départ de la recherche est une question. Qu'est ce qui fait que, aujourd'hui en France, le constructeur d'un bâtiment résidentiel collectif préfère le voile porteur au poteau porteur ? Autrement dit, pour reprendre les termes des professionnels du Bâtiment : pourquoi préfère-t-on les structures linéaires aux structures ponctuelles ?[1]

Pour répondre à cette question j'ai choisi de procéder à une étude de comportements d'acteurs, d'une part, et de reconstituer le cheminement de la diffusion du voile banché, d'autre part. C'est cette dernière démarche qui est ici présentée, cheminement dont le schéma ci-après est la synthèse.

1. Je me réfère ici uniquement aux voiles “ épais ” et non aux voiles minces, ou coques, utilisés uniquement pour les toitures ou pour les parois de certains ouvrages spéciaux (tours de refroidissement de centrales nucléaires). Je traite donc des voiles verticaux employés en bâtiments résidentiels. Connus comme voiles de refend, ils ont cependant d'autres usages que les murs de refends, comme les façades et les pignons. Il s'agit de voiles de béton, béton généralement armé, et généralement coulé sur place (à l'exception, dans certains cas, des voiles de pignon). Nous ne désignons pas sous le terme “ voile ” les murs préfabriqués en usine, appelés “ panneaux ”.

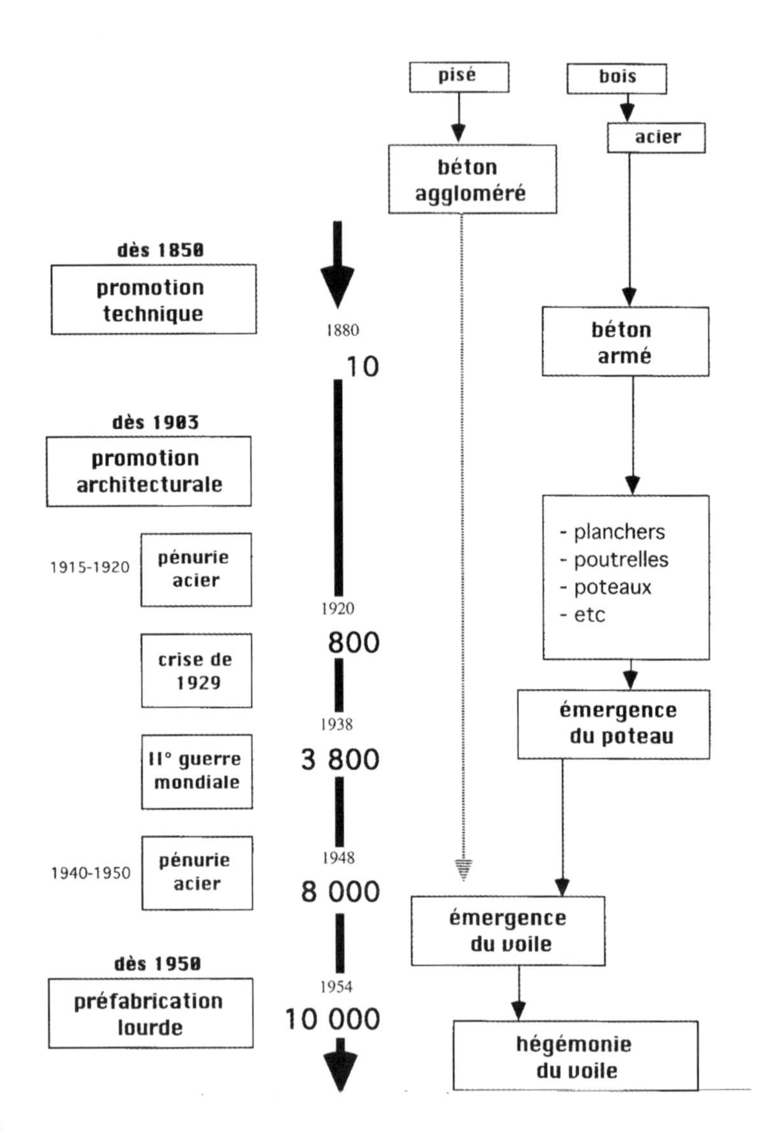

Ce schéma donne les principaux paramètres impliqués dans une diffusion du béton en France. Diffusion qui conduit à l'hégémonie du voile dans le logement collectif. Il met en parallèle trois dimensions :

1) les facteurs sociaux, politiques, économiques et culturels conditionnant la diffusion du béton armé ;

2) l'évolution de la production de ciment ;

3) le mode de diffusion du béton armé.

QUELQUES PRÉCISIONS SUR LA DIFFUSION DU BÉTON DANS LE RÉSIDENTIEL

Le premier emploi important connu de béton, dans un bâtiment résidentiel, date de 1835 avec la maison de l'architecte Lebrun. En 1853, c'est la maison de Coignet à Saint-Denis. Joseph Tall améliore le procédé Coignet en mettant notamment au point des coffrages démontables et réutilisables, employés en 1867 pour une cité ouvrière avenue Daumesnil (Paris) : c'est là sans doute le plus ancien exemple d'utilisation de voile en logement social.

Aussi bien Lebrun, que Coignet et Tall, mettent en oeuvre du béton aggloméré.

Ce n'est qu'en 1892 qu'est construit le premier immeuble en béton armé. Et il ne s'agit pas vraiment d'un immeuble résidentiel : c'est le siège social de Hennebique (rue Danton, Paris).

En somme, au XIXe siècle, on n'a pratiquement pas de bâtiments résidentiels dont le gros oeuvre est réalisé en béton, armé ou pas.

En revanche, on assiste à un développement de l'emploi du béton, d'abord aggloméré puis armé, pour des parties d'ouvrages : en fondations ; puis dans les planchers et murs.

Pour la superstructure, la diffusion du béton prend plutôt la forme d'une substitution du métal, métal qui lui-même était venu substituer le bois pour ce qui est des poteaux, poutres, poutrelles. Ceci est particulièrement notable dans les planchers où, de simple garni au départ, le béton acquiert progressivement, une fois qu'il est armé, une fonction structurelle (dalle de compression ou hourdis).

Puis, grâce au pouvoir liant du béton, s'entame un processus de substitution (des poutrelles d'acier notamment), qui se transforme en partie en processus d'expulsion de tout ce qui n'est pas béton armé. Là où l'emploi d'un profilé d'acier n'est pas fondamental, ce profilé est réduit, si l'on peut dire, à un fer à béton et englouti par le béton (à partir du moment où il devient plus intéressant d'employer du béton armé).

Par conséquent, en étant schématique, le chemin de diffusion du béton dans les immeubles résidentiels ne se situe pas dans le droit fil du béton aggloméré et de la banche à pisé. Il se situe plutôt dans le droit fil de l'acier et du bois, c'est-à-dire du poteau et de la poutre.

On m'a signalé l'emploi de la maçonnerie banchée dans le pavillonnaire parisien entre 1880 et 1939[2]. Mais il ne semble pas que l'on puisse établir une connexion directe entre le pavillonnaire et le logement collectif en termes de diffusion du béton armé, pour pouvoir comprendre l'émergence du voile de béton armé.

2. M. Francis Pierre, président de l'Association internationale pour la Promotion des engins techniques et historiques comme objets d'art et de culture.

Bien sûr, si l'on considère la structure horizontale, on peut constater un essor du béton monolithique aussi bien dans le logement collectif que dans le pavillonnaire : la dalle de béton armé. En revanche, pour ce qui est de la structure horizontale on assiste, dans le logement collectif, à une émergence du poteau et non pas du voile, plutôt dans l'entre-deux-guerres. Le poteau porteur est employé dans les murs porteurs des immeubles résidentiels principalement parce qu'il autorise une diminution de l'épaisseur des murs. Il s'emploie en particulier dans les murs de refend, c'est-à-dire les murs porteurs internes autres que les murs de façade et de pignon.

Le fait que l'emploi du poteau se développe dans les murs de refend paraît très intéressant : pourquoi ? Parce qu'aujourd'hui, en France, " voile de refend " est le terme le plus usité, pour désigner le voile en béton armé banché.

Ce qui m'amène à préciser ce que j'entends par " hégémonie du voile " dans le logement collectif. Car, en termes de façades et de pignons on ne peut pas parler d'hégémonie. C'est, en revanche, pour les murs de refends que le recours au voile banché de béton armé est pratiquement systématique.

La promotion du béton : d'objet technique à objet architectural

Au XIXe siècle le béton est un objet technique : il n'est pas le support d'une architecture spécifique.

Notons qu'à l'époque il n'y a pratiquement pas d'ingénieurs dans le Bâtiment. Tout au moins, l'ingénieur-concepteur n'y est pas connu, à l'inverse des Travaux Publics.

La figure qui émerge alors est celle de l'ingénieur-entrepreneur de béton. Le plus connu est Hennebique, mais il y a aussi Coignet, Considère et bien d'autres.

A l'époque, le béton fait l'objet de brevets : béton aggloméré puis béton armé. Le béton armé n'est pas proposé comme simple matériau mais comme procédé.

Hennebique se lance dans un travail impressionnant de promotion du béton armé : il licencie ses procédés, ouvre des bureaux de représentation, démarche les architectes et décideurs, organise des visites de chantiers, des colloques, édite une revue…

Les entrepreneurs de béton armé ont leurs propres brevets et/ou exploitent une ou plusieurs licences. Cependant tous, ou presque, ont une pratique systématique du démarchage. Jusqu'au début du XXe siècle, les entrepreneurs de béton armé vont être les acteurs principaux de la promotion du béton.

Précisons que breveter du béton armé, en caricaturant à peine, c'est comme breveter le beurre : les copies se multiplient et les brevets aussi. En outre, ces brevets tombent peu à peu dans le domaine public. Or, le moment où les principaux brevets tombent dans le domaine public coïncide avec le moment où les architectes relaient les ingénieurs-entrepreneurs de béton armé dans leurs efforts de promotion de ce matériau.

C'est le moment où le béton acquiert le statut d'objet architectural. C'est-à-dire que le béton est associé à une architecture qui est non seulement nouvelle, mais est spécifique au béton armé, et met explicitement celui-ci en valeur.

Les premières réalisations remontent aux frères Perret, avec l'immeuble de la rue Franklin à Paris (1903) et le garage Ponthieu (1905). Au niveau théorique, le premier projet concerne la cité industrielle de Tony Garnier (1901-1904), que l'on peut considérer comme un véritable catalogue d'utilisation architecturale du béton. On trouve aussi dans l'ouvrage de Le Corbusier de 1923, *Vers une nouvelle architecture*, divers projets en béton, dont certains datent de 1914.

Mais surtout, dès 1915, Le Corbusier idéalise, sans toutefois l'inventer, l'ossature Domino ; c'est-à-dire l'ossature poteau-dalle, alors que celle bois-métal ne connaissait que l'ossature poteau-poutre[3]. On peut même supposer qu'il s'agit d'une structure nouvelle et spécifique au béton, sauf peut-être si l'on remonte très loin dans l'histoire de l'architecture de pierre ; structure que l'on associe à une architecture.

Il est de plus en plus accepté que le XIXe siècle prend fin en 1914. Mais les architectes qui viennent d'être cités sont des précurseurs du Mouvement Moderne, qui prend véritablement son essor dans l'entre-deux-guerres. Par conséquent, je suppose que ce n'est qu'au XXe siècle que le béton devient objet architectural. Plus exactement, ce n'est que dans les années 1930 que le béton se constitue pleinement en objet architectural, époque où, en tant qu'objet technique, il connaît déjà une expansion certaine.

D'autant plus que, si l'on revient au schéma présenté plus haut, au sortir de la première guerre mondiale, la production d'acier connaît des difficultés dont peuvent tirer parti les partisans du béton. Il faut insister sur le fait que les professionnels de l'acier ne pratiquaient pas dans le Bâtiment le démarchage systématique qui caractérise les entrepreneurs du béton. Bien plus, l'introduction de la poutre métallique dans le Bâtiment n'est pas loin de relever du hasard : la grève des charpentiers des années 1840 qui amena à utiliser des rails de chemin de fer en guise de poutres.

De même, si l'on s'intéresse aux industriels du ciment, on détecte un important effort de promotion commerciale. Effort que l'on peut repérer dès la naissance de cette industrie, en 1830 avec Pavin de Lafarge. Celui-ci ne produit plus dans le seul objectif de répondre à la demande, mais essaye de l'anticiper et surtout de la créer. D'autres ingénieurs-entrepreneurs suivent la même démarche : Dupont et Demarle en 1846, Vicat fils vers 1850.

3. Sur la base de nos lectures le plancher, structure monolithique horizontale, prend directement appui sur les poteaux. En fait, l'appui se fait par l'intermédiaire de poutres, mais celles-ci ne sont pas visibles, étant noyées dans la dalle.

Ce souci de promotion commerciale n'est assurément pas spécifique à l'industrie cimentière française, mais, s'étant manifesté très tôt, il figure au nombre des facteurs contribuant à cette spécificité française, qui est l'emploi massif du voile banché dans le résidentiel collectif après la seconde guerre mondiale.

Ces efforts concomitants des cimentiers et des entrepreneurs de béton d'abord, puis des architectes, contribuent à ce, que dans les années 1930, le décollage du béton dans le Bâtiment paraît assuré :

- le marché est créé, notamment en termes architecturaux ;
- on dispose d'une production de ciment relativement abondante, puisqu'elle est presque multipliée par cinq entre 1920 et 1938, passant de 0,8 à 3,8 millions de tonnes.

Pourtant, en revenant au schéma comparatif exposé plus haut, on constate que le décollage du béton coïncide dans le Bâtiment à un essor du poteau et non pas du voile.

1930-1960 : DES MURS À POTEAUX DE BÉTON ARMÉ AUX MURS EN BÉTON ARMÉ

Si, en France dans le résidentiel collectif, le poteau de béton émerge dans l'entre-deux-guerres, en particulier dans les murs de refends, c'est le mur monolithe de béton armé qui s'impose progressivement après la seconde guerre mondiale. La question qui se pose est alors : pourquoi l'essor du poteau porteur avorte-t-il au profit du voile porteur ?

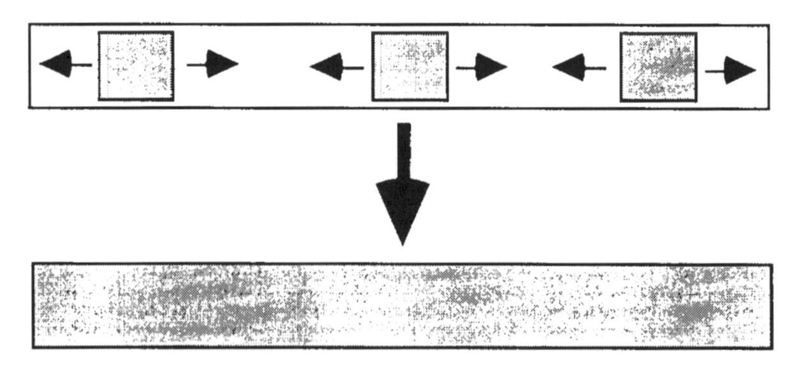

piste 1 : le voile est une dilatation du poteau

Le schéma illustre une première voie d'exploration : à force de construire des murs constitués de poteaux en béton armé entre lesquels on disposait des briques, des entrepreneurs ont fini par se dire qu'il était plus intéressant d'utiliser uniquement du béton armé, notamment pour des questions de gain de temps, la construction d'un mur en béton armé requiert un nombre de manipulations beaucoup moins élevé qu'un mur en maçonnerie classique. Cette piste était d'autant plus séduisante que ce basculement poteau-voile était de plus en plus susceptible de se produire à mesure que l'usage du parpaing comme matériau de maçonnerie de remplissage se répandait, et que le prix relatif du ciment baissait (en raison de l'augmentation de sa production). Toutefois, cette hypothèse se révèle insuffisante. Une seconde hypothèse est nécessaire. Elle n'est pas incompatible avec la première, mais se révèle plus forte : le voile résulterait d'une application à la construction " traditionnelle " à murs porteurs de matériels et de méthodes des Travaux Publics.

Ces matériels et ces méthodes sont nouveaux pour la branche du Bâtiment d'alors, et utilisent le béton armé. Ils se développent dans la branche des Travaux Publics avec les politiques des grands travaux des années 1930 ; grands travaux que relaieront les travaux d'ouvrages défensifs et de reconstruction de ponts après la seconde guerre mondiale. Il faudrait, ici, cerner pourquoi les entreprises de Travaux publics sont venues au Bâtiment. Toujours est-il qu'à une exception près (Bouygues) les majors du BTP en France seront tous issus des Travaux Publics.

Remarquons que l'effort de guerre ne mobilisera pas que les entreprises de Travaux Publics. Les entreprises de Bâtiment sont elles aussi réquisitionnées, et sont amenées à couler du béton et à se frotter aux méthodes et matériels des Travaux Publics. En outre, une importante main d'oeuvre sera formée au béton, notamment au travers du STO (service du travail obligatoire), pour la construction de blockhaus. Ces facteurs ne sont certainement pas à négliger, pour comprendre certaines réticences à l'utilisation du béton parmi la main d'oeuvre et les entreprises.

Par ailleurs, l'état de la production cimentière a joué un rôle sensible dans l'avènement du voile de béton banché dans le logement collectif. D'une part, parce que la production de ciment est en hausse constante depuis le début du XXe siècle, alors que la production sidérurgique connaît une croissance en dents de scie (guerre, crise, guerre, reconstruction). D'autre part, parce qu'au sortir de la guerre, l'industrie cimentière est dans une situation pratiquement opposée à celle que connaît la sidérurgie. En effet, c'est une pénurie d'acier que l'on enregistre en France jusqu'au début des années 1950, et non de ciment. L'industrie cimentière, qui a fortement crû pour les besoins de la guerre, peut craindre, au sortir du conflit mondial, de se retrouver à terme en situation de surcapacité.

En résumé, c'est l'épisode de la préfabrication lourde qui consacre le voile de béton armé. Pas sous la forme que l'on connaît aujourd'hui (voile coulé en place) mais sous la forme de panneau de béton armé. Mais l'utilisation de ce dernier dans le logement collectif déclinera assez vite au profit du voile coulé en place, d'abord au moyen de coffrages-tunnel, puis avec les coffrages-outils couramment utilisés aujourd'hui en France.

L'épisode de la préfabrication lourde (années 1950 et 1960) en France paraît être le moment-clef du passage de l'utilisation du poteau au voile en béton armé dans le logement collectif. De nombreux paramètres explicatifs peuvent être mis en avant. Cependant trois d'entre eux paraissent primordiaux : l'épisode de la préfabrication lourde et l'avènement du voile résulteraient d'une triple conjonction. Conjonction entre : la disponibilité d'une main d'oeuvre et l'existence d'entreprises formées aux matériels et méthodes du béton monolithique ; un Etat désireux de construire vite et beaucoup, parce qu'il est engagé dans un vaste processus de modernisation (planification française) ; une bonne disponibilité en matériaux constitutifs du béton. Sur ce dernier point, une question se pose : le ciment aurait-il été plus facile d'accès en France que dans d'autres pays européens à la même époque ?

Une problématique pour une histoire des infrastructures urbaines à travers le cas de Montréal (1845-1960)

Robert Gagnon - Dany Fougères - Michel Trépanier

Aqueducs, réseaux d'égouts et lignes de Métro constituent des infrastructures urbaines les plus structurantes et les plus complexes. Or, on a trop tendance à oublier que le développement de ces systèmes centralisés dans les villes industrielles a aussi entraîné de multiples transformations de la société urbaine. Notre programme de recherche a eu, conséquemment, pour objectif de reconstruire l'histoire de la mise en place du système d'égouts et d'aqueducs de la ville de Montréal, tout comme la construction de son Métro.

Bien que notre objet soit éminemment technique, notre problématique refuse de l'appréhender dans le cadre étroit d'une histoire internaliste de la technologie. Nous voulons avant tout rendre compte de la multiplicité et de la diversité des facteurs à la fois politiques, scientifiques, techniques et sociaux qui, souvent indissociables, ont donné à ces infrastructures montréalaises ses caractéristiques particulières.

C'est pourquoi nous proposons d'analyser l'espace social extrêmement hétérogène et métissé dans lequel se sont prises les décisions relatives à la construction de ces infrastructures.

Bien que la technologie ait joué un rôle crucial dans le développement des villes, ce n'est que tout récemment que les historiens ont commencé à étudier cet aspect de l'histoire urbaine[1]. Les travaux de Joel Tarr et Louis Cain sur le développement des infrastructures urbaines aux Etats-Unis, ceux de Georges Knaebel, Philippe Cebron de Lisle et Gabriel Dupuy pour les villes européennes et les travaux de Christopher Hamlin sur la construction du réseau d'égouts de Londres, ont montré que ce ne sont pas des facteurs économiques et des innovations techniques qui déterminent pour l'essentiel le développement de

1. Pour une bibliographie exhaustive sur le sujet voir : A. Durkin Keating, *Invisible Networks. Exploring the History of Local Utilities and Public Works*, Floride Krieger Publishing Co. 1994.

ces infrastructures urbaines[2].

Comme le souligne l'un d'eux *the preferences and perceptions of different actors such as business leaders, politicians, and public health and engineering professionals in a particular city at a particular time may be more important in the timing of the city-building process than a generalized set of forces that relates to all cities*[3]. Bref, c'est plutôt l'interaction de facteurs scientifiques, techniques, sociaux, institutionnels, économiques et politiques qui donne aux infrastructures sa forme finale. Les sociologues de la technologie ont d'ailleurs montré comment les ingénieurs sont souvent confrontés à des considérations de natures très différentes ainsi qu'à des acteurs sociaux eux aussi très diversifiés[4]. On sait, à présent, que les ingénieurs ont développé toute une série de rôles et d'habiletés qui vont bien au-delà de ceux liés à la conception technique proprement dite. Ce courant récent de la sociologie nous fournit une grille d'analyse pertinente pour comprendre et interpréter certains aspects de notre étude. Jusqu'à maintenant, l'analyse des processus de décisions qui conduisent à la construction de ces équipements et qui rendent compte de leurs caractéristiques a surtout été l'apanage des sociologues[5] ou des politicologues[6]. Or, le rôle extrêmement important du temps et de la contingence dans ces processus en font des objets tout désignés pour les historiens[7].

2. J.A. Tarr, " Building the Urban Infrastructure in the Nineteenth Century : An Introduction ", *Infrastructure and Urban Growth in the Nineteenth Century*, Chicago, Public Works Historical Society, 1985, 61-85 ; J.A. Tarr , " The Separate vs. Combined Sewer Problem : A Case Study in Urban Technology Design Choice ", *Journal of Urban History*, V, 3 (mai 1979), 308-339 ; J.A. Tarr, " The Evolution of the Urban Infrastructure in the Nineteenth and Twentieth Centuries " in R. Hanson (ed.), *Perspectives on Urban Infrastructure*, Washington D. C., Committee on National Urban Policy, Commission on Behavorial and Social Sciences and Education, National Research Council, National Academy Press, 1984, 4-66 ; J. Tarr et G. Dupuy, *Technology and the Rise of the Networked City in Europe and America*, Philadelphie, Temple University Press, 1988 ; L. Cain, " Raising and Watering. Ellis Sylvester Chesbrough and Chicago's First Sanitation System ", *Technology and Culture*, XIII (juillet 1972), 353-372 ; G. Knaebel, " Bielefeld. Genèse d'un réseau d'égouts 1850-1904 ", *Les annales de la recherche urbaine*, 23-24 (juillet-décembre 1988), 90-102 ; C. Hamlin, " William Dibdin and the Idea of Biological Sewage Treatment ", *Technology and Culture*, XXIX, 2 (avril 1988), 189-218 ; F. Caron, J. Dérens, L. Passion, P. Cebron de Lisle, *Paris et ses réseaux : naissance d'un mode de vie urbain XIXe-XXe siècles*, Paris, Hôtel d'Angoulême-Lamoignon, 1990 ; G. Dupuy, " La science et la technique dans l'aménagement urbain ", *Les annales de la recherche urbaine*, 6 (janv. 1980), 3-18.

3. J.A. Tarr et J.W. Konvitz, " Pattern in the Development of the Urban Infrastructure " dans H. Gillette Jr et Z.L. Miller (eds), *American Urbanism. A Historical Review*, New York, Greenwood Press, 1987, 196.

4. Nous pensons, notamment, à B. Latour, *Science in Action : How to Follow Scientists and Engineers Through Society*, Cambridge, Havard University Press, 1987, 132-144, et, surtout, à Y. Gingras et M. Trépanier, " Constructing a Tokamak : Political, Economical and Technical Factors as Constraints and Resources ", *Social Studies of Science*, XXIII, 1 (1993), 5-36.

5. M. Trépanier, *L'aventure de la fusion nucléaire. La politique de la Big Science au Canada*, Montréal, Boréal, 1995 et D. MacKenzie, *Inventing Accuracy : A Historical Sociology of Nuclear Missile Guidance*, Cambridge, MIT Press, 1990.

6. G.T. Allison, *Essence of Decision : Explaining the Cuban Missile Crisis*, Boston, Little Brown, 1971.

7. Voir par exemple l'excellente monographie de A. Hermann, J. Krige, U. Mersits et D. Pestre, *History of CERN*, Amsterdam, North-Holland, 1987 et 1990, Vol. I et II.

Les égouts

En ce qui concerne la mise en place d'un réseau intégré de conduites d'égouts à Montréal, c'est entre 1850 et 1920 que nous situons sa genèse et son développement.

A l'instar de plusieurs grandes villes industrielles en Amérique du Nord et en Europe[8], Montréal s'engage, au cours de la deuxième moitié du XIXe siècle, dans la construction de grandes infrastructures urbaines. L'industrialisation de la ville, l'accroissement de sa population, les épidémies de choléra et de typhus, les conceptions médicales de l'époque (théorie des miasmes), ne sont que quelques-uns des facteurs qui incitent les élus municipaux à se préoccuper de la santé publique et à envisager la construction d'égouts souterrains pour évacuer les eaux usées[9]. Or, l'ampleur des budgets nécessaires à la construction d'un réseau d'égouts, tout comme l'impact que cette infrastructure a sur le milieu, font qu'une multitude d'acteurs sociaux participent aux débats qui s'instaurent sur la santé publique, l'intervention des gouvernements locaux dans la vie des citadins ou le rôle de la technologie dans l'environnement urbain, dans les dernières décennies du XIXe siècle[10]. Les ingénieurs, en particulier, commencent à s'imposer comme des experts incontournables. L'ingénieur sanitaire et l'ingénieur municipal deviennent des acteurs centraux avec lesquels les hommes d'affaires, les élus municipaux et les médecins devront composer[11]. Il s'agira donc de cerner les forces et les faiblesses des différents acteurs et d'examiner l'ensemble des stratégies qu'ils mettent en oeuvre pour convaincre les différentes instances décisionnelles. Ainsi, nous pourrons comprendre pourquoi un système d'égouts unitaire, qui reçoit aussi bien l'eau de ruissellement que les eaux usées, a été privilégié au détriment d'un système séparé ou, encore, comment le caractère technique de ces infrastructures a servi de point d'appui aux ingénieurs pour investir différents comités municipaux et jouer un rôle accru dans la gestion de la ville. Cette montée des experts s'explique également en la rapportant au courant réformiste qui traverse les grandes villes nord-américaines et qui gagne Montréal à la fin du siècle dernier. Ce courant vise, en effet, à rationaliser l'administration municipale et à s'attaquer aux questions de nature sociale ou environnementale.

8. J.A. Tarr, J. McCurley III, F.C. McMichael, T. Yosie, " Water and Waste : A Retrospective Assessment of Wastewater technology in the United States, 1800-1932 ", *Technology and Culture*, 2 (1984), 226-263.

9. M. Farley, O. Keel, C. Limoges, " Les commencements de l'administration montréalaise de la santé publique ", dans P. Keating et O. Keel (éds), *Santé et société au Québec*, Montréal, Boréal, 1995, 85-114.

10. M. Trépanier, " L'eau, la technique et l'urbain : l'ingénieur n'est jamais seul dans l'univers des infrastructures urbaines " dans L. Pothier (ed.), *L'eau, l'hygiène publique et les infrastuctures*, Groupe PGV, Diffusion de l'archéologie, 1996, 65-83.

11. S.K. Schultz et C. McShane, " Pollution and Political Reform in Urban America : The Role of Municipal Engineers, 1840-1920 " dans M.V. Melosi (ed.), *Pollution and reform in Amercian Cities, 1870-1930*, Austin et Londres, University of Texas Press, 155-172.

Le service d'aqueduc

Le service d'alimentation en eau potable de Montréal a été marqué, quant à lui, par une réévaluation sérieuse des rôles respectifs du secteur public et du secteur privé : en 1845, achat et municipalisation du service ; en 1928, achat et municipalisation de la Montreal Water and Power Co. ; et finalement aujourd'hui, l'actuelle remise en question du service montréalais à la lumière des nouveaux rapports qui se nouent entre les secteurs public et privé. Le caractère récurrent des remises en question de la composition publique/privée du service d'aqueduc d'une part, et de son profil technique, d'autre part, indique que nous ne sommes pas ici en présence d'une loi universelle incontournable — surtout immuable — mais plutôt d'un espace de négociation spécifique.

Le service d'eau montréalais a connu une histoire qu'il partage, dans ses grandes lignes, avec celle des autres grandes villes occidentales. Mais il a aussi une histoire bien à lui, avec ses rythmes au sein desquels les défis et les enjeux se succèdent à intervalles irréguliers, où les acteurs et les facteurs, les contraintes et les opportunités, s'entremêlent pour en faire un espace de négociation spécifique, où les résultats du processus demeurent en partie contingents. L'histoire du service montréalais est donc directement inscrite dans celle de son lieu d'appartenance, la ville.

La place des secteurs public et privé dans la gestion du service et le profil technique qui caractérise le service d'eau potable tiendraient selon nous, d'une construction plus ou moins planifiée, à la jonction du technique et du social. Les transformations que connaît le service au fil du temps témoigneraient de la rencontre et de la hiérarchisation de multiples facteurs (politique, technique, économique, juridique, *etc.*), de l'état des rapports de force entre les intervenants et du caractère contingent dans lequel s'inscrivent les évolutions en cours. Le service d'eau potable serait donc un “ lieu d'arbitrage ” entre ce qui relève à la fois pleinement du social et du géo-technique (la dotation en ressources hydrauliques, la résistance des matériaux, les connaissances de l'époque, *etc.*).

Le Métro

L'histoire du Métro de Montréal, elle aussi, offre un bel exemple d'une infrastructure où l'interaction des facteurs techniques, sociaux, politiques et économiques est déterminante. Conçu au début des années 1960, le Métro de Montréal a comme particularité technologique que les voitures sont montées sur des pneus. Elles roulent donc pneumatique sur fer plutôt que fer sur fer. Outre le fait que cette technologie améliore la qualité du roulement et accroit ainsi le confort des usagers, cette innovation s'explique en partie par le désir de l'administration municipale de se servir du Métro pour projeter à l'extérieur du Québec et du Canada une image de modernité, de progrès technologique et de savoir-faire industriel en prévision de l'Exposition universelle que Montréal devait accueillir en 1967.

Dans l'histoire du Métro interviennent une multitude d'acteurs. Tout d'abord une équipe d'ingénieurs et de gestionnaires du Bureau de transport métropolitain qui conçoit le projet et en supervise la construction. Pour ces ingénieurs, ce projet est un défi technologique majeur qui leur fournit l'occasion de développer une expertise de premier plan en génie des transports. Partenaire financier du projet, le gouvernement du Québec y voit un équipement de transport en commun à vocation intermunicipale ou régionale. Il intervient donc dans le tracé des lignes et se préoccupe des coûts importants qu'entraîne le choix de la technologie pneumatique. Quant à l'administration municipale montréalaise, ses objectifs sont multiples. Comme nous l'avons souligné, elle voit dans le Métro un équipement collectif de prestige susceptible d'améliorer sa visibilité et son image. Finalement, les entreprises privées qui participent au projet développent elles aussi de nouveaux créneaux d'expertise et, pour plusieurs d'entre elles, la construction du Métro sera l'occasion de se doter d'un savoir-faire théorique et pratique dans le domaine des transports. Ce dernier point donne d'ailleurs un intérêt supplémentaire à l'histoire du Métro de Montréal puisqu'à travers celle-ci on peut déceler l'origine et l'émergence d'une expertise québécoise en ingénierie des transports qui, aujourd'hui, constitue un secteur industriel important et regroupe des entreprises qui se distinguent autant sur le plan technologique que par leurs succès sur les marchés d'exportation.

LES SCIENCES DANS LE PROCESSUS D'EXPLOITATION DES RESSOURCES NATURELLES AU BRÉSIL AU XIXe SIÈCLE[1]

Heloisa Maria BERTOL DOMINGUES

INTRODUCTION

L'objectif de ce travail a été d'étudier les branches des sciences naturelles qui, au XIXe et au début du XXe siècle, se sont révélées les plus déterminantes dans le processus d'exploitation des ressources naturelles au Brésil. L'analyse documentaire a été centrée sur la vision de la nature, qui apparaît comme un élément fondamental dans les changements de ligne scientifique.

Depuis les temps immémoriaux, les ressources naturelles font l'objet d'une exploitation par l'homme. Cependant, ce n'est que bien plus tard que les connaissances scientifiques qui pouvaient faciliter l'exploitation se sont systématisées. Au Brésil, les sciences se sont institutionnalisées au moment où, à la fin du XVIIIe siècle, les intérêts politico-économiques se sont conjugués avec la pensée philosophique des Lumières. A cette époque, alors que le Brésil était encore une colonie portugaise, les premières institutions scientifiques, visant à l'exploitation des ressources naturelles ont été créées et intégrées dans la politique du gouvernement.

Au XIXe siècle, après l'indépendance du pays en 1822, la nature brésilienne a cessé d'être un simple objet de la politique économique pour se transformer en symbole de la nation (une nation qui se glissait dans le monde capitaliste des Etats nationaux). Dans ce sens, la nature du pays a été célébrée par les Romantiques brésiliens. Dans le même temps, elle a fourni des arguments aux initiatives qui tentaient de relativiser les sciences et la nature. L'agriculture était la base économique du pays. La diversité de la flore, de la faune et des

1. Je voudrais tout d'abord remercier le MAST-CNPq pour le financement de cette recherche ainsi que le FAPERJ pour son appui. De même, je remercie tout spécialement le Professeur Michel Paty qui a eu la gentillesse de relire cet article et de me suggérer d'y apporter nombre de corrections linguistiques. Les fautes qui subsistent me sont imputables.

minéraux fut regardée comme riche des potentialités que la nature " romantique " offrait comme marque de sa singularité par rapport aux autres pays.

Dans le contexte intellectuel et politique du XIXe siècle brésilien, les sciences naturelles ont été conçues comme un moyen de conquérir toute la nature. Cependant, on ne peut pas oublier que les sciences gardent des spécificités théoriques et pratiques, dont l'analyse peut être faite à partir de leur insertion dans la culture la plus générale du pays. Mais, il faut compter aussi sur le " champ scientifique " dans son ensemble où les divisions internes — surtout quand il s'agit de choisir une théorie — déterminent les positions de chacun et peuvent limiter les travaux des scientifiques. Cela signifie que les positions des scientifiques et leurs choix théoriques peuvent donner une direction à la pratique scientifique et ainsi la déterminer à un moment donné. C'est ce qu'on peut appeler le style ou la tradition scientifique d'un groupe ou d'un pays[2]. Au Brésil, quand on pense à la question du rapport entre les sciences et le processus d'exploitation du milieu, il est très difficile de laisser de côté cette double imbrication sociale des sciences : dans la société et le champ scientifique dans son ensemble.

Au XIXe siècle, les sciences naturelles ont été structurées dans des institutions fondées par le gouvernement (colonial), dont les plus importantes, qui existent encore aujourd'hui, sont le Jardin botanique de Rio de Janeiro (1808) et le Musée National (1818). Pendant cette même période, des branches des sciences naturelles se sont également développées dans le cadre de sociétés scientifiques qui se créaient, telles que la Société d'Encouragement de l'Industrie Nationale (1825), l'Institut Historique et Géographique brésilien (1838), la Société Vellosianna (1851), l'Institut Imperial Fluminense de l'Agriculture (1860). Après 1870, est en outre apparue l'Ecole des Mines d'Ouro Preto dans la province de Minas Gerais (1874), qui s'est fixé pour objectif de développer l'exploitation des ressources naturelles du pays. Parallèlement, d'anciennes institutions se sont profondément restructurées. Le Musée national a ainsi vu son organisation interne modifiée. Quant à l'Observatoire astronomique, créé en 1827 et jusqu'alors rattaché à l'école militaire avec une activité limitée aux seules missions didactiques, il gagna en autonomie à partir de 1871, tout en renforçant sa prédisposition pour la connaissance de la nature. Le décret du gouvernement qui modifia ses attributions précisa qu'il appartiendrait à l'Observatoire de " renseigner le Gouvernement sur toutes les questions d'astronomie, de géodésie, de géographie et de navigation importantes à l'usage de tous ceux qui pouvaient s'intéresser aux sciences "[3]. D'un autre côté, innombrables ont été les expéditions organisées tant par des Brésiliens

2. M. Paty, *L'analyse critique des sciences ou le tétraèdre épistémologique*, Paris, L'Harmattan, 1990, chap. IV.

3. H. Morize, *Observatorio Astronomico*, Fac simile MAST/Salamandra, 1987, 67.

que par des étrangers, dans le but d'exploiter scientifiquement l'intérieur du pays, dont la plus grande partie était encore inhabitée et même inconnue, ou plutôt habitée et connue des seuls Indiens, eux aussi interprétés par la culture dominante comme des " objets naturels ".

Dans les dernières décennies du XIXe siècle, quand la conjoncture sociale et politique du pays a commencé à changer et que le pouvoir de l'Empereur a perdu sa force, les sciences naturelles ont elles aussi commencé à se diversifier et à aborder de nouvelles problématiques. Les changements n'ont donc pas été seulement institutionnels, ils ont été plus larges, puisque l'optique épistémologique même des sciences de la nature a connu un changement très remarquable. C'est dans cette nouvelle façon de voir la nature que se situent les bouleversements subis par les institutions.

Pendant ce temps, si on peut dire que la conjonction des intérêts scientifiques et politiques a été déterminante pour structurer le champ des sciences naturelles au Brésil, principalement quand il s'agissait de l'exploitation de la nature, on ne saurait omettre d'en dire autant sur l'interprétation théorique des objets dits " naturels " (végétaux, animaux, pierres et/ou hommes). Il s'agissait toujours d'étudier l'espace, mais celui-ci était vu de façon différente. Et dans ce sens, on peut dire de l'interprétation de la nature qu'elle a connu deux moments différents sans que, néanmoins, l'un n'exclue l'autre. D'un autre côté, il est très difficile de préciser ce qui, dans ce cas, a été la raison déterminante du changement, de la science ou de la politique qui l'accompagnait.

LES RESSOURCES NATURELLES ONT ÉTÉ EXPLOITÉES POUR ELLES-MÊMES

Dans un premier temps, l'interprétation des ressources naturelles se fit de la même manière que celle des sciences naturelles en général : en termes de collection, de classification et d'acclimatation. C'était le temps de la " mise en ordre de la nature ", quand on nommait les espèces selon les familles et les genres, dans le seul but de connaître les plantes, les animaux et les pierres exploitables et commercialisables[4]. Au Brésil, les jardins botaniques et le musée d'histoire naturelle ont les premiers pris leur essor, en étant intégrés au contexte économique, que dominait l'agriculture. Dans ce sens, la botanique a été pendant longtemps la science la plus forte et la mieux structurée.

En même temps s'est développée l'étude systématique de la minéralogie, dont l'intérêt économique était déjà ancien. On ne doit en effet pas oublier que le pays avait été depuis le XVIIe siècle une des plus grandes colonies exportatrices d'or vers l'Europe. Mais, du point de vue institutionnel au XIXe siècle, la pratique scientifique de la minéralogie était restreinte au Musée National où un laboratoire de chimie avait été installé pour ces études depuis 1824. Pour la botanique, outre l'institution spécifique qu'était le Jardin Botanique de Rio de

4. M. Foucault, *Les mots et les choses*, Traduction port., Lisboa, Martins Fontes, 1966.

Janeiro, il y avait des jardins provinciaux et une section au Musée National. Son importance dans le pays était telle que, en 1839, la Société d'Encouragement de l'Industrie Nationale fit publier, sur l'ordre de l'Empereur, un livre intitulé *Manual do Agricultor Brasileiro*, où la botanique était en bonne place. Ce livre était écrit par un journaliste, Carlos Augusto Taunay, avec la collaboration de Ludwig Riedel, un botaniste, directeur de la Section de Botanique et d'Agriculture du Musée National depuis sa création, en 1842, et jusqu'à sa mort en 1861. Riedel, d'origine allemande, était venu au Brésil dans les années 20 comme botaniste de l'expédition scientifique organisée par le consul russe Langsdorf. Dans le *Manuel de l'Agriculteur*, Riedel a publié, en portugais, une grande liste de plantes brésiliennes avec leurs classifications respectives et l'explication de leur utilisation populaire. L'auteur du *Manual*, Taunay, exaltait la botanique en disant qu'elle était l'aînée des branches de l'agriculture. Pour un pays qui se glissait dans la " marche de la civilisation " comme producteur agricole cela signifiait que la botanique était la science la plus importante à développer à ce moment-là. Cependant, la pratique de la botanique faisait partie du champ scientifique dans son ensemble.

Les spécialistes brésiliens étaient depuis longtemps en rapport avec les naturalistes étrangers. Le premier directeur du Jardin Botanique de Rio de Janeiro après l'indépendance, Leandro do Sacramento, avait obtenu son doctorat à l'Université de Coïmbra. Ensuite, à Rio de Janeiro, il entretint une correspondance avec, entre autres, Auguste de Saint-Hilaire, qui reconnut la qualité de ses travaux et les mentionna à l'Académie des Sciences de Paris. Il fut le premier professeur du cours de botanique et d'agriculture créé à Rio de Janeiro et fut membre de sociétés scientifiques dans de nombreux pays. De plus, ses travaux ont été publiés dans des revues spécialisées en Europe et dans d'autres pays de l'Amérique.

Au Jardin Botanique de Rio de Janeiro, les activités d'acclimatation des plantes exotiques destinées à développer l'agro-exportation obligeaient les directeurs de cette institution à connaître et à discuter ses classifications. D'un autre côté, s'amorça un mouvement d'échange entre des espèces indigènes et des plantes exotiques envoyées par de nombreux pays et institutions scientifiques étrangers. Ainsi, par exemple, le successeur de Leandro do Sacramento, Bernardo Serpa Brandão, a laissé un catalogue de classification riche de plus de 400 espèces de plantes cultivées au Jardin Botanique et qui servaient autant aux autochtones qui entreprenaient de les cultiver qu'aux étrangers qui souhaitaient les échanger. Ces activités d'échanges d'espèces apportaient l'avantage de mettre les Brésiliens en rapport avec beaucoup d'institutions scientifiques du monde, elles mettaient aussi en évidence la place que les travaux des Brésiliens occupaient dans le champ scientifique international. Ainsi, s'ils ont pu bénéficier parfois de la reconnaissance dans le milieu scientifique, ils ont maintes fois vu leurs travaux ignorés par la concurrence, ce qui suscita des protestations de leur part.

Le gouvernement brésilien, par contre, en ignorant ces concurrences et les controverses qu'elles ont pu susciter, a démontré son intérêt pour le développement de la connaissance de la flore du pays. A partir des années 40, il finança l'ouvrage *Flora Brasiliense* dirigé par l'allemand von Martius, directeur du Jardin Botanique de Munich. Martius avait visité le pays en compagnie de Spix entre les années 1817 et 1821 avec l'appui de la princesse Leopoldina. A cette époque, il se mit en contact avec des Brésiliens, parmi lesquels beaucoup collaborèrent ensuite à son oeuvre, qui est considérée jusqu'à maintenant comme l'une des plus complètes sur la flore du Brésil[5]. Cependant, quelques-uns de ses collaborateurs, comme par exemple Freire Allemão (directeur du Musée National entre les années 1866 et 1874), renonçèrent à envoyer des données botaniques, en disant que Martius manquait de respect pour leur travail[6].

Les Brésiliens, par contre, s'organisaient dans le but d'étudier eux-mêmes l'histoire naturelle de leur pays et, en 1850, sous la direction du même Freire Allemão, ils créèrent la Société Vellosiana, dont les membres se recrutaient exclusivement parmi les Brésiliens s'intéressant à l'étude de la flore, de la faune et de la minéralogie[7]. Pour atteindre cet objectif, une de leur premières décisions fut de définir la position de la Société par rapport au champ scientifique international et d'adopter une terminologie commune pour la classification et l'identification des espèces. Ainsi, ils décidèrent que pour les travaux minéralogiques et chimiques le système à adopter serait celui de Dufresnoy ; la nomenclature, celle de Berzelius et les notations cristallographiques, de Naumann. Pour la botanique, ils se restreindraient le plus possible aux classifications et à la terminologie adoptées par Endlicher dans son *Genera Plantarum.* La zoologie suivrait Cuvier. Ils publiaient dans la revue de la Société ces travaux de classification en portugais aussi bien qu'en latin, démontrant ainsi qu'ils voulaient faire connaître les produits de la terre brésilienne, suivant les conventions taxonomiques internationales mais tout en utilisant la langue vernaculaire.

L'importance des différentes branches des sciences naturelles développées au Brésil à cette époque, telle qu'elle apparaît au travers des travaux, est inégale. Ainsi, la zoologie est encore mal connue, faute d'études existantes. Mais, si l'on examine les registres des archives du Musée national, on s'aperçoit que la plus grande partie des collections constituées pendant ce temps porte sur

5. Le contrat de Martius avec le gouvernement brésilien a duré au-delà de sa mort jusqu'en 1905, quand le gouvernement républicain a résolu d'arrêter le financement. Martius, quant à lui, fut si lié au milieu intellectuel du Brésil qu'en 1847 il participa à un concours de l'Institut Historique et Géographique, qui avait pour but de présenter un plan pour écrire l'histoire du pays, et le gagna.

6. J. Saldanha da Gama, " Biografia e Apreciação dos Trabalhos do Botânico Brasileiro Francisco Freire Allemão ", *Revista do Instituto Histórico e Geográfico Brasileiro*, (1875), 103.

7. *Sociedade Vellosiana*, Rio de Janeiro, Biblioteca Guanabarense, 1851-1855.

l'ornithologie, ce qui permet déjà de conclure à un grand intérêt pour les oiseaux du pays.

Les travaux sur les minéraux, bien que moins nombreux que ceux de botanique, furent significatifs dans les premiers temps du XIXe siècle. On rappellera malgré tout que la première collection du Musée National fut une collection de minéraux classifiée selon le système de Werner. Les pierres précieuses et la recherche des filons d'or ou de fer ont donné lieu à beaucoup de travaux de minéralogie au Musée National pendant les premieres années de son existence : on y analysait des pierres, et principalement des échantillons de minéraux venus de divers endroits du pays, envoyés par des voyageurs étrangers qui étaient obligés de faire ces dépôts, ou par des savants établis localement dans les diverses régions. Le Brésil vivait encore son âge de l'or et des échantillons de ce métal, aussi bien que de diamants, étaient envoyés au Musée. Beaucoup de travaux sur la minéralogie du pays ont été aussi développés à la Société Vellosiana. Parmi les travaux des sociétaires, se distinguent ceux de Fréderico Cezar Burlamaqui, qui a précédé Freire Allemão à la direction du Musée National[8].

Le principe de la minéralogie s'apparentait à celui de la botanique. Il consistait à étudier les pierres en elles-mêmes pour remonter ensuite à leur composition chimique : c'est dans ce but qu'avait été créé le laboratoire de chimie au Musée National en 1824. Dans la première moitié du XIXe siècle, la pratique courante en chimie était d'effectuer des analyses de composants de minéraux ou de végétaux ; la chimie non organique et les travaux réalisés dans ce laboratoire servirent autant à l'avancée de la science qu'à son enseignement (à l'Ecole Militaire, où se formaient les ingénieurs, on utilisait le laboratoire du Musée pour donner la formation nécessaire en sciences naturelles aux élèves).

A cette époque-là, les ressources naturelles les plus étudiées au Brésil, dans le cas de la botanique, étaient les arbres producteurs de bois pour la construction des navires et la menuiserie ; les arbres producteurs de fibres pour la fabrication des tissus, et aussi des plantes nouvelles qui appartenaient à la culture traditionnelle des Indiens, et qui commençaient en même temps à entrer dans le marché économique. Parmi elles, on retiendra celles qui produisaient des matières colorantes, des boissons comme le guaraná, des huiles, des cires et des gommes, dont l'exemple le plus connu est celui de l'*hevea brasiliensis*, arbre producteur du caoutchouc. Dans le cas de la minéralogie, la plus grande partie des travaux portait sur les pierres précieuses telles que le diamant ou le cobalt ; ils portaient aussi sur les marbres et sur les métaux comme le fer, l'or et, à un moindre degré, sur le cuivre, l'étain, le plomb et le charbon[9]. Autour

8. Sur l'institutionnalisation des travaux de minéralogie et/ou de géologie au Brésil voir S.F. de M. Figueroa, *Em Busca do Eldorado*, São Paulo, FFLCH/USP, 1992 (Thèse de doctorat).

9. Documents des Archives Historiques et Scientifiques du Museu Nacional, Dossiers 01, 02, 03.

de 1848 commencèrent à apparaître au Musée National des commandes d'analyse d'échantillons de combustibles fossiles comme le charbon minéral et les schistes. C'était déjà le signe du début d'une autre période pour les recherches de ressources naturelles[10].

Le désir de connaître et d'enrichir toujours plus les catalogues de classification des plantes, des animaux ou des pierres amena les naturalistes à s'enfoncer dans les forêts, les rivières, et les montagnes jusqu'alors complètement inconnues. Chaque fois, ils s'éloignaient davantage des centres urbains, ce qui contribua certainement à modifier le regard porté sur la nature.

On peut dire néanmoins que, jusqu'aux environs des années 1850, la recherche pour elle-même des ressources naturelles a déterminé les espaces à explorer. A cette époque, les naturalistes cherchaient à connaître tout simplement les plantes nouvelles ou celles qui ressemblaient à celles déjà connues. Ou alors, ils cherchaient l'or, les diamants, *etc.*, à l'intérieur des terres, en suivant les chemins étroits, ouverts par d'autres naturalistes ou même par les Indiens.

Dans la seconde moitié du XIXe siècle, le gouvernement brésilien a développé une politique visant ce que j'ai appelé une " recolonisation " ou une " intériorisation du pays ", par rapport à l'exploitation des ressources naturelles et au peuplement. Le peuplement était un facteur important pour la question de l'exploitation des ressources naturelles : il n'y a pas d'exploitation de la nature sans bras pour exploiter ou pour permettre la commercialisation. Ainsi le temps durant lequel les ressources naturelles ont déterminé l'exploitation de l'espace a donné lieu à un autre dont l'espace détermina son exploitation.

L'ESPACE A DÉTERMINÉ L'EXPLOITATION DES RESSOURCES NATURELLES

Peut-être pourrait-on dire qu'après 1850, la préoccupation ancienne de " mettre de l'ordre " dans la nature a cédé la place à celle de " comprendre le désordre de la nature ".

Dans la seconde moitié du XIXe siècle a commencé à s'intensifier au Brésil le mouvement abolitionniste dont le premier pas fut la loi de 1850 mettant fin au trafic d'esclaves. Le gouvernement créa une législation qui permettait de vendre des terres (il faut rappeler que jusque-là un simple document de possession ou une donation permettait à quiconque d'être propriétaire). La terre fut transformée en bien de capital, et l'offre de main-d'oeuvre se trouva menacée. S'intensifièrent alors les intérêts politiques, économiques et scientifiques tournés vers l'exploitation de l'intérieur du pays. C'était dans des régions encore inconnues que l'on trouvait non seulement de nouvelles richesses naturelles, mais aussi le potentiel de main-d'oeuvre indigène qui, en outre, connaissait le milieu.

10. *Idem*, Pasta 3.

Au fur et à mesure que la frontière d'exploitation des ressources naturelles s'éloignait, pour produire et distribuer ces produits, il fallut créer des moyens de transport, voies de navigation ou chemins de fer, qui permettraient de transporter les produits de l'intérieur vers les marchés consommateurs. En même temps, il devint nécessaire de peupler les régions inhabitées. C'est ainsi qu'une nouvelle optique épistémologique de la nature commença à s'imposer. Dans ce contexte, d'autres sciences, comme la géographie, la géologie, l'astronomie et l'anthropologie, commencèrent à prendre la place de celles qui étaient seulement taxinomiques. La connaissance de l'espace, avec toutes les caractéristiques conditionnant la mise en valeur des ressources naturelles de toutes sortes vint à occuper le premier rang. L'important devenait de connaître l'endroit correspondant aux conditions géologiques ou climatologiques qui permettait l'apparition des différentes ressources de la nature.

La géographie a gagné son importance parce qu'elle était la science qui, non seulement dessinait les chemins de l'intérieur, mais situait les différences physiques de chaque endroit. D'un autre côté, la géographie dépendait de l'astronomie qui connut aussi un énorme développement avec l'étude des latitudes, des altitudes et des climats. En même temps, la géologie, avec l'étude du sol, du sous-sol et de tout ce qui le compose vit augmenter ses études au-delà des simples analyses de minéraux. La pratique de ces sciences augmenta énormément au Brésil après 1850 grâce aux initiatives politiques qui ont engendré l'intériorisation. Par là même, certaines régions du pays considérées comme les plus riches ont été valorisées comme objet d'étude.

Les théories du déterminisme géographique ont été transformées en arguments pour justifier l'importance de l'espace et la pénétration par les naturalistes de l'intérieur le plus éloigné : les voyages scientifiques ont été grandement soutenus à ce moment-là[11]. De plus, les voyages aux endroits inconnus et la possibilité de les peupler entraîna le développement de l'ethnographie qui, peu à peu, se sépara de l'anthropologie. L'ethnographie se proposait de comprendre la culture des indigènes, ses valeurs, ses habitudes, ses pratiques matérielles, *etc*. Ce qu'on a appelé " anthropologie " était une branche des sciences naturelles formée par une association de l'archéologie et de l'anatomie avec les études des fossiles et la mesure des crânes, tout en invoquant comme conditionnement de la culture le milieu physique.

Un des premiers indices du changement de la vision de la nature dans la pratique des sciences naturelles au Brésil s'observe à l'occasion de l'organisation, en 1859, de la Commission Scientifique du Gouvernement, en vue d'exploiter le nord du pays. Cette Commission devait être composée uniquement de Brésiliens, et rassembler toutes les branches des sciences naturelles importantes pour l'exploitation de la région. Elle fut structurée en cinq

11. H.M.B. Domingues, " As Ciências Naturais e a Construção da Nação Brasileira ", *Revista de História*/USP, 135 (1996), 41-60.

sections : Botanique, Astronomie et Géographie, Géologie et Minéralogie, Zoologie et Ethnographie.

A cette époque, la botanique, la zoologie et l'ethnographie faisaient partie des spécialités du Musée National, et la botanique, comme on l'a vu, possédait une institution spécifique, le Jardin Botanique. L'ethnographie était devenue en 1847 une section de l'Institut historique et géographique brésilien et formait avec la géographie les fondements scientifiques de la construction d'une histoire nationale. L'astronomie se maintenait encore comme une activité didactique, liée à l'Ecole Militaire.

Par contre, dans la structure de la Commission Scientifique, l'astronomie apparaît déjà comme un des fondements les plus importants de la géographie. Il semble qu'elle gagna à la Commission Scientifique plus d'importance qu'elle n'en avait institutionnellement dans le pays. Les finalités assignées par cette commission à l'astronomie étaient doubles. Tout d'abord les membres de cette section devaient faire les études astronomiques et topographiques concernant la détermination de la position géographique des points les plus importants du territoire exploré. Et dans ce sens, ils devaient étudier la localisation des villes, déterminer les latitudes et l'altitude méridienne (pour déterminer les latitudes géodésiques ils utilisaient des chronomètres, des mesures de distances de la Lune, *etc.*). L'astronomie devait faire aussi des travaux d'investigation sur la physique générale du globe, à travers le calcul des moyennes de latitudes, chercher les différences de températures et de pressions barométriques qui pourraient influencer les résultats obtenus dans des circonstances données. Elle devait observer les altitudes méridiennes des astres, en les observant pour deux états hygrométriques de l'atmosphère, gardant les températures et les pressions barométriques à peu près constantes. D'un autre côté, les instructions données à la section d'astronomie stipulaient que, comme la Commission se trouvait au Cearà une région toujours touchée par la sécheresse, les commissaires, en accord avec ceux mandatés par la Commission de géologie devraient en profiter pour faire des sondages afin de découvrir des traces susceptibles de guider l'ouverture de puits artésiens[12].

Dans la même Commission Scientifique, la géologie gagna aussi d'autres attributions qu'elle n'avait pas dans les institutions. La Section de Géologie et de Minéralogie était chargée principalement de déterminer le potentiel des gisements trouvés, les types de minéraux, et aussi les fossiles. Cette section devrait classifier les minéraux et dire s'il s'agissait de combustibles, de marbres, de ciments, de charbons, ou de bitumes qui pouvaient être distillés pour la production de naphte, *etc.* Dans la partie de géodynamique, la Section devrait examiner les roches, leur composition chimique, leurs dimensions, leurs directions, les formes de fossiles, faire des dessins des montagnes, *etc.*,

12. Les études sur la sécheresse dans la région nord-est du pays ont été intenses au XXe siècle, néanmoins la question posée dès le XIXe siècle reste toujours sans solution.

et enfin faire les plans géologiques de la région. Cette section devrait aussi faire une collection d'échantillons des différents sols, pour analyser le degré hygroscopique de chacun.

Tandis que les sections de géographie et d'astronomie localisaient les endroits où se trouvaient les végétaux ou les roches, la section de géologie et minéralogie spécifiait la composition des roches et de chaque terrain. Le directeur de cette section affirmait : " Quels que soient ces minéraux cristallisés, on devra chercher le plus grand nombre possible d'individus parfaits, en visant à établir la série la plus complète possible des combinaisons cristallographiques. Si ce sont des minéraux décomposés, on recherchera leurs matrices respectives pour obtenir les individus dans leur forme primitive, en étudiant en même temps les causes des décompositions actuelles ". La géologie commençait à être dissociée de la minéralogie, et regardait le sol et le sous-sol comme faisant partie d'une histoire de la Terre.

Dans la nouvelle vision de la nature qui s'imposait, la géologie analysait les minéraux à partir de l'endroit où ils se trouvaient et pourtant, dans les institutions scientifiques du pays, la pratique de la minéralogie était demeurée inchangée. La section correspondante du Musée National, malgré sa dénomination de Section de géologie reçue dès sa création en 1842, ne procédait qu'à des analyses minéralogiques[13]. De même, dans les cours officiels au milieu du siècle, l'enseignement de la minéralogie se limitait encore à des classifications de minéraux et à l'étude de leur composition chimique[14]. Ce n'est qu'à la Société d'Encouragement de l'Industrie Nationale que l'on peut percevoir un changement dans la pratique de la géologie à cette époque. En effet, quand la discipline y a été introduite en 1857, elle s'est associée à la chimie, dont les études commençaient aussi à changer, et où l'on commençait à pratiquer de la chimie organique[15]. D'autre part, l'Institut Historique et Géographique faisait paraître beaucoup d'articles évoquant la géologie. Pour faire l'histoire du Brésil on se basait sur " l'histoire de la terre " et on situait l'origine du pays dans l'Antiquité de la " civilisation "[16].

Au cours de la décennie 1860, les mouvements à l'intérieur du pays s'intensifièrent et, parmi les travaux résultant des voyages réalisés, on distingue ceux de l'astronome français Emmanuel Liais et ceux du géologue américain Charles Hartt. Emmanuel Liais, après avoir voyagé dans le sud du pays fut

13. L'organisation du Musée National, créé en 1842, comprenait les sections d'Anatomie Comparée et Zoologie, de Botanique, Agriculture et Arts Mécaniques, de la Minéralogie, Géologie et Sciences Physiques ; Numismatique et Arts Libéraux ; de l'Archéologie, Usages et Coutumes des Nations Modernes (Archives Historiques du Musée National, Dossier 02, Doc. 124). Sur la dominance de la Minéralogie voir aussi S.F de M. Figueroa, *op. cit.*, 1992.

14. S.F de M. Figueroa, *op. cit.*, 1992.

15. H.M.B. Domingues, *A Sociedade Auxiliadora da Indústria Nacional e as Ciências Naturais no Brasil Império. Notas Técnicas*, MAST/CNPq, 002/1996.

16. H.M.B. Domingues, *A Noção de Civilização na Visão dos Construtores do Império*, Niterói, RJ, Depto. d'Histoire/UFF, 1990 (Dissertação de Mestrado).

engagé par contrat par le gouvernement brésilien pour explorer la région de la Vallée du Rio São Francisco en Minas Gerais, pendant les années 1863 et 1864. Charles Hartt, le géologue de la Commission Thayer dirigée par Agassiz, avait parcouru une grande partie de la côte du pays à partir de Rio de Janeiro vers le Nord, dans la région amazonienne et aussi dans la région du São Francisco, peu après Liais. Ils furent ensuite tous deux chargés de diriger des institutions scientifiques du pays. Liais fut nommé en 1871 directeur de l'Observatoire National. Hartt devint chef de la Section de Géologie du Musée National en 1876, quand il organisa la Commission Géologique de l'Empire, dont les études étaient vouées principalement à l'agriculture[17].

Le contrat de Liais sur l'exploitation de la Vallée du São Francisco prévoyait un rapport sur la géographie physique de la région pour une collection cartographique et une analyse comparative entre cette région et celles de Rio de Janeiro, Paraná et Pernambuco. Ce travail devait être une *Introduction Générale à la Géologie de l'Empire*[18]. Son rapport devait évoquer des gisements de minéraux, leur nature, et les moyens de les exploiter ; il devait aussi décrire les divers climats de la partie orientale du Brésil, ses phénomènes météorologiques, considérer la flore et la faune en ce qui concernait la paléontologie et la géographie (générale, botanique et zoologique). De plus, il devait faire des considérations sur la statistique agricole et sur les systèmes de culture les plus appropriés aux diverses régions parcourues par sa Commission. Enfin, ce rapport devait comprendre une description des nouvelles méthodes pour dessiner géographiquement la région, dans le but d'augmenter la précision et de faciliter les voyages[19].

Liais a publié officiellement le rapport du voyage dans la Vallée du São Francisco, dans un livre qui s'intitule, paradoxalement, *Traité d'Astronomie Appliquée et Géodésie Pratique - Comprenant l'Exposé des Méthodes suivies dans l'Exploration du Rio São Francisco et précédé d'un rapport au Gouvernement impérial du Brésil*[20]. Mais, bien que le rapport officiel figure dans cette édition du *Traité d'Astronomie* (le rapport ne figure pas dans toutes les éditions de ce livre), Liais a publié les résultats des observations de ce voyage, et des autres qu'il avait faits auparavant dans le sud du pays, dans deux autres livres : *L'Espace Céleste*[21] et *Climats, Géologie, Faune et Géographie Botanique du Brésil*, publiés aussi par ordre du Gouvernement Impérial du Brésil selon ce

17. Sur ce sujet voir S.F de M. Figueroa, *op. cit.*, 1992.

18. 2e contrat signé entre E. Liais et le Gouvernement le 9 mars 1864 (Dossier 52, Archives de la Société des Sciences Naturelles et Mathématiques de Cherbourg, France. Des copies de certains articles de ces Archives m'ont été fournies par Antonio Augusto Videira, que je remercie).

19. *Idem.* Liais est resté au Brésil comme directeur de l'Observatoire jusqu'en 1881, quand il est rentré en France et a été remplacé par le belge Louis Cruls.

20. Paris, Garnier Frères, Libraires Editeurs, 1867.

21. E. Liais, *L'Espace Céleste ou Description de L'Univers accompagnée de récits de voyages entrepris pour en compléter l'étude*, Paris, Garnier Frères, Libraires Editeurs, 1881 (2e ed.). La première édition de ce livre date de 1865.

qu'on peut lire sur la page de garde[22]. D'après les termes mêmes employés dans les titres de ses ouvrages, on devine quelle importance la description de la nature prise dans son ensemble a acquise.

Ces livres contiennent la marque caractéristique des sciences naturelles de l'époque et laissent voir clairement que les pratiques scientifiques étaient en relation les unes avec les autres. Ainsi, le *Traité d'Astronomie* montrait l'étroite relation entre la pratique de l'astronomie et la géographie. Ce livre qui avait pour finalité de divulguer la pratique d'un astronome est partagé en trois parties : dans la première, il s'agit d'expliquer les instruments astronomiques et leur méthode d'utilisation ; la deuxième partie traite des positions géographiques et des coordonnées des astres. Dans cette partie, Liais traite de l'astronomie en général, discute la détermination de l'heure, la détermination des latitudes par les observations des altitudes, la détermination des longitudes par l'observation de la Lune et par l'observation dans l'azimuth, *etc*. La troisième partie du livre, *Géodésie appliquée à la Géographie*, démontre les liens étroits entre cette science et l'astronomie. Dans cette partie du *Traité d'Astronomie,* Liais parle de la détermination des différences de longitude de deux points, ou de l'évolution de triangles géodésiques. Il fait aussi des considérations sur la mesure des dimensions de la Terre et sur les opérations avec la boussole, les tracés des routes, *etc*. C'est dans sa conclusion qu'il souligne l'importance de l'astronomie pour la géographie en disant que son " traité d'astronomie contient tous les renseignements nécessaires aux explorateurs pour préparer leurs plans géographiques avec le plus grand degré de précision possible ". Ce qui illustre les besoins du géographe à l'égard de l'astronomie pour dessiner la nature.

D'un autre côté, dans le livre *Climats, Géologie, Faune et Géographie Botanique du Brésil*, Liais, au chapitre sur la géologie, souligne que : " la distribution des végétaux à la surface du globe est surtout réglée par les climats, si on l'observe non seulement du point de vue des températures moyennes et extrêmes mais encore par l'humidité et par la distribution des pluies en suivant les saisons ". La dite acclimatation gagna ainsi plus de systématisation scientifique grâce au travail de ceux qui, comme lui, se consacraient à l'astronomie, parce que les latitudes, les longitudes, aussi bien que les altitudes, étaient pour eux des indicateurs des conditions de l'acclimatation végétale ou animale. Au Brésil, constate Liais, la distribution des espèces est conditionnée par l'altitude des surfaces dans les zones tropicales et intertropicales où l'humidité et la distribution des pluies jouent un rôle très fort. Et, à côté de l'action principale de l'humidité, la question de la température dépend de la distance à l'Equateur et de l'altitude au-dessus de niveau de la mer. A la fin du XIXe et au début du XXe

22. E. Liais, *Climats, Géologie, Faune et Géographie Botanique du Brésil*, Paris, Garnier Frères, Libraires Editeurs, 1872.

siècle, l'étude du climat a revêtu une importance considérable au Brésil et s'est avérée l'une des activités les plus remarquables des astronomes de l'Observatoire National[23]. Cependant, pour Liais, la détermination géographique n'était que partielle parce que, selon lui, on ne pouvait pas considérer le climat comme, par exemple, un élément déterminant des différences de couleurs des races et il doutait même de son influence sur les végétaux comme le voulaient beaucoup de ses partenaires. Il considérait, tout en rappelant Darwin, que les différences raciales des espèces végétales, qu'on ne pouvait pas négliger, étaient soumises à d'autres lois encore inconnues, liées au phénomène de la reproduction[24].

En évoquant aussi les principes de la géologie de Charles Lyell, dont on trouve le livre dans son édition de 1845 (la dernière date de 1860) à la bibliothèque de l'Observatoire de Rio de Janeiro, Liais interprétait les ressources naturelles à partir de leur milieu. Cependant il le faisait en se basant sur des théories précises. Pour lui, la géologie, par exemple, devait observer les dépôts les plus anciens des terrains en prenant en compte les failles survenues et qui mêlaient fréquemment les dépôts inférieurs avec les couches plus superficielles. Donc, en donnant des descriptions des gisements, il disait qu'il était difficile de dater les fossiles au Brésil, il croyait néanmoins qu'ils pouvaient être encore plus anciens que ceux déjà trouvés. Il observait de plus que dans les couches d'huile très inférieures on trouve des traces de plantes fossiles[25]. Le problème, disait-il, c'est que les gisements du Brésil sont encore inconnus. Il l'affirmait au contraire d'autres qui n'acceptaient pas l'idée qu'au Brésil les terrains étaient si anciens puisqu'on trouvait encore les traces des cultures les plus primitives.

Dans le même raisonnement interprétatif, on trouve les travaux de Charles Hartt sur la géologie du Brésil. Lui aussi interprète la géologie brésilienne en suivant les plus nouvelles théories d'évolution de la terre[26]. En 1870, Hartt publia le livre *Géologie et Géographie Physique du Brésil*, où il décrit les résultats de ses études à l'Expédition Thayer et d'autres qu'il avait faites sur la côte du Rio de Janeiro, Bahia, Pernambuco et baie d'Abrolhos où il avait étudié les récifs de coraux. Hartt décrit aussi la géologie de Santa Catarina, Rio Grande do Sul e du Nord, Alagoas, Sergipe, Paraíba, Piauí, en se fondant sur

23. Liais a publié entre autres études sur les climats : " Les Sensations de la Chaleur " dans la *Revista Brasileira* en 1898. Son successeur à la direction de l'Observatoire National, avait une colonne dans le *Jornal do Comercio* où il a publié *L'Etude sur le Climat du Brésil* et son successeur, Henri Morize, publia lui aussi un travail très détaillé sur le *Climat du Brésil*.

24. E. Liais, *L'Espace Céleste*, *op. cit.*, 191.

25. E. Liais, *Climats, Géologie, Faune et Géographie du Brésil*, *op. cit.*, 200.

26. Aux Cours Publics créés au Musée National en 1876, Hartt diffusait les idées évolutionnistes qu'orientait la théorie de Lyell et Darwin (M. Romero Sá, H.M.B. Domingues, " O Museu National e o Ensino das Ciências Naturais no Brasil no século XIX ", *Revista da Sociedade Brasileira de História da Ciênca*, 15 (1996), 79-88).

les travaux d'autres naturalistes[27]. Dans ce livre, Hartt, comme l'avait fait Liais pour la région du San Francisco, décrit les aspects physiques des endroits en expliquant les différences des altitudes des terrains, des conditions d'accès et finalement, il parle des disponibilités en ressources naturelles aussi bien que du peuplement. Hartt a comparé ses observations avec d'autres travaux identiques comme par exemple ceux de Fernando Halfeld (1851), celui de Liais pour la Vallée du São Francisco, ou celui de Nathaniel Plant pour la région sud du pays.

Hartt analyse les formations géologiques tout en situant les ressources naturelles qui y correspondent. Ainsi, il présente les études géologiques de Plant qui situait la région du charbon de Rio Grande do Sul. Quant à l'or, il parle de son " occurrence dans les roches métamorphiques anciennes, dans des argiles de " drifts ", dans les sables alluviaux des pierres lavées dérivées des dégâts des roches ". Il décrit encore les formations qui produisent l'or de la meilleure qualité : " schistes argileux traversés par des veines de quartz aurifères, les roches itacolumito ", en les situant géographiquement[28]. Comme l'avait fait Liais, mais à propos de la flore et de la faune, Hartt analyse les dépôts d'huiles trouvés au sud du pays par Plant comme étant des ressources en minerais.

Au Musée National, les analyses sur les matériaux fossiles ont commencé aux environs des années 1850, quand y sont arrivés beaucoup d'échantillons de roches bitumineuses. A cette époque, les savants ont conclu à l'existence de gisements d'huile de carbures d'hydrogène à Bahia, qui furent utilisés pour la fabrication de gaz destinés à l'illumination publique[29].

A partir de 1876, l'étude du sous-sol brésilien fit l'objet d'un enseignement institutionnalisé, avec la création de l'Ecole des Mines de Ouro Preto. Celle-ci répondait à la volonté de l'empereur Pedro II de rechercher les meilleures conditions pour l'exploitation des ressources naturelles du Brésil. A cette fin, il avait convié le directeur de l'Ecole des Mines française, Daubrée à venir au Brésil institutionnaliser l'enseignement de la géologie. Daubrée déclina l'offre pour lui-même, invoquant les contrats qu'il avait avec son gouvernement, mais recommanda son ex-élève Henri Gorceix, qui accepta l'invitation. L'Ecole put ainsi être créée. Son principal but était de former des ingénieurs capables d'exploiter les gisements de fer trouvés nombreux dans la région. A la suite de cela, un grand nombre d'industries furent créées dans cette zone, par des élèves

27. C. Hartt, *Geologia e Geografia do Brasil*, São Paulo/Rio, Cia. Ed. National, 1941 (Traduction port., Col. Brasiliana, 1941).

28. *Idem*, 573.

29. Un envoi du 31 mars 1851 adressait au Musée National des échantillons dans lesquels on pensait trouver du charbon, des bitumes et du gaz (Arquivo Administrativo, Histórico-Científico do Museu Nacional - AAHC-MN). Les rapports sur les gisements de fossiles bitumineux ou des bois fossiles qui contenaient des carbures d'hydrogène se trouvent dans l'Aviso du 26 juillet 1858 et dans celui du 31 juillet 1858-Dossier 06, Doc. 22 (AAHC-MN)

et même par des professeurs de l'Ecole[30]. Il ne faut pas oublier l'importance de l'exploitation de fer à l'époque de la construction des grandes machines pour les industries.

On peut, peut-être, dire qu'à un certain moment, les études scientifiques sur les ressources naturelles sont sorties du sol pour pénétrer dans le sous-sol, lorsque les fossiles et les métaux ont pris plus d'importance. Cependant, à l'inverse de la botanique ou de l'astronomie, la géologie s'est institutionnalisée par un enseignement dans l'Ecole des Mines d'Ouro Preto, tandis que la recherche a dû attendre la période républicaine pour recevoir une consécration officielle. Ce n'est en effet qu'en 1906 qu'a été créé le Service Géologique et Minéralogique du Brésil.

Les changements qu'ont subi les pratiques scientifiques du pays ont impliqué une nouvelle façon d'interpréter la nature. Tout d'abord on regardait la nature à travers les objets naturels eux-mêmes. Ensuite, la nature a été conçue dans son ensemble : les ressources naturelles aussi bien que les peuples ont été vus comme des produits de l'espace dont ils étaient issus et dont la connaissance provenait de la géographie, de l'astronomie ou de la géologie qui incluait aussi l'archéologie et la paléontologie.

L'INTERPRÉTATION DE L'ESPACE : LES TEMPS, LES HOMMES ET LES RESSOURCES NATURELLES

L'interprétation de l'espace, faite sous les feux croisés de nouvelles théories d'interprétation de la terre, de son évolution, aussi bien que du développement biologique et social des hommes, a engendré un rapport temps/espace déterminant de la nouvelle conception de la nature. La théorie de l'évolution de la terre de Charles Lyell (1830) a trouvé ses partisans parmi les astronomes, géographes et géologues du pays. Et la théorie la plus révolutionnaire, celle sur l'origine des espèces de Charles Darwin, apparue en 1859, a touché plus au moins tout le milieu intellectuel brésilien. Les gens se sont déclarés pour ou contre. Ces théories, non seulement contribuent au changement dans la façon d'interpréter la nature, mais aussi orientent les pratiques scientifiques dans les institutions s'intéressant aux sciences naturelles du pays. En même temps elles ont engendré des conceptions de rapport espace/temps.

On a vu qu'au Brésil l'exploitation des ressources naturelles, après le milieu du XIXe siècle, a été conditionnée par la connaissance détaillée de la géographie et de la géologie du pays. La recherche du sous-sol a commencé à attirer l'attention sur les fossiles humains et animaux. Les analyses des scientifiques se ressentent de leurs positions théoriques, ce qui s'observe tantôt dans les lettres d'accompagnement des envois d'échantillons, tantôt dans leur correspon-

30. J.M. Carvalho, *A Escola de Minas de Ouro Preto*, Relatório FINEP, 1986 (Arquivo MAST).

dance ordinaire, où se discerne une vision catastrophiste de la terre. Dans telle lettre, on demande d'analyser les petits fossiles et les terrains où ils se trouvent, afin de les dater : " c'est dans les couches les plus basses des terrains que l'on trouve les formes les plus inférieures des animaux parce que sont les plus anciennes ; celles des hommes sont les plus récentes ". Et l'auteur continuait, " on sait que le mouvement de la mer et les catastrophes naturelles concourent à garder les fossiles à l'intérieur de la terre ". Cette vision de la géologie, manifestée en 1855, est celle d'un naturaliste engagé par contrat par le Museu Nacional – le français Jacques Brunet. Elle est complètement différente de celle dont parlait Liais.

Pour Liais ou Hartt, le temps était en rapport avec l'espace mais il n'était pas déterminant. Liais, dans son livre *L'Espace Céleste*, a très bien remarqué cela : " l'intellect par lequel le temps se manifeste directement à notre être, grâce à la propriété de la mémoire de graver l'ordre de succession des idées, ne renferme en lui-même aucun moyen de mesurer la durée. Nous cherchons alors notre unité dans les phénomènes physiques du monde extérieur, et nous l'empruntons au mouvement. Mais où est la preuve que des phénomènes successifs et en apparence identiques répondent vraiment à des intervalles de temps égaux, c'est-à-dire en tout semblables ? (...) L'idée de mesurer l'espace renferme implicitement l'idée de temps, car il faut supposer le transport successif d'une unité, et l'idée de mesure du temps contient l'idée de l'espace. L'espace peut donc lui-même varier avec le temps dans sa constitution intime, comme le temps peut varier progressivement suivant les régions de l'espace. Des rapports et seulement des rapports... "[31].

Ces théories qui conditionnaient les travaux scientifiques sur l'exploration des ressources naturelles ont donc donné des directions différentes aux travaux institutionnels réalisés à la fin du XIXe siècle au Brésil. Par exemple, dans le Musée National on s'en tenait plutôt à l'anthropologie, partagée entre les pratiques de la paléontologie et de l'anatomie humaine tout en s'efforçant de donner un âge à l'homme américain. Tandis que à l'Observatoire National on avait embrassé la question de distinguer les endroits les plus riches pour l'exploitation des ressources naturelles en déterminant la géographie géologique, botanique et climatique du pays.

L'espace vu comme élément déterminant du temps géologique et social a créé une image d'un milieu physique très riche mais, en même temps, a forgé la représentation d'un peuple très primitif qui n'avait presqu'aucune possibilité de surmonter ses limites culturelles. C'est ce qui permet de dire que les sciences ont donné les contours de ces images. Cela a coïncidé avec l'époque de l'impérialisme, quand l'Europe luttait pour le " partage " du reste des territoires du monde. A partir des sciences naturelles, on construisait l'image du pays. Celle-ci constituait un fondement pour toute la politique économique que l'on

31. Liais, *L'espace céleste*, *op. cit.*, 12, Introduction.

menait. C'était le moment où on a trouvé les racines historiques de certains problèmes encore actuels, comme ceux de la marginalité ou de la dépendance[32]. C'est ainsi que les sciences naturelles ont cimenté les idées en marche de " sous-développement ".

32. P. Petitjean, " Entre a Ciência e a Diplomacia : A Organização da Influência Cientifica Francesa na América Latina, 1900-1940 ", org. Hamburger *et al.*, *A Ciência nas Relações França-Brasil (1850-1950)*, São Paulo, EDUSP/FAPESP, 1996, 89-120.

A Study of Controlled Nuclear Fusion in Japan Concerning Arguments about the Policies of Research Organization

Eisui UEMATSU - Sigeko NISIO

In Japan, nuclear fusion research is recognized not only as technology but also science. Studying nuclear fusion as a science means specifically basic research into plasma physics aiming at controlled nuclear fusion. When study into nuclear fusion began in Japan, the thrust of research policy was mainly toward basic research into plasma, and this led to the establishment of the Institute of Plasma Physics at Nagoya University.

The study of nuclear fusion in Japan began in 1955. In October, a workshop on nuclear reactions in stars, held at the Yukawa Institute for Theoretical Physics (YITP) of Kyoto University, included the first public discussion of the need for research to establish whether the nuclear reactions in stars could be reproduced on the Earth. As a result, a workshop on very-high-temperature phenomena was held at YITP in April 1956. This meeting discussed ways to produce very-high-temperatures by electric discharges or shock waves and the basic items required for understanding the high-temperature state, such as the statistical mechanics of ionic gas, electric conduction, magneto hydrodynamics, and star models. Also reported was the situation in other countries.

Meanwhile, knowing of the plan of the April workshop on very-high-temperature phenomena, the Welding Engineering Department of the Faculty of Engineering at Osaka University decided to begin nuclear fusion research with the co-operation of the Physics Department of the Faculty of Science. Because a study group within the Welding Engineering Department had already been conducting an experiment to converge an arc by magnetic fields in order to obtain a high temperature arc, the Osaka University group remodelled this equipment to conduct experiments in large electric current discharges. In June 1956, these experiments were open to every researchers in Japan and Cho-kouon Kenkyukai (research group of very high temperature phenomena), chaired by Kodi Husimi of Osaka University, was formed.

Between that time and the end of 1957, spurred on by the meetings and the experiments, theoretical and experimental studies also began at other universities such as Tohoku University, Nagoya University, the University of Tokyo, and Nihon University as well as at governmental research organizations such as the Electro-Technical Laboratory (ETL). Small research groups were formed in various districts in Japan.

Then Kakuyugo Kondankai (nuclear fusion discussion group) of the researcher-initiated organization was established in February 1958. The Kakuyugo Kondankai played important roles as a common forum of the Japanese fusion community until Purazuma-Kakuyugo Gakkai (Japan Society of Plasma Science and Nuclear Fusion Research) was established as a corporate body in April 1983. At the start of the Kakuyugo Kondankai a symposium on nuclear fusion was held in February 1958. And publication of the transaction of the forum Kakuyugo-Kenkyu (nuclear fusion research) was commenced by the editorial office placed at Nihon University in July 1958.

Meanwhile, the Japan Atomic Energy Commission (JAEC) took up the controlled fusion research. JAEC was established in the Prime Minister's Office in order to discuss a national policy for atomic energy in January 1956. The Science and Technology Agency (STA) was set up in May 1956 and given the task of managing Japan nuclear research and development and of providing staff and other support for JAEC. In February 1957 JAEC instigated Kakuyugo-Hanno Kondankai (assembly for discussion of the nuclear fusion reaction), with Hideki Yukawa of Kyoto University as chairman and held the meetings in February and October of 1957. The assembly was, however, dissolved after the October meeting. JAEC then established Kakuyugo Senmonbukai (special panel on nuclear fusion research), again with Yukawa as chairman, in April 1958 and discussed the nuclear fusion research program. The following March, the Senmonbukai submitted its recommendation to JAEC that a double policy be implemented : (a) that basic research into plasma be supported (Plan A) and (b) that a medium-sized experimental fusion device be constructed (Plan B). Goro Miyamoto of the University of Tokyo, after having attended at the second International Conference on the Peaceful Uses of Atomic Energy, had voiced this double-featured plan at the Senmonbukai. Following to the recommendation Kakuyugo Kenkyu Iinkai (fusion research committee, abbr. Yamamoto Iinkai), with Kenzo Yamamoto of Nagoya University as chairman, was set up, and Plan B was intensively discussed at the first meeting of the Yamamoto Iinkai in April 1959. In June of the year an interim report of the Yamamoto Iinkai which proposed a mirror device and a stellarator device as the plasma confinement devices of the Plan B was submitted to the Senmonbukai.

While, however, the Senmonbukai was working on the original idea of Plan B, in October 1958 the Kakuyugo Kondankai requested the Science Council of Japan (JSC) to form a special committee to discuss the nuclear fusion research program in universities. At the end of the same month JSC decided to promote the nuclear fusion research and submitted this recommendation to the director general of STA. JSC decide to set up Kakuyugo Tokubetsu Iinkai (a special committee for nuclear fusion, abbr. Yutokui), with Kodi Husimi as chairman, in April 1959. And the preparatory committee for setting up the Yutokui held a symposium on the above Plans A and B in May, when the first meeting of the Yutokui was held.

Broadly, Plan A had aimed at promoting basic research in order to foster new ideas, train researchers, establish a study group to construct an experimental fusion device based on the new ideas, and put together a core organization to plan for the future. Plan B was to construct a medium-size fusion device able to maintain a temperature of about ten million degrees Celsius, working with a budget of about one billion-yen. However, as no nuclear fusion researches anywhere in the world had gone beyond basic research, researchers were unable to devise a theoretical basis for the promotion of Plan B.

The recommendation of JSC gave rise to a debate about policy, the so-called AB argument, among researchers about to be involved in the study of nuclear fusion. This argument was not simply a choice between two plans but one about whether or not Plan B should be implemented at the same time as Plan A, since Plan A, being basic research, did not need any special support for its promotion. The AB argument was immensely significant, as it was the starting point of nuclear fusion researches in Japan, and it was taken up by JSC. At the fourth meeting of the Yutokui, held on 2 August 1959, promotion of the Plan A was decided and at the same time the Yutokui decided to set up a preparatory study group for establishing of an institute of plasma science. It was finally decided not to submit a budget for Plan B at the fifth meeting of the Yutokui held jointly with the Kakuyugo Senmonbukai on 10 August 1959. This meant that Japan decided to promote only the Plan A, that is, basic research into plasma, and subsequently the Institute of Plasma Physics was established in 1961 at Nagoya University, a national university under the jurisdiction of the Ministry of Education (MOE). The Institute was later dissolved to become in May 1989 the National Institute for Fusion Science under the direct control of MOE.

It is of interest that while researchers involved in the early stage of the study of nuclear fusion wanted the sound promotion of basic research into nuclear fusion in general, many of them had an antipathy toward excessive power belonging to any particular governmental organization, such as JAEC or STA. They firmly rejected any idea that STA should support the controlled fusion research, for they had already had a negative experience of government initiated nuclear fission research for energy purpose. Consequently, they decided to do away with the influence of STA on the Kakuyugo-Hanno Kondankai, the first discussion group formed within JAEC, dissolving it and organizing their own democratically-based nuclear fusion discussion group, the Kakuyugo Kondankai, in February 1958. In this general environment, the Plan B was reconsidered in terms of its theoretical basis as well as its proposed planning and funding, and was finally abandoned by the Kakuyugo Senmonbukai of JAEC in the summer of 1959. The Senmonbukai was dissolved with the determination. Thereafter until 1966, JAEC took no formal action regarding any nuclear fusion research. JAEC established the second Kakuyugo Senmonbukai in 1967 and decided to promote the Plan B in 1968. As the consequence of the decision the research and development of the nuclear fusion were designated as Special Comprehensive Research Project of Atomic Energy. From this time, nuclear fusion research in Japan entered the new era and moved to a direction of big science.

The research project was promoted by assigning the tokamak device as the main device. The Japan Atomic Energy Research Institute (JAERI) was in charge of promotion of this project and RIKEN (The Institute for Physical and Chemical Research) took charge of development of relating technologies. ETL undertook the research of the high beta plasma device as the supplementary device. JAERI and RIKEN are corporations having special status under the control of STA, and ETL is a national laboratory under the direct control of the Ministry of International Trade and Industry (MITI).

In the early stage of the nuclear fusion research under consideration a large part of the budget for the fusion research was accounted for managed by a Grant-in-Aid of MOE and the Trust Fund for Peaceful Uses of Atomic Energy of STA. In Japan, the atomic energy budget is managed by STA. But it cannot be used for universities. The nuclear fusion researches in universities were conducted only within the budget of MOE. In 1957 the Grant-in-Aid of MOE for Institutional Research to Osaka University was the only budget allotted to the fusion research. From 1958 the Grant-in-Aid for Comprehensible Research and several themes of the Grant-in-Aid for Institutional Research were approved and budget of nuclear fusion research became rather plentiful. The " Integral research of nuclear fusion " of the Grant-in-Aid for Comprehensive Research was allocated for three years and it expired in 1960. From 1961 the Grant-in-Aid for Comprehensive Research was allocated under the title of " Basic research of nuclear fusion ". The Trust Fund for Peaceful Uses of Atomic Energy was allotted to the fusion research during the five years from 1958 to 1962. The main item is the " Research on fusion reaction " of ETL, and the remainder was allotted to some companies as Toshiba, Hitachi, and Mitsubishi Atomic Power Industries, and JAERI. In JAERI the preliminary study of nuclear fusion had been started already from 1958, but after the allotment of this Trust Fund JAERI started the nuclear fusion research in real earnest.

Today, administratively two organizations, MOE and JAEC-STA manage the nuclear fusion research in Japan, with separate budgets.

BIBLIOGRAPHY

1) For historical surveys on the nuclear fusion research in Japan, for examples, Kenro Miyamoto, " Kakuyugo wo mezashita purazuma no kenkyu " (Plasma physics and controlled fusion in Japan), *Butsuri* (Proceedings of the Physical Society of Japan), 51, n° 8 (1996) 549-556 ; Kenzo Ymamoto, *Kakuyugo no 40nen-Nihon ga susumeta kyodai kagaku (Fifty years of nuclear fusion research-big science in Japan)*, Tokyo, ERS Shuppan, 1977.

2) For history of the early nuclear fusion research in Japan, see Satio Hayakawa, Kazue Kimura, " Kakuyugo kenkyu kotohajime (1) (2) (3) " (Beginning of the nuclear fusion research in Japan), *Kakuyugo Kenkyu* (Transaction of the Kakuyugo Kondankai), 57, n° 4-6 (1987) 201-214, 271-279, 346-378 ; Sigeko Nisio, Yoshinosuke Terashima, " Nihon ni okeru shoki no kakuyugo kenkyu ni kansuru shiryo " (Material for the history of early nuclear fusion research in Japan), *Butsurigaku-shi Noto* (Historical notes in the physical sciences), 1 (1991), 47-52 ; Sigeko Nisio, " Satio Hyakawa's letter to Hideki Yukawa (19 Sept. 1955). A Trigger for controlled Nuclear Fusion Research in Japan ", *Historia Scientiarum*, 1, n° 3 (1992), 221-222.

Car Industry, Technology, and Society in the USA
The Fight against Air Pollution between 1950 and 1990

Joop Schopman

Every technology has an impact on the society in which it functions. This influence will not be restricted to intended effects but it will also bring some which are considered harmful by members of that society. As a consequence, internal social forces will try to counter (or at least to contain) these latter ones. That means at each stage of a technology the society has to find an equilibrium between its positive and negative aspects.

To this general " law " the introduction of automotive vehicles has been no exception.

This paper concentrates on one of its negative side-effects : the environmental degradation. It describes efforts by politics and car industry in the U.S.A. after the second World War to counter this particular, negative aspect of a successful automotive history. First, a story will be told about initiatives by the federal government to counter the negative impact. The resulting legislation provoked reactions from the side of the industry. The consequent balance of power between government and industry led to a situation which was stable from 1977 till 1990.

Building up of legal pressure

Ever bigger sales

After the end of the second World War the U.S. industry could return to the production of consumer goods, as the automotive industry successfully did : the number of new cars grew steadily. Detroit where most of the automotive industry was concentrated, employed a substantial number of people. But the success came at a prize ! The very success, the enormous amount of cars on the roads, hastened the appearance of shadow sides such as the degradation of the natural environment. The accumulation of the exhaust gases of the fossil fuel burning vehicles became a source of increasing concern : public health appeared to be at risk. These anxieties translated into social pressures which were felt in the political centers. Being aware that the problem was an interstate one, the federal government decided in 1954 that action should to be taken.

In January 1955, president Eisenhower appointed a special committee in the department of Health, Education and Welfare (HEW) to deal with pollution. It was far from clear what action should be taken. Therefore the U.S. Congress decided in its *Air Pollution Control Act* of 1955 to authorize " a program of research and technical assistance to obtain data and to device and develop methods for control and abatement of air pollution by the Secretary of [HEW] and the Surgeon General ". For this it provided five million dollars per year from 1955 to 1960. Several sources of pollution, " the inevitable results of our increasing mechanization and industrialization ", had to be studied. Early on scientists " generally concurred that a major source of fumes which are ingredients of smog and pollution in most cities is the motor vehicle "[1]. Hearings in 1958-1960 by the Health and Safety Subcommittee made it clear that progress had been made in air pollution control, in particular with regard to motor vehicle exhausts but not enough. So the committee could favourably report on a new bill (public law 86-493) " requiring increased emphasis on research into the motor exhaust problem "[2]. The Surgeon General was asked to pay special attention to the effects on human health. How much the latter was at the focus of the public attention can be seen in the speech by president John Kennedy on February 7, 1963. He then mentioned the relation of pollution with an aggravation of heart conditions and with an increased susceptibility to chronic respiratory diseases ; estimated damage 11 billion dollars. Not surprisingly, the same argument can be found in the public law of that year which mentioned as one of the motives for the law : " the increasing use of motor vehicles, has resulted in mounting dangers to the public health and welfare "[3].

Study alone will not do the job

The 1963 law required that the Secretary (HEW) would semi-annually report on " measures taken toward the resolution of the vehicle exhaust pollution problem [...] ". One of the things which should be reported, were " criteria on degree of pollutant matter discharged from automotive exhaust ", and suggestions for further legislation[4]. The next time the Congress in Washington issued a public law (89-272) on this topic, the tone had changed somewhat : a " National Emission Standards Act " was mentioned. The establishment of standards was required, but the formulation of the procedure con-

1. *U.S. Code Congressional and Administrative News* (for short : *US Code*), 87th Congress 2nd session (1962), 2818-9.

2. *US Code*, 88th Congress 1st session (1963), 1261.

3. *United States Statutes at large* (for short : *US Statutes*), 77 (1963), 392.

4. *US Statutes,* 77 (1963), 399 Sec 6.

tained several snags : " The Secretary shall by regulation, giving appropriate consideration to technological feasibility and economic costs, prescribe as soon as practicable standards, applicable to the emission of any kind of substance, from any class of new motor vehicles or new motor vehicle engines [...] "[5]. (emphasis j.s.).

Why did this change of tone happen ? The story so far suggests that progress was being made. Research was getting results ; the industry was busy, but with what ? Outsiders got the impression that it was not making a serious effort to control emissions. The industry should have been aware that public pressure was building up because as a reaction to its inertness Californian politicians had started to formulate emission standards for the 1966 car models. It should be no surprise that California took the lead. It had to do something against the smog in Los Angeles. Because California itself did (and does) not have a car industry this legislation was a clear message in the direction of Detroit. California formed such an enormous market that Detroit could not afford to ignore this signal.

This *fait accompli* confronted the Detroit car manufacturers with a range of problems. First, they had to decide to make special " editions " of their cars for the Californian market, or to adapt all their models to the Californian standards. They decided for the first option. The next question was how to fulfil these Californian norms. The industry chose for a " systems engineering " approach[6]. Chrysler e.g. succeeded to burn fuel more completely by changes in the carburettor, distributor and a spark adjustment. The extra costs for this Cleaner Air Package (CAP) were 13-25 $. An other solution was to install a pump which would add air to the hot exhaust gases to burn them more thoroughly ; costs 50-75 $[7]. The choice of the automotive industry for a dual market was not welcomed in Washington. The Subcommittee of the U.S. Senate on air and water pollution made it clear that " since the industry advises it would not voluntary provide automobiles meeting Cal's standards on a nationwide basis it is up to the Congress to act "[8]. The U.S. Senate wanted that all new 1968 models would be equipped with exhaust control devices. It gave the Secretary the power to set the norms and even to go for an earlier introduction. The automotive industry began to realize that Washington meant serious business. The car industry hurried to indicate its willingness. At a House hearing Barr of the Automotive Manufacturers Association (AMA) stated that all new cars could be equipped with an exhaust control system " if Congress decided they were warranted after considering the many factors involved "[9]. The House

5. *US Statutes, 79* (1965), 902 Title II, Sec 202 (a).

6. J. Dunne, " Smog control confusion ", *Automobile industries* (for short : AI), *131,* n° 3 (1964), 19.

7. *AI, 131,* n° 3 (1964), 19 ; Dunne, " Chrysler's anti-smog unit ", *AI, 131,* n° 12 (1964), 41.

8. *AI, 131,* n° 12 (1964), 42. *US Code,* 95th Congress 1st session (1977), 1310.

9. *AI, 133,* n° 2 (1965), 114.

could only follow the Senate's initiative because the general feeling was that air pollution is a serious problem which " threatens to grow and worsen in direct proportion to upward trend of urban and economic growth and technological progress "[10]. The politicians decided to go ahead. They could hardly have acted differently. The data collected so far " present impressive evidence of damage to health resulting from air pollution ". Together they presented a disturbing portrait of a major health menace. So " the conclusion was inescapable that air pollution was a major factor in the development of many diseases affecting millions of Americans ", and that situation was steadily worsening.

Congressional anger

Thus, officials and citizens on the grassroots level throughout the U.S. had every reason to be seriously aroused over the threat of air pollution 60 % of which came from cars[11]. Some measures had already been taken : the law of 1963 had given the Secretary of HEW the power to set the pollution maxima. He had set the 1968 norms of 6 g.p.m. for hydrocarbons (HC), 52 g.p.m. for carbon monoxide (CO) but none for nitrogen oxides (NO_x), or 6/52/- ; later tightened for 1970 to resp. 3.9/43/-[12]. For some commentators " the industry had shown itself willing and able to make the modifications for its lucrative Californian market "[13] but for many the industry had waited too long, and then acted too reluctantly. " The problem of air polluted by automobiles has become, in our view, so massive that it cries out for drastic measures. We are not impressed by the ails of the automotive industry that meaningful improvements in their products pose insurmountable cost and engineering problems. We listened to the same complaints back in 1967, when Congress agreed to permit California to depart from national auto emission norms in setting and enforcing more stringent controls. The industry demonstrated then that it has the expertise and the know-how to make just about any change for the better when the public demand is great enough "[14]. Some accused the manufacturers : " they have stalled research which might have helped to clean our skies, except when the Government pressure has been brought to bear "[15]. In an effort to turn the tide, the Secretary HEW proposed tighter standards for the 1975 models : 0.5/11/0.9, and for 1980 resp. 0.25/4.7/0.4[16].

10. *US Code*, 89th Congress 2nd session (1966), 3473.

11. *US Code*, 91st Congress 2nd session (1970), 5360-1.

12. *AI, 138*, n° 3 (1968), 80. *Federal register part II, 33* (108), (June 4, 1968), 8303-8324.

13. *US Code*, 90th Congress 1st session (1967), 1938 ; resp. 1955.

14. Comment by Deerlin, Ottinger and Tiernan on House Report 17255, in *US Code*, 91st Congress 2nd session (1970), 5372.

15. Statement by Anderson (Cal) in the U.S. House, Dec. 18, 1970. *Legislative history of the clean air amendments of 1970* (Washington, 1974), 119.

16. *Federal register part II, 35* (28), (Feb. 10, 1970), 2791.

Not impressed the Congress demanded “ to speed up, expand, and intensify the war against air pollution. ” It wanted that “ the air we breathe throughout the nation is wholesome again ”[17]. As one representative put it : “ the divine right of every citizen to breathe clean air ” must be respected. “ By 1970, it had become apparent that motor vehicles emissions would not be lowered to levels sufficiently to protect public health unless Congress specifically established emissions standards and set schedules for attaining such standards. In doing so, Congress abandoned the administrative “ technical and economic feasibility ” approach ”[18]. It gave the Administrator [head of a new agency, the Environmental Protection Agency, for short : EPA] clear indications : “ (A) The regulations [...] (a) applicable to emissions of [CO] and [HC] from light duty vehicles and engines manufactured during or after model year 1975 shall contain standards which require a reduction of at least 90 per centum from emissions [...] allowable under the [...] year 1970 ”. (B) Similarly it wanted in 1976 a 90 % reduction of NO_x compared with 1971. It realized that its norms would be tough for the industry. So, it allowed the individual manufacturer to request a suspension for one year. But then interim norms must be indicated which “ shall reflect the greatest degree of emission control which is achievable by application of technology which the Administrator determines is available, giving appropriate consideration to the cost of applying such technology within the period of time available to manufacturers ”. Congress made it also clear that this law would not be the last word in this. It wanted to know how far pollution can be reduced. Therefore “ [t]he Administrator shall undertake to enter into appropriate arrangements with the National Academy of Sciences to conduct a comprehensive study and investigation of the technological feasibility of meeting the emission standards required to be prescribed by the Administrator ”[19]. For the same reason it also wanted that special low-emission vehicles would be developed. The *Clean Air Amendment* of 1970 (public law 91-604, also called after its main promoter, the “ Muskie ” law) has become a landmark in environmental legislation for many years to come. For some its norms were not going far enough. They proposed an amendment which would give the industry a chance to clean up their internal combustion engines. “ If the existing engines could not measure up, they should gradually be phased out, starting with the dirtiest, largest horsepower ” ; they should disappear if they can not perform better than steam or turbo. The proposers were convinced that “ despite the protestations of Detroit [other systems] can be mass produced at little or no additional cost to consumer ”[20]. Another amendment proposed to

17. *US Code*, 91st Congress 2nd session (1970) ; House report 91-1146 ; 5356.
18. *US Code*, 95th Congress 1st session (1977), 1310.
19. *US Statutes, 84* (1970), 1960, resp. 1961 5(C), and 1962.
20. *US Code*, 91st Congress 2nd session (1970), 5372.

adopt the Californian norms : why should people in New York not enjoy Los Angeles norms ?

These political initiatives underlined the public demand for stringent measures. It wanted vehicles which produce as little pollution as possible. But there is no trace of evidence that it considered to give up its newly acquired tool of mobility. Health was important but apparently not the ultimate value, at least not for the majority of the population. So the car industry might have been pushed in the corner, it also learned from this that it had a secure place in the American market place for a long time to come. It had not to take too seriously the threat of the return to steam engines, but it had to act. Apparently the society was willing to pay a prize (i.e. some degree of pollution) for its mobility but it wanted the health-prize to be as low as possible.

Reactions by the industry

Industry follows the developments

From the mid 60s the automotive manufacturers realized that they had no choice. They knew that they had not done enough to counter the pollution produced by their products[21], and that the public was in no mood to have much consideration for their arguments. So they complied with the Californian norms. That first step was not too difficult for the industry. Chrysler fulfilled the 1966 Californian norms by its CAP. Other manufacturers used an air pump to add air to the hot exhaust gases. These technologies allowed the car manufacturers to satisfy the federal 1968 norms as well. Through their speaker Mann (AMA) they promised Congress that all 1968 models would satisfy the CO and HC norms : the industry would bring emissions below these norms. Congress looking for the lowest emissions possible, pressured Mann to find out what really could be done by the industry. But it did not want to inform Congress about its technical abilities.

Although the automotive industry had moved too late, and initially too reluctantly, one should realize the problems they were facing. Polluting is not difficult but to find out what pollutes, can be far from obvious. Then methods have to be invented to measure the degree of pollution. For this devices had to be developed and build to measure the presence of sometimes minuscule amounts of pollutants. They had to be calibrated to be usable on all locations and circumstances. All this required an expertise which so far the car industry did not have[22]. What came closest were the departments which tried to opti-

21. Ford II quoted in : *AI, 148,* n° 3 (1973), 19 : he personally feels that the government " forced us to do a lot of things we should have done before ".

22. Beckman Instruments Inc. in California produced " smog mobiles ", trailers type laboratories to measure CO, aldehyde, HC, oxidants, particle matter. *AI, 137,* n° 8 (1967), 39.

mize the fuel consumption of the engines. Even they were, given the low cost of fuel, more interested in developing more powerful engines than in more efficient ones. But the emission research led to a close co-operation with the oil industry, a connection which proved to be extremely useful when emission norms became stricter[23].

There was no reason why the downwards trend should stop with any particular number. Many people were upset about the risk for health and environment ; they did not experience much improvement. The evasive attitude from the side of the industry gave them the feeling that the industry could do better but did not really want to. Moreover, from the scientific-technological side no serious objections could be brought forward against lower norms because no sufficient knowledge was available[24]. To study the different aspects of smog, a new program was started by a Co-ordinating Research Council for the AMA and the American Petroleum Institute. It would give attention to such topics as : the effect of CO on humans ; the impact of aromatic HC on lungs ; the result of combinations of pollutants[25].

The political initiatives had changed the industry. Not only had it " taken more and more the character of a regulated industry with constantly greater attention to safety features, reduction of pollution [...] "[26], it also was running out of conventional answers. It had improved existing technologies to meet the new Californian measures with encouraging results : reduction of HC and CO by catalysts ; special carburettors ; and devices which reintroduced exhaust gases to diminish NO_x. Only GM's Oldsmobile had been working on a different line : an engine ignition spark timing system which reduced NO_x and HC[27]. The lack of time to develop new answers could have been mitigated by a co-operation between the automotive manufacturers and there existed several forms of co-operation. But in the eyes of the Johnson Administration these were not intended to speed up things but to make a common front in delaying them. So on January 10, 1969 it sued the big-four for " conspiracy to delay development and installation of air pollution devices ". It came to settlement notwithstanding the opposition by several members of Congress.

In fact, the automotive manufacturers had come to acknowledge their social duties, their environmental responsibility. They were spending half a billion

23. E.g. Chrysler had a program with Standard Oil to reduce emissions by cars : *AI, 137,* n° 6 (1967), 64.

24. N.M. Lloyd, *AI, 144,* 10 (1971), 7 : " Detroit hardly deserves all the credit for the anti pollution progress to date. Without consumer pressure and the resulting government regulation, it is doubtful that we would have come even this far in the quest for cleaner air. But the hitch has been that federal standards have not been related to technology ".

25. *AI, 138,* n° 7 (1968), 68 ; and n° 10, 132.

26. J. Geschelin, *AI, 139,* n° 7 (1968), 59.

27. *AI, 140,* n° 11 (1969), 21 ; Cfr. resp. : *AI, 139,* n° 5 (1968), 220 ; *AI, 140,* n°10 (1969), 35.

dollars yearly to find solutions for the air pollution. But they had given the initiative away. The social and political machinery as moving. That was a painful experience for the industry : their arguments were no longer heard by a nation in " near-panic ". The industry experienced the legal norms as hostile. Was the purpose of the politics an extermination of the automotive industry ? " The latest controls could mean auto industry's death "[28] which it was not willing to accept without a fight.

With their back against the wall

For a long time the automotive industry kept a low profile for which it had many reasons but it was not willing to be regulated out of existence[29]. So facing the " Muskie law " the industry tried to come back in the driver's seat with no strategy left unused although some of them were incompatible.

One strategy used in different forms was a denial of irresponsibility. By (re)writing history the industry tried to convince the public opinion it did not deserve such negative publicity. Terry (Chrysler) e.g. defended his company while acknowledging that in a competitive world the government had to intervene for social and environmental reasons. To satisfy these government regulations is new for the automotive industry which needed external help for problems outside its competence. This, however, did not mean that Chrysler had not been active in controlling pollution. He then presented *the Chrysler story* to recall the industry's contribution. When in 1953 the nature of Los Angeles smog became known, Chrysler recognized cars as part of the problem, and immediately wanted to do something about it. Thanks to our research the government could issue norms in 1960. In 1961 we developed and produced a control for crankcase fumes. After much research the CAP was so far ready that lasting devices could be produced for California in 1966. Now, in 1970 we have a vapour saver program to reduce the amount of evaporating fuel which will be available nation-wide in 1971. As a result, even before all the old models have disappeared from the roads, the condition of the air in 1980 will become as it was in 1940. In short, the automotive industry has not been as passive as often presented.

The opposite strategy denied the responsibility of the automotive industry by attacking the data on which the political authorities based their norms. It quoted a Berkeley study : cars were responsible only for 12 % of the pollution, and not for 61 % as the EPA had claimed[30]. The CO norm was based on a 4 year old study, a " false alarm ". The industry complained that notwithstanding

28. *AI, 142,* n° 11 (1969), 21 ; J.M. Callahan, *AI, 143,* n° 8 (1970), 17.

29. N.M. Lloyd, " AI in the offensive after having kept a low profile for a long time ", *AI, 144,* n° 2 (1971), 3.

30. J.M. Callahan, *AI, 144,* n° 4 (1971), 17 and *145,* n° 7 (1971), 19.

the disputed status of the data EPA had not been willing to change its norms[31].

Blaming others. Sometimes the reactions of the industry sounded like these of an angry child : the government was the real culprit which created distrust between industry and public[32]. Detroit faced an alliance of interests such as suppliers, insurance's. Everybody was against it even its own employees ; its Union of Automobile Workers (UAW) together with 6 conservation groups had written letters to every U.S. senator asking for an abolishment of the internal combustion engines after 1975 and to replace them by steam engines[33] if they remained a health threat. A variant of blaming others was the reproach that EPA had pushed the industry to a premature response[34]. In fact, as EPA replied, the automotive manufacturers had several options to meet the norms : " They were not forced into using catalysts. It was their option because they felt this was the only way they could do it ". So, when in June 1973 GM announced its' lifetime catalytic converters' this meant support for EPA's lonely position. Clean Air Act of 1970 had been a gamble ; nobody knew whether its standards would be realizable. Sometimes one had to push a technology.

On the counterattack

Another strategy by the industry was to fight the norms head-on : to portray them as illegal ; as against the economic interest of the US economy ; or as anti-American.

First, the 1965 law explicitly stipulated conditions for norms. Thus if the industry thought that the " demands [...] aren't practically attainable "[35] then legal action seemed to be an obvious reaction. The industry considered this. Apparently it was not fully aware of the mood in Washington : the government wanted to bring the industry to court. Luckily the automotive manufacturers did not pursue this legal move ; they even decided not to appeal against EPA's 1976 norms although it had to acknowledge that the NO_x norm was too strict given the lack of data about its health effects[36]. Economy/employment was a second argument. When Muskie's law tightened HEW's 1975 norm to 98 %

31. J.M. Callahan, *AI, 145,* n° 10 (1971), 17 and 35. According to the original study by Beard and Wertheim of Stanford University, Nov. 1966, 2 % CO would damage human health. This result could not be duplicated. EPA's own study did not find negative effects with 13 %, only some indications for 15 %.

32. J.M. Callahan, *AI, 143,* n° 6 (1970), 17 : they keep saying that car is responsible for 60 % of the pollution. *AI, 146,* n° 1 (1972), 17. E.g. HEW reports 82 % failure of exhaust controls.

33. D.A. Colburn, *AI, 153,* n° 1 (1975), 11 ; Callahan, *AI, 143,* n° 4 (1970), 21.

34. J.M.Callahan, *AI, 144,* n° 10 (1971), 21. Why go for a " most disadvantageous " technology asked Gottesman, if with more time, a more promising technology can be developed ? *AI, 148,* n° 6 (1973), 17.

35. J.M. Callahan, *AI, 152,* n° 11 (1975), 25 ; resp. *AI, 144,* n° 4 (1971), 17.

36. R.J. Fosdick, *AI, 149,* n° 11 (1973), 17 ; J.M. Callahan. *op. cit.*, n° 10, 9 : " We have overreacted ". *Id.*, *op. cit.*, 45. CO norms not based on scientific evidence. Nature is a larger source of CO than traffic.

reduction, Iacocca countered : the " clean air act will might stop car production by 1975 ". This argument became quite powerful when the threat of unemployment became a reality[37], in the mid 70s a quarter of a million people had already lost their job. Pushing the norms meant a threat of a decline in US well-fare and its entire way of life[38]. To strengthen their point the manufacturers tended to exaggerate the costs of emissions control devices. Another powerful argument when oil became scarce was that a relaxation of the standards would save 30 % fuel[39].

A third powerful group of arguments were ideological. There was the slogan of a complete freedom threatened : " The drivers right to and ability to go anywhere at any time can no longer be taken for granted "[40]. The industry was also effectively playing on the American aversion of regulation and big government : Governmental institutions only served themselves. Politicians had no idea of technology. They thought that everything was available at push-button ; we got a fully controlled economy " run by a bunch of socialists "[41]. Not surprisingly their norms didn't serve the public. We, the industry, had to protect the public interest. " We got to guard against abuses ". Detroit wanted Washington from its back in the interest of the common people. The public demonstrated that it did not like regulation : the government was to blame for the drop in sales, " the public rejection of the government designed car ".

Similarly dangerous were the ideas from the side of EPA and others, that if the exhaust gases could not be cleansed sufficiently, one had to think about drastic changes in the modes of transportation ; one had to prepare the nation for a change transportation habits, thus in US lifestyle[42] which had been too auto-minded. Another argument was that by the imposition of these norms the industrial landscape was changing. The smaller companies could not meet the requirements at this pace. This threatened their existence, and thus the U.S. entrepreneurial future.

Technological answers

The automotive industry realized that words alone would not the job. Although it confined itself to improvements of existing technologies, these appeared to be far from easy because the conditions for the solution of one problem often aggravated other problems or created even new ones. E.g. the exhaust gas recirculation (EGR) system required a higher than usual tempera-

37. *AI, 146,* n° 10 (1972), 22. Ford will have to put 800.000 people on the street.

38. Ford II quoted in : *AI, 148,* n° 3 (1973), 19 ; J.M. Callahan, *op. cit.*, n° 5, 17.

39. J.M. Callahan, *AI, 148,* n° 1 (1973), 17 ; Cole (GM) : pollution norms must change given the energy shortage, in *op. cit.,* 20

40. N.M. Lloyd, *AI, 143,* n° 5 (1970), 33.

41. Ford II cited by J.M. Callahan, in : *AI, 148,* n° 1 (1973), 51.

42. J.M. Callahan, *AI, 144,* n° 8 (1971), 41 ; resp. *AI, 152,* n° 3 (1975), 21 ; and *AI, 148,* n° 3 (1973), 17 ; J.B. Pond, *AI, 144,* n° 10 (1971), 21 quotes Middleton (EPA).

ture of the engine ; otherwise the dirt in the exhaust got into the engine itself. That would be catastrophic for the functioning of modern engines which had been made so precise to use fuel as economically as possible.

In particular, the industry realized legal standards could not be met without a device borrowed from the chemical industry, the catalyst (or cat) which uses special chemicals to reduce the toxic elements in the exhaust gases. But these chemicals don't work efficiently for every toxic element. That means to reduce all pollutants a car would need several cats ; thus it would become heavier and use more fuel. In addition, cat's chemicals will not work for ever. Particularly the lead in exhaust gases has a disastrous effect on the cat's lifetime ; an argument to try get rid of the lead in petrol. Statements by companies such as Ford that if lead-free petrol became available there would be cars to use it (as early as 1971), convinced the oil refineries that it would be economic to provide lead-free petrol for the market. Again things have two sides : a lowering of the octane meant an increased use of petrol. Nevertheless, EPA demanded that from 1 July 1974 at least one type of lead-free petrol would be available everywhere so that the cat could be used. As a result, lead-free petrol, rating 91 octane, was then available after a crash program by the oil industries, and an investment of 3-4 billion $[43]. Another problem for cats was the high temperatures in the stream of the exhaust gases. That required the use of expensive materials but that would push the extra cost per car too high. Consumers might want cleaner air, did they want to pay for it, and if so, how much ? Research concentrated on finding cheaper material[44] which can work in the whole temperature range from 0 - 2000 °F. An additional problem was that both extremes in temperature formed a threat. If it became too high, the cat would be destroyed. If it was too low, the cat would malfunction, i.e. the car would produce more pollution. The latter situation is unavoidable : every engine is cold as it starts. Begin 1973, GM still claimed it would not be able to meet the emission criteria mandated in the Muskie bill. Even after testing 223 catalysts from 13 potential suppliers it could not find any appropriate catalyst and they seemed to be crucial to do the 1975 job[45]. Ford II confirmed that nobody in the industry knew how to meet the 1975 norms. EPA was not impressed. It did not even want to postpone the norms for one year at the industry's request. It acknowledged that there were difficulties, but, in its opinion, the industry had not demonstrated that such technology did not exist. Indeed, within a few months GM changed its opinion : its cat did even save some energy, lasted for

43. J. Geschelin, " Political and social pressure have combined to make this a crash program ", *AI, 142,* n° 8 (1970), 83. Thereby ignoring the effect which the replacing anti-knock additives might have on human health.

44. R.J. Fosdick, *AI, 147,* n° 9 (1972), 18 : Questor Inc. claimed to have developed a cat which did not require exotic material, and met the 1976 norms for leaded and lead-free petrol.

45. J.M. Callahan, *AI, 144,* n° 10 (1971), 20.

more than 50.000 miles and cost 150 $[46].

Other answers

An obvious way to reduce pollution should have been the adoption of *smaller cars*. The Americans have a tradition of much larger cars. Only for a time smaller cars became fashionable in the USA but this was not to last.

An extremely pragmatic solution, hardly technological at all, was the so called " 2 car strategy "[47]. Many families owned two cars of which one was used to go to work in the " dirty " cities. For this a clean, even an electric, car could be used. A second larger and dirtier car was for out of town activities with the whole family ; thus for use in the cleaner air areas.

As already stressed the technological solutions remained mainly restricted to modifications of the existing technology, although one heard from time to time that *radical different type of engines* should be developed ; data from developments in Japan and Europe were very promising.

Unexpected help

When the pressure on and in the industry was rising help came from two unexpected sides. The first was an hesitation within EPA to continue with the cat. There were indications that its negative side-effects might be worse than its usefulness : cats were reported to emit platinum and sulphur particles. The question was whether the assumptions of that troubling study were extreme or on the safe side ? Most other research found less sulphur. EPA found itself caught between different interests : that of the automotive industry and that of a concerned public. The only way out was to start a study on its own but the results would not be available before March 1974, too late to stop the introduction of the 1975 cats. Moreover, that spring would be the cut-off point for any other counter measures such as the reduction of sulphur in fuel[48]. Given the advantages of cats, one should be reluctant to abandon them. EPA decided to suspend the statutory HC and CO standards for 1977, and to extend the 1975 (49 states) norm.

The second form of help came from the oil producing Arabic countries. After the start of the Kippur war on October 6, 1973, the U.S.A. decided to support Israel. The OPEC countries retaliated with an oil embargo. Although it did not last long, the world had learned a new word : energy crisis. Then the U.S. government wanted to postpone the introduction of the Muskie norms. Indeed, Congress decided to do that. Another consequence of the embargo, the rising energy prizes, threw the economies of all Western countries in disarray.

46. R.J. Fosdick, *AI, 146,* n° 11 (1972), 20 ; *AI, 149,* n° 2 (1973), 17.
47. J.M. Callahan, *AI, 153,* n° 4 (1975), 15 ; D.A. Colburn, *AI, 150,* n° 10 (1974), 23.
48. J.M. Callahan, *AI, 152,* n° 11 (1975), 25 ; resp. D.A. Colburn, *AI, 149,* n° 11 (1973), 20.

The following recession led to massive unemployment. That provided the automotive industry with ammunition for their case. In the eyes of the automotive industry the prize for the stricter norms for cars exhaust gases had been more expensive cars. It was senseless to push prices in a time of a deepening recession and increasing unemployment.

THE ACTUAL BALANCE OF POWER

One must acknowledge that the opposition by the industry had been effective. So far the introduction of stricter norms had been postponed three times, first by the Administrator, then by the legislator, and again by the Administrator[49]. EPA even proposed a freeze for these HC and CO norms from 1979 till 1982 for which permission by Congress is required. California, however, did not hesitate to tighten its 1977 standards to 0.4/9/1.5 (only CO remained at 9 g.p.m.)[50].

How did the changes influence the positions of the other partners in this conflict ? As one could expect the *UAW* made a U-turn. It now took position against regulation which was blamed for a loss of employment in 1975 of 150-200.000 people : employment had become now more important than the environment[51]. The captains of the automotive industry could not agree more. According to Iacocca, there existed a " hell of a danger " that " hundreds of thousands of people will be regulated out of job " if the costs of cars continued to rise. Employment should now be our first priority. Moreover, the industry had attained a lot the last five years. For GM e.g. the car was no longer an environmental factor.

The question was will Congress concur with this conclusion. Some thought there will be more regulation to come. This was quite feasible : technologies now existed allowing to set even lower standards. But not only the captains of industry were convinced that priorities had to change, many in the political arena shared this position : the political climate had changed. The 94th Congress proved unable to reach agreement before the end of its term. In February 1977, a reconsideration had been made on president Carter's request. Administrator Costle had brought together a special Task Force to examine the auto emission question and to formulate a new proposal. This task force found that a study composed for the previous administration contained unrealistic assumptions and serious flaws in analysis. It had ignored the urgency of a short term NO_x norm, instead of an annual one, and also its role in oxidant formation. The previous study — did not point out clearly there would be serious deterioration of air quality at any NO_x emission standard above 0.4 g.p.m.

49. *US Code*, 95th Congress 1st session (1977), 1311.
50. *AI, 152,* n° 6 (1975), 7 ; resp. R.J. Fosdick, *AI, 152,* n° 8 (1975), 9.
51. *AI, 153,* n° 5 (1975), 9.

EPA's new proposal for the president intended to be a compromise (1979-80 norms : 0.41/9/2 ; for 1981 : 0.41/3.4/1) which " incorporates a measure designed to provide incentives to industry to develop inherently low-emission, high-fuel economy automobiles ".

The reactions of the Congress were mixed. Predictably, part of it reacted negatively. It sided with the industry : health arguments must " be balanced with the urgent national need to secure energy savings and consumer savings with regard to the use, purchase, and maintenance of automobiles ". Jobs had to be secured to maintain a strong American economy. It considered the norms in particular for CO and NO_x as unnecessary strict. They also used the occasion for an attack of the position of the EPA. Another commentator stated : we all want clean air, but this was not the way to do it. It was harmful to our economy and other important national goals. It brought higher costs for the consumers, and did not treat all manufacturers equally. Moreover, " the last thing our economy needs is another increase in paperwork generated by government regulations ".

But there were supporters as well. These stressed that the emission reduction schema was a compromise. " We believe protection of public health would demand, and technology would allow, a more dedicated commitment to speedy attainment of the emissions standards set in the 1970 Clean Air Amendment ". That applied in particular to NO_x. Its 1970 norm had been postponed to 1983 and even then it was not clear what would happen. They acknowledged that " quantifiable evidence on health effects is indeed imperfect "? but ignorance was no reason for laxer norms, but for more study. " In fact, GM and Ford had already produced full-size test cars that meet the statutory standards not only for NO_x but for HC and CO as well ". Remaining problems were the 50.000 miles and the mass production test. Moreover, past experience suggested that the industry would not vigorously pursue commercial development of equipment to meet the NO_x standard unless the intent of the Congress in this regard was clear[52].

The Congress agreed on 1.5/15/2 for 1978-9 ; on 0.41/7/2 for 1980, and 0.41/3.4/1 for 1981. After this the issue of car emissions seemed to have lost its urgency. It would take some time before pollution could return on the political agenda ! No further discussions in the US Congress have been reported till 1990.

52. *US Code*, 95th Congress 1st session (1977), 1077-1577.

UN ENFOQUE ACTUAL DE LA TECNOLOGÍA

Claudio KATZ

En los últimos años se han desarrollado los estudios sociales de la tecnología, expandiendo un tipo de investigaciones que tradicionalmente estuvo restringido al campo de la economía. Este reciente escenario teórico nos permite ampliar el análisis de la visión marxista a los nuevos terrenos de la sociología y la filosofía de la innovación.

En un ensayo anterior[1] establecimos una contraposición entre el enfoque económico marxista y las interpretaciones neoclásicas, keynesianas, schumpeterianas y evolucionistas. Ahora realizaremos un análisis equivalente, pero referido a escuelas y autores representativos de los estudios sociales de tecnología. Estas visiones son culturalistas (Pacey, Basalla), deterministas (Ogburn, Bell), críticas (Winner), espiritualistas (Ellul), sistémicas (Gille), constructivistas (Bijker, Pinch, Callon, Latour) y racionalistas (Bunge, Quintanilla).

Estudiando la tecnología como una fuerza productiva social y observando como es caracterizada en los estudios sociales recientes, clarificaremos la especificidad del marxismo en relación a las otras corrientes.

DEFINICIONES Y FUNCIONES

La tecnología se define como el conocimiento científico aplicado a la producción. Este conocimiento se materializa en objetos — máquinas y artefactos — o en sistemas de gestión y organización de la producción. Tanto estos instrumentos y acciones como el conocimiento necesario para desarrollarlos, constituyen para el marxismo fuerzas productivas, es decir medios para realizar una actividad económica, en un modo de producción históricamente específico.

1. C. Katz, " La concepción marxista del cambio tecnológico ", *Revista Buenos Aires. Pensamiento económico,* 1 (1996).

Con mayor precisión, corresponde definir a la tecnología como una fuerza productiva social, ya que en su caso “ el conocimiento aplicado a la producción ” esta configurado directamente por las normas de funcionamiento del sistema capitalista. Debido a este condicionamiento, la tecnología actúa en dos dimensiones. Por un lado permite el cumplimiento de una finalidad práctica útil y por otra parte contribuye a la valorización del capital. Viabiliza de esta forma la creación de valores de uso, que operan en el mercado como valores de cambio. Tomando una clasificación de Dussel[2] se puede denominar a la primera función “ tecnología en general ” y a la segunda “ tecnología como capital ”.

El cambio tecnológico en el capitalismo plasma estes dos características. Es la modificación introducida en la actividad económica por nuevos productos, procesos de trabajo y formas de organización de la producción, que corresponden a necesidades y posibilidades técnicas (“ tecnología en general ”) y a los principios de beneficio, competencia y mercado (“ tecnología como capital ”). Una invención es un descubrimiento que cumple los requisitos de utilidad (primer aspecto) y una innovación es su aplicación productiva, cuando satisface las exigencias de rentabilidad (segundo aspecto).

El marxismo analiza a la tecnología en los dos terrenos. Distingue cuales son los rasgos universales del conocimiento que aporta, que tienen instrumentación similar en cualquier régimen social. Pero también describe cuales son las características de este saber, asociadas al uso capitalista. Cuando Marx estudió por ejemplo, los tres componentes de la máquina (fuerza motriz, mecanismos de transmisión y máquinas-herramientas), analizó la aplicación de una “ tecnología en general ”. Investigó un conjunto de rasgos técnicos, que si bien aparecieron con el capitalismo naciente no obedecen exclusivamente a las exigencias de este modo de producción. En cambio cuando estudió a la máquina como instrumento a través del cual se genera plusvalía relativa, realizó un análisis de la “ tecnología como capital ”, ya que aquí trató la utilización del artefacto como vehículo de la acumulación. La misma máquina presenta por lo tanto este doble carácter : es un medio para fabricar más y mejores bienes y es una herramienta para reforzar la explotación de los trabajadores asalariados. La tecnología es un conocimiento aplicado para satisfacer ambas finalidades.

Estudiar a la tecnología como un fenómeno social, significa tomar en cuenta sus aspectos autónomos y dependientes del capital. Implica considerar que el proceso de valorización fija la forma concreta que asume la tecnología, pero sin eliminar su carácter universal y sin convertirla en un patrimonio exclusivo de este régimen social. Lo que impone la vigencia de la ley del valor es la adaptación de las innovaciones al objetivo prioritario del lucro.

2. E. Dussel, “ Estudio preliminar de los cuadernos ”, en C. Marx, *Cuaderno tecnológico histórico*, Puebla, UAP, 1984.

La efectividad de una tecnología se mide en el capitalismo con el doble parámetro del cumplimiento de una tarea en el proceso de producción y la creación de beneficios para los empresarios. Cualquier sistema informático, máquina textil o tipo de gestión administrativa presenta siempre una función universal y un papel particular en la valorización del capital. Aclarar esta determinación simultánea es el objetivo del concepto " fuerza productiva social ".

FACTIBILIDAD TÉCNICA Y VIABILIDAD ECONÓMICA

Todas las corrientes de interpretación de la tecnología aceptan la existencia de los dos componentes que subraya el marxismo, pero los analizan como el " rasgo técnico " y el " aspecto económico ". En la tecnología converge siempre la factibilidad del primero con la viabilidad del segundo[3]. Pero esta contextualización es simplemente descriptiva. Se limita a distinguir entre la función útil y la función rentable de la tecnología, personificadas en la figura del ingeniero y el economista y concretadas en el diseño técnico y la estimación de rentabilidad, respectivamente.

El marxismo vá más allá de esta constatación, porque analiza como está condicionado y determinado el desenvolvimiento de la tecnología por las leyes del capital. Señala con esta caracterización un patrón de funcionamiento, que las restantes teorías solo ilustran fragmentariamente y puntualiza contradicciones, que son ignoradas por estas concepciones. Solo observando a la tecnología como una fuerza productiva social se puede entender el carácter convulsivo e incierto que adopta el cambio tecnológico, explicar la predilección empresaria por innovaciones que refuerzan el control patronal del proceso de trabajo, o comprender las causas del desaprovechamiento sistemático de tecnologías de gran utilidad social pero baja promesa de beneficio. Ignorando las leyes del capital resulta imposible analizar los efectos más debatidos del cambio tecnológico actual, como la masificación del desempleo, el estancamiento de los salarios o la expansión de la pobreza.

Particularmente en la visión racionalista los componentes " técnico y económico " son estudiados como el criterio interno y el externo de evaluación de la tecnología. Para Quintanilla[4] la eficiencia constituiría el parámetro interno y se mediría por el grado de control que se alcanza de un proceso, o por la capacidad para gobernar cierta propiedad en función de un obietivo buscado. Ingenieros, técnicos y expertos serían los responsables de esta evaluación. La evaluación externa, en cambio, estaría condicionada por la competencia y por el " imperativo de la innovación ". Aquí prevalecerían los criterios

3. Ver : N. Rosenberg, " The impact of technological innovation ", in N. Rosenberg, R. Landau, *The Positive Sum Strategy*, Washington, National Academy Press, 1986.

4. M. Quintanilla, *Tecnología : un enfoque filosófico,* Buenos Aires, Eudeba, 1991.

económicos, políticos, sociales y culturales fijados por cada sociedad, para medir el impacto y la idoneidad de una tecnología.

Esta distinción entre criterios internos y externos podría parangonarse con la diferenciación entre " tecnología en general " y " tecnología como capital ", que establecen los marxistas. En ambas interpretaciones la eficiencia se evalúa con parámetros objetivos e independientes de las metas del capital, mientras que la evaluación externa está sujeta al principio de la maximización del beneficio. Quintanilla incluso acepta la primacía de la evaluación externa sobre la interna en la dinámica del cambio tecnológico, pero a diferencia de los marxistas no reconoce en las leyes del capital las normas definitorias de este predominio externo.

Pero ademas, al presentar a la " sociedad " como una entidad homogénea que fijaría las reglas de la tecnología, Quintanilla se olvida que las clases sociales no resuelven en común la introducción o el descarte de las innovaciones. El mismo error cometen los culturalistas, cuando definen los criterios políticos o culturales que adoptaría la " sociedad " para configurar a la tecnología. Ignoran que las decisiones de innovación las adoptan solamente los capitalistas, guiados por el principio de la valorización del capital. Para entender este fenómeno se requiere analizar la tecnología como una fuerza productiva social.

Con este concepto se pue de comprender además, la existencia de un conflicto estructural entre la " eficiencia técnica " y la " evaluación social externa ". En lugar de una armonización espontánea entre ambos rasgos, la dinámica de la acumulación produce un choque entre la optimización técnica y la maximización del beneficio, que genera la sobreproducción y el sub-empleo de los recursos en el capitalismo.

Tecnología y técnica

Los autores racionalistas afirman correctamente que un rasgo distintivo de la tecnología es su carácter científico. Sitúan en esta característica la diferencia con la técnica, que se desenvuelve con métodos pre-industriales a través de los oficios[5]. La tecnología, en tanto sistema de acciones orientados a transformar objetos en forma eficiente, exige un grado de conocimiento organizado muy superior a la destreza técnica. Requiere no solo " saber hacer ", sino también " saber como hacer ". Este aprendizaje se logra con la utilización de los conocimientos científicos, que no estaban disponibles cuando prevalecía el uso de la técnica empírica. La tecnología es " la técnica que pasa por la ciencia, se asocia al laboratorio y se utiliza en la fábrica "[6].

5. Ver : J.J. Saldañas, " Historia de la ciencia y la tecnología. Aspectos metodológicos ", en E. Martinez, *Ciencia, tecnología, y desarrollo*, Caracas, Nueva Sociedad, 1994.

6. J.-J. Salomon, *Le destin technologique,* Gallimard, 1992, cap. 3 y 11.

Históricamente la tecnología surgió de la técnica con la profesionalización del ingeniero y el abandono de la actividad artesanal. La sustitución de una destreza subjetiva por una labor formalizada marca el punto de ruptura entre una y otra disciplina[7]. Pero a esta caracterización debe añadirse el sometimiento de la tecnología a las reglas del capital. Este dato es sustancial.

La subordinación a los criterios de la rentabilidad se trasladó a la enseñanza técnica y bajo este nuevo principio se desarrolló la sistematización, la escolarización y la transmisión de los nuevos conocimientos. Ingeniería, management y finanzas comenzaron a estudiarse conjuntamente en la formación de los " ingenieros de las cosas y de los hombres ", que actuarían como directivos de empresa. El enfoque racionalista acepta que la tecnología es un conocimiento científico, pero ignora que su aplicación a la producción se desenvuelve bajo el comando del capital.

Al definir a la tecnología como una fuerza productiva social se evita otro tipo de error, propio del enfoque sistémico, que extrapola las características contemporáneas del cambio tecnológico a cualquier modo de producción. Se presenta en este caso a la técnica y a la tecnología como equivalentes. Para Gille[8] por ejemplo, la noción " sistema técnico ", sería válido para la antiguedad, el medioevo, o el capitalismo. Como no establece ninguna diferencia cualitativa entre la " concatenación de estructuras y conjuntos técnicos " de los " sistemas " egipcio-mesopotámico, clásico-renacentista o industrial y contemporéneo, no se entiende cual es la lógica que guiaría el pasaje de un modelo a otro. Los sistémicos simplemente constatan que el agotamiento de un régimen favorece su reemplazo por el siguiente.

Pero esta simple contrastación entre " sistemas " no permite deducir las peculiaridades de la innovación. La tecnología se explica por el componente científico, indicado por los racionalistas y por la influencia del proceso de valorización, señalada por los marxistas.

Solo distinguiendo a la tecnología de la técnica se puede ademas evitar el típico uso ahistórico de ambas nociones, que hacen los economistas neoclásicos. Especialmente cuando hablan del " progreso técnico ", parecen referirse a acontecimientos que navegan fuera del tiempo y del espacio. A diferencia de la innovación, el " progreso técnico " es una abstracción, concebida para modelos imaginarios y carentes de contenido empírico.

También los autores culturalistas[9] que analizan los impulsos psicológicos, lúdicos o irracionales, que subyacen en el proceso innovador, tienden a identi-

7. Ver : A. Picon, " Le dynamisme des techniques ", *L'Empire des techniques*, Paris, Seuil, 1994.

8. B. Gille, " Science et technique ", dans B. Gille, *Histoire des Techniques et Civilisations*, Paris, 1978.

9. Ver : B. Jacomy, " Le faire savoir des savoir faire ", *L'Empire des techniques*, Paris, Seuil, 1994.

ficar técnica con tecnología. Se internan en un campo de gran interés para la sociología de la tecnología, pero sin ubicar histórica y socialmente al cambio tecnológico. La subjetividad de los innovadores debe encuadrarse en los dos parámetros que distinguen a la tecnología de la técnica : procedimientos científicos y leyes del capital. Para esta contextualización, la noción " fuerza productiva social " resulta irremplazable.

Ciencia y tecnología

El análisis de la tecnología como un campo de conocimientos, que estudia " el diseño de los artefactos, el planeamiento de su realización, así como su operación, ajuste, mantenimiento y monitoreo, mediante la aplicación del conocimiento científico ", es relativamente reciente[10].

En las visiones tradicionales de los años 50 y 60 se consideraba a la tecnología como una simple aplicación de la ciencia, carente de una dimensión intelectual propia, relevante y separada de la ciencia. En gran medida las interpretaciones culturalistas pusieron a fin a esta desjerarquización de la tecnología[11]. Rechazaron el " modelo unilineal ", que atribuía total primacía a los descubrimientos en relación a las aplicaciones y demostraron que esta última secuencia, no es de ninguna manera única ni predominante. Por el contrario prevalece una interacción entre las preguntas teóricas que formula la ciencia y las soluciones prácticas que encuentra la tecnología. La ciencia y la tecnología constituyen dos ámbitos separados por instituciones, reglas y tipos de conocimientos, que se influyen mutuamente sin preeminencia de uno sobre otro. Esta diferencia se expresa en la aparición de una sociología y una filosofía de la tecnología delimitada de sus equivalentes tradicionales en la ciencia.

En la tecnología se estudia cómo y porqué se desarrollan objetos útiles con finalidades prácticas, mientras que en la ciencia se analizan diversas teorías con el objetivo de alcanzar la verdad. Existe una diferencia de propósitos entre " conocer por conocer y conocer para hacer ". A la tecnología le interesa la aplicabilidad y se desenvuelve por medio de la creación de artefactos, mientras que la ciencia se desarrolla a través de la publicación de artículos. Por eso Price[12] dice que la primera es " papirofóbica " y la segunda es " papirocéntrica ".

Particularmente los culturalistas se han preocupado por subrayar estas diferencias, batallando contra la idea que la tecnología constituye una simple aplicación de la ciencia.

10. Ver : M. Bunge, *Treatise on basic philosophy,* Dordrecht, D. Reidel Publishing, 1985, vol. 7, part 2, chap. 5.

11. Ver : R. Laudan, " Natural alliance forced marriage ? ", *Technology and Culture,* vol. 36, 2 (1995).

12. D.S. Price, " The difference between science and technology ", *Science beyond Babylon,* Yale University Press, 1975.

Algunos autores como Perrin[13], destacan incluso que el conocimiento teórico depende de los instrumentos disponibles y que la investigación científica está condicionada por desarrollo de los objetos. En la misma linea, Basalla[14] reivindica el papel del pensamiento visual sobre el verbal y la gravitación de la acción práctica sobre las intuiciones teóricas.

En el enfoque marxista estas características distintivas de la tecnología se explican por su mayor grado de conexión con las leyes del capital. La ciencia es también una fuerza productiva, pero más indirecta y menos configurada por los requerimientos sociales de la acumulación. Por eso mantiene mayor autonomía de las exigencias inmediatas del proceso de valorización. Lo que distingue la actividad de preguntarse por la validez de una teoría de la acción de construir un prototipo, es la mayor influencia del principio de rentabilidad sobre esta última labor.

Tradicionalmente se ha destacado que la ciencia pura estudia las propiedades de un fenómeno, la ciencia aplicada los transforma en objetivos humanos y la tecnología concreta su aplicación productiva. En esta división de tareas también resulta evidente que las reglas de la competencia y el mercado se refuerzan cualitativamente en los últimos eslabones de la cadena[15].

Sin embargo, la creciente mercantilización actual de la ciencia pone un legítimo signo de interrogación en torno a la distinción tradicional entre ciencia y tecnología. Lander[16] por ejemplo afirma que con la consolidación de los laboratorios en las grandes compañías, la privatización de la universidad y el estricto control comercial de las patentes, la ciencia asume objetivos utilitarios, mientras que la tecnología se desliza hacia problemas más teóricos. Vessuri[17] considera que esta " cientifización de la tecnología e industrialización de la ciencia " tiende a reunificar la ciencia pura con la tecnología industrial.

La fusion actual de las " tecno-ciencias " es muy visible en algunas ramas, como la ingeniería química o la biología molecular y en ciertas industrias, como la farmaceútica, en dónde las reglas de costo-beneficio gobiernan todas las etapas de la investigación. Pero este tipo de convergencia no se ha generalizado a todos los sectores, ni se ha consumado en todas las industrias. El surgimiento de nuevas disciplinas téoricas recrea además, la diferenciación entre ciencia y tecnología. Lo que fusiona el laboratorio se vuelve a desdoblar en la investigación ulterior. Aunque la linea demarcatoria entre ciencia y tecnología

13. J. Perrin, " Pour une culture technique ", *L'Empire des techniques*, Paris, Seuil, 1994.

14. G. Basalla, *Evolución de la tecnología,* Crítica, 1991.

15. Ver : J. Feibleman, " Pure science, applied science and technology. An attempt an definition ", in C. Mitcham, R. Mackey, *Philosophy and technology. Readings in the philosophical problems of technology*, New York, London, The Free Press, 1983.

16. E. Lander, *La ciencia y la tecnología como asuntos políticos*, Caracas, Nueva Sociedad, 1994 (Cap. 5).

17. H. Vessuri, " Distancias y convergencias en el desarrollo de la ciencia y la tecnología ", en C.A. Di Prisco, E. Wagner, *Visiones de la ciencia*, Caracas, Monte Avila, 1992.

se ha vuelto más borrosa, la delimitación continúa siendo útil para analizar dos tipos singulares de actividad. Numerosos procesos científicos deben ser actualmente incluidos dentro del concepto " fuerza productiva social ", porque han quedado sometidos a las leyes del capital de manera equiparable a la tecnología. Pero en términos generales, continua siendo válida la diferenciación conceptual entre las dos disciplinas.

El criterio delimitativo que postula el marxismo es el grado de dependencia del proceso de valorización. Este parámetro es el único que permite evitar clasificaciones abstractas y servir al entendimiento del cambio tecnológico.

Ideología y cultura

Caracterizar a la tecnología como fuerza productiva social permite entender su especificidad en relación a la técnica y a la ciencia. Pero además, sirve para explicar exactamente en qué sentido la tecnología es un fenómeno social, al poner de relieve como influye en su configuración concreta la ideología de la clase dominante y la dinámica de la lucha de clases.

La tecnología no es un instrumento neutral del progreso. Los capitalistas la utilizan para maximizar sus beneficios, imponiendo incluso este principio en el propio diseño de los artefactos. Como fuerza productiva social, la tecnología tiene parcialmente incorporados los valores y las normas ideológicas de las clases dominantes.

Mientras que el rasgo universal de la " tecnología en general " está a la vista en un martillo, en un torno, o en una computadora, el componente social de la " tecnología como capital " se evidencia en la cadena de montaje taylorista, en las innovaciones orientadas a descalificar al trabajador o en la reciente " flexibilización " del trabajo, dirigida a aumentar la competitividad de una economía mediante el aumento de la tasa de plusvalía. Las relaciones de producción capitalistas están contenidas en todos los aspectos de la tecnología que contribuyen al aumento de la explotación. Aquí aparece nítidamente el carácter de clase que tiene el uso de las innovaciones y también la ideología que justifica esta instrumentación.

La precencia de normas sociales y culturales en la tecnología es un hecho aceptado por los analistas de la innovación. Pero el problema radica en precisar cuales son los patrones determinantes de estos valores. Para el marxismo la respuesta esta en los intereses materiales de las clases capitalistas, que impone la adaptación de la tecnología a los principios de rentabilidad, mercado y explotación. Una expresión típica de esta ideología es la creencia que el uso de una tecnología se optimiza asegurando las mayores ganancias.

Prestar atención a la raíz social del uso de tecnología, no es de ninguna manera incompatible con reconocimiento de otras características culturales de innovación, que están desligadas total o parcialmente de esta función clasista. El marxismo no excluye, ni desvaloriza esta dimensión, sino que subraya la primacía de la utilización capitalista en la explicación de las formas que adopta la innovación. Sin lugar a dudas existen culturas tecnológicas nacionales, regionales, sectoriales y particulares. Esta realidad no está en debate. El problema es si el país, la comunidad, o la empresa, constituyen puntos de partida más satisfactorios para el análisis social de la tecnología, que las necesidades y los intereses de las clases dominantes. El marxismo no descarta la primera alternativa, pero jerarquiza a la segunda.

Los estudios recientes de la " tecnología como una cultura " han servido para ilustrar como se emparentan los particularismos culturales con los tipos de innovación. Por ejemplo en las investigaciones de Pacey[18] o de los autores de la revista *Technology and Culture*[19], se describen los fundamentos culturales de ciertas habilidades, formas de organización laboral, o tipos de maquinarias. Pero estos retratos solo explican parcialmente el origen de los éxitos y los fracasos tecnológicos y no dan cuenta debidamente del propósito primordial de la innovación. Solo restituyendo el contenido de clase al análisis de las innovaciones puede alcanzarse una comprensión global del cambio tecnológico.

Los patrones de novedad y selección tecnológica nunca son primordialmente culturales, como cree Basalla. Estos criterios están antecedidos y condicionados por los intereses materiales de las clases sociales. Constatar la influencia decisiva que ejerce la pertenencia a una u otra clase, no supone " ningun reduccionismo ". El lugar que ocupa en la producción cada agrupemiento social es determinante de la concepción que detenta de la tecnología.

El materialismo histórico interpreta las culturas tecnológicas partiendo de la categoría modo de producción. Esta noción ordena históricamente y explica lógicamente, las distintas normas y valores que configuran a la tecnología.

Pero el rasgo decisivo que se puede analizar por medio de la categoría modo de producción es la explotación del trabajo asalariado y esta característica queda totalmente diluida en el enfoque culturalista. Entre consideraciones sobre la " vitalidad innovadora " de una cultura o la " ausencia de ímpetu tecnológico " en otra comunidad, la explotación desaparece o queda relegada a un lugar secundario. Esta omisión significa dejar de lado, la principal motivación y finalidad de la innovación en el capitalismo.

A diferencia del culturalismo, que desestima la importancia del contexto económico para subrayar la relevancia de la cultura, el marxismo no encuentra ninguna oposición entre ambas explicaciones. Partiendo del concepto modo de producción capitalista esta antinomia desaparece, ya que la propia categoría establece una correspondencia entre los valores ideológicos preponderantes y la dominación de las clases sociales que los propaga. El modo de producción conceptualiza ambas características, evitando un dilema ocioso en torno a la

18. A. Pacey, *El laberinto de ingenio,* Barcelona, Gilli, 1980.

19. Ver reseña general en : Hector Ciapuscio, *El fuego de Prometeo,* Bueno Aires, Eudeba, 1994.

primacía de las normas culturales o el mercado, en la determinación del cambio tecnológico. La burguesía ejerce dominación ideológica del proceso de innovación porque es propietaria material de los recursos tecnológicos. Esta conclusión surge de un enfoque unitario, superador de la dicotomía culturalista.

El significado social de la innovación

Para el marxismo el cambio tecnológico es un fenómeno social, porque depende de las leyes del capitalismo y porque se desarrolla en torno de los conflictos entre clases sociales. Un estudio social de la innovación que desconozca estas confrontaciones es un contrasentido.

Marx[20] inaguró la interpretación social del cambio tecnológico, describiendo el efecto del maquinismo sobre la opresión de los trabajadores. Los marxistas han continuado este enfoque, analizando el efecto opresivo del aumento de la productividad en diversos períodos, circunstancias y países. En la actualidad esta visión es particularmente pertinente para interpretar la generalización del desempleo o el incremento de la explotación, a escala internacional.

La valorización del capital induce formas muy diversas de expropiación por medio del cambio tecnológico. La extracción de plusvalía en la fábrica continúa siendo la principal característica de este proceso. Pero la diferenciación entre empresarios e inventores añade un tipo de confiscación, particular del proceso innovador. La fractura entre los creadores que descubren y los aplicadores que " rentabilizan " una tecnología, dá lugar a una apropiación de rentas tecnológicas en favor de los aplicadores y en desmedro de los creadores.

La división de tareas entre hombres dotados para la creación de nuevas tecnología y personalidades capacitadas para organizar su aplicación introdujo una diferenciación central en la historia de la tecnología, desde el contraste inicial entre Watt, Boulton y Arwright. Mientras que Watt concibió la máquina de vapor desinteresándose de su comercialización, Boulton se ocupó de negociarla y patentarla. Arwright convirtió posteriormente este instrumento en un medio de explotación fabril[21]. Esta fractura entre inventores e innovadores no responde solamente a dos tipos psicológicos diferentes, ni a dos funciones complementarias del proceso económico, que Schumpeter creía indispensables para el buen funcionamiento y continuidad del capitalismo. Se trata de una forma de consumar la extracción, transferencia y realización de plusvalía.

En el capitalismo monopólico la confiscación estructural de conocimientos se institucionaliza, a través de la absorción del inventor independiente por los

20. C. Marx, *El Capital*, México, Fondo de Cultura Económica, 1973, t. 1, sección 4ta.

21. Ver : F.M. Scherer, " Invención e innovación en la aventura de la máquina de vapor Watt-Boulton ", en M. Kranzberg, W. Davenport, *Tecnología y Cultura*, Barcelona, Gili, 1978.

laboratorios de las grandes compañías. Siguiendo la secuencia que privó de la tierra al campesino durante la acumulación primitiva, que degradó la calificación del artesano en la revolución industrial, y que quitó al obrero taylorista su control del proceso de trabajo durante la segunda revolución tecnológica ; el capital conforma un " proletariedo científico " que pasa a formar parte de la masa de trabajadores expropiados. Convierte la labor de los trabajadores calificados en una nueva fuente de plusvalía, y somete la aplicación de sus conocimientos a las necesidades de lucro de las grandes compañías.

Noble[22] describe como esta confiscación se consumó en Estados Unidos al comienzo de la época monopólica, a través del sistema de patentes, que impuso una apropiación de los derechos de los empleados sobre cualquier innovación realizada en la empresa. Este avasallamiento se convirtió a partir de ese momento en expropiación organizada de los asalariados científicos y técnicos.

Las interpretaciones no marxistas eluden estudiar el antagonismo creado dentro de la empresa con la generalización del un " proletariado científico ". Estas escuelas consideran que estudiar el cambio tecnológico como un fenómeno social, consiste en sustituir el análisis " trascendentalista de la innovación " (un acto de inspiración personal del inventor) por la descripción de las condiciones histórico-sociales, que rodean al cambio tecnológico. Pero no basta desmistificar las " simplificaciones románticas " para entender el fenómeno. Se necesita conceptualizar cuales son los conflictos sociales, que tienen como denominador común al cambio tecnolúgico y este análisis involucra ante todo a la lucha de clases.

La comprensión social del proceso innovador requiere entender tres tipos de conflictos. La oposición entre capitalistas y trabajadores frente al aumento de la tasa de explotación que acompaña el cambio tecnológico, el choque entre empresarios rivales en el mercado por la apropiación de la renta tecnológica ; y la contradicción entre el beneficio y necesidades sociales, que opone a los capitalistas y consumidores como consencuencia de la subordinación de la utilidad social de los nuevos productos al imperativo de la ganancia.

La innovación dá lugar a una gran variedad de confrontaciones, que atraviesa a numerosos grupos y sub-grupos. Esta diversidad solo pue de conceptualizarse satisfactoriamente con el análisis de las clases sociales. Entre ellas se dirigen los choques y se dilucida el efecto que tendrá el uso de las nuevas tecnologías. La contraposición de intereses sociales es un sub-producto visible de la innovación, en todas las esferas de la reproducción del capital.

Estudiar el cambio tecnológico a través de la lucha de clases es una alternativa opuesta al análisis corriente del " impacto social de la tecnología ". En lugar de indagar la adaptación de los comportamientos sociales al uso de nuevos artefactos, el marxismo parte de los conflictos existentes en la producción,

22. D. Noble, *El diseno de Estados Unidos,* Madrid, Ministerio de Trabajo y Seguridad Social, 1967.

sin aceptar ningún principio de armonización de la vida social a las exigencias de la tecnología.

Partiendo de la lucha de clases se puede interpretar con realismo cuales son los intereses en juego en la " configuración de la tecnología ". Pero el mismo fenómeno estudiado como una " negociación entre actores relevantes ", pierde toda relevancia. Es que el término social tiene para el marxismo un significado totalmente distinto, al asignado por los autores deterministas, constructivistas y sistémicos. Estas corrientes discuten si " tecnología impacta a la sociedad ", si por el contrario la " sociedad configura a la tecnología ", o si prevalece un proceso de " co-evolución " entre ambo fenómenos[23]. La total indefinición del término " social " en este debate dificulta la comprensión del problema.

Lo " social " es para el determinismo sinónimo de un cambio de hábitos derivado del surgimiento de nuevos artefactos (la radio es el ejemplo clásico). Para el constructivismo revelaría la influencia que tuvieron distintos grupos en la conformación final del objeto (la bicicleta es un ejemplo típico). En un caso lo " social " alude a transformaciones humanas derivadas de la tecnología y en el otro a la influencia que tienen los distintos grupos en la forma que asume un objeto. La visión sistémica por su parte, intenta presentar una " co-evolución " conciliatoria de las dos interpretaciones anteriores. En ninguno de estos casos, lo " social " hace referencia a las normas de funcionamiento del régimen social que define al cambio tecnológico, ni tampoco sirve para evalúar los conflictos de este sistema. Sin leyes del capital y sin lucha de clases, los estudios empíricos de estas escuelas no permiten una comprensión del proceso innovador.

Un proyecto emancipatorio

La dinámica del cambio tecnológico hay que explicarla partiendo de movimiento contradictorio que imponen las leyes del capital al proceso innovador. La competencia obliga a introducir nuevas tecnologías, pero la rivalidad por la obtención del máximo beneficio termina generando sobreproducción y crisis. Se trata de un conflicto específico del capitalismo, que domina en la tecnología contemporánea. La ley del valor opera como un movimiento dual de revolución y asfixia periódica de las fuerzas productivas.

Esta interpretación es muy diferente a la imagen racionalista del " progreso técnico ", como una evolución ascendente de la humanidad, garantizada por la simple aplicación productiva de los descubrimientos científicos. Afirmar que la innovación está simultáneamente sujeta al desarrollo y a la crisis implica definir un patrón de comportamiento objetivo del cambio tecnológico, que no depende de las " normas y los valores " culturalistas. Esta dinámica es por otra parte totalmente independiente de " la negociación para configurar cada artefacto ", que indagan los constructivistas, cuando estudian puntualmente

23. Ver : D. Vinck, *Sociologie des sciences,* Paris, Armand Colin, 1995, cap. 6, 7.

cada innovación, desconociendo las tendencias generales del proceso de valorización.

Para el marxismo los efectos nocivos del cambio tecnológico no tienen causas transhistóricas, ni se originan en la naturaleza del hombre, o en la rebelión de su alma, como creen los espiritualistas. Es más útil indagar las raíces sociales de las " desgracias tecnológicas " que buscar en el pesimismo conservador una respuesta al optimismo acrítico de los apologistas de la tecnología.

El marxismo se opone a las tres variantes más corrientes del determinismo tecnológico : considerar que la innovación configura fatalmente el comportamiento social (Ogburn), suponer que la tecnología se " autodirige " luego de haber escapado al control humano (Ellul), y creer en la " supremacía del conocimiento " sobre los distintos tipos de organización social (Bell).

Las categorías fuerzas productivas y relaciones de producción dan cuenta de los condicionamientos históricosociales, que limitan el alcance del cambio tecnológico a un radio de posibilidades. Pero estas nociones establecen claramente que son los hombres, estructurados en torno a las clases sociales y en acuerdo a sus intereses y convicciones, quienes definen siempre la dirección del proceso innovador. Este curso no viene pre-establecido por determinaciones tecnológicas, sino que es un resultado de las confrontaciones sociales.

Una vez reconocida la dinámica objetiva del cambio tecnológico es posible formular propuestas emancipatorias, para alcanzar una utilización socialmente provechosa de la tecnología. El marxismo recoge la rica tradición de innovadores que concibieron sus creaciones sin finalidad de lucro, buscando favorecer el bienestar general y contribuir a la emancipación social. " Inventar para cooperar, no para competir " es el principio que históricamente impulsó la conducta de numerosos innovadores utopistas, cooperativistas, románticos, sindicalistas, y por supuesto socialistas y verdaderos comunistas.

Esta misma actitud también caracteriza la acción espontánea de la mayoría de los técnicos, ingenieros y científicos, que inventan bajo el impulso de la curiosidad y la propensión natural a mejorar las formas de trabajo[24]. Este impulso no mercantil al cambio tecnológico, que apunta al logro de mayores satisfacciones en la vida social y laboral, ha estimulado además la búsqueda contemporánea de tecnologías alternativas a las dominantes bajo el capitalismo.

En el terreno del cambio tecnológico el proyecto socialista se nutre tanto de las tradiciones de resistencia a la explotación, como de las propuestas de emancipacion social. La defensa de los derechos de los trabajadores contra las " flexibilizaciones laborales " y los proyectos de un nuevo uso de la tecnología están indisolublemente ligados. Constituyen dos momentos de un mismo pro-

24. Además del mencionado texto de Noble, ver J. Christie, " Morris Cooke and energy of America ", in C. Pursell, *Technology in America*, Massachusetts, MIT Press, 1981.

ceso de ruptura del desenvolvimiento tecnológico, con las exisencias de la ganancia. Otro tipo de sociedad y de tecnología pueden erigirse en base a este principio de emancipación de la tiranía del beneficio.

La mirada desde el observatorio

La tecnología ha comenzado a estudiarse como una disciplina de las ciencias sociales, situada en el límite con las ciencias naturales. No es asimilable a la ingeniería, a la física o a la química, pero depende significativamente de los conocimientos de estas y otras disciplinas. La estrecha relación con las ciencias naturales diferencia a la tecnología de las ciencias sociales clásicas, como la economía, la sociología o la política.

La tecnología es un área de investigación rodeada por disciplinas " duras " y " blandas ". Se asienta en las matematicas, se nutre de las humanidades e influye sobre el arte. Se encuentra en la " frontera caliente " de las ciencias sociales con las ciencias naturales. Recibe criterios de análisis a la economía o la sociología, pero no está totalmente regida por este tipo de principios.

La definición marxista de la tecnología como una " fuerza productiva social " tiene en cuenta esta determinación múltiple y esta diversidad de vasos comunicantes de la tecnología y otras ramas del saber. La distinción entre " tecnología en general " y " tecnología como capital " permite justamente a delimitar, los aspectos universales de la tecnología derivados de su cercanía con las ciencias naturales, de la innovación como proceso social, interpretable con los parámetros de las ciencias sociales.

Algunos autores culturalistas conceptualizan este doble carácter distinguiendo a la " tecnología en un sentido restringido " de la " tecnología en un sentido abarcativo ".

Para Pacey[25] se trata de una situación equivalente a la medicina, en dónde también resulta conveniente discriminar entre la " ciencia médica universal " y la " práctica médica específica ". La primera sintetiza principios de validez general, mientras que la segunda se refiere a aplicaciones de diverso tipo, dependientes de condiciones organizativas y culturales particulares.

Pacey introduce esta clasificación para realzar el aspecto humano de la tecnología, que a su juicio está totalmente relegado en los enfoques racionalistas y deterministas. Describe como en la práctica tecnológica se manifiestan las normas culturales de cada sociedad. Las categorías marxistas " tecnología en general " y " tecnología como capital " cumplen una función semejante, con la ventaja de situar estas determinaciones socio-históricas en las leyes del capital. A su vez el concepto " modo de producción " permite observar, no solo las relaciones funcionales de la tecnología con los otros campos del pensamiento,

25. A. Pacey, *La cultura de la tecnología,* México, Fondo de Cultura Económica, 1990 (Cap. 1 y 2).

sino también la forma precisa en que el capitalismo desenvuelve estas conexiones.

La capacidad del marxisimo para comprender a la tecnología surge de la aplicación de estos instrumentos analíticos a los estudios empíricos. Pero este entendimiento es a su vez posible, por la opción social que asume esta concepción en defensa de los intereses de los trabajadores. El marxismo busca desligarse explícita y concientemente de las conveniencias de las clases dominantes y esta decisión le permite situarse en una ubicación privilegiada del conocimiento de la tecnología. Puede jerarquizar el problema de la explotación y situar las contradicciones de la innovación en el marco de la transitoriedad histórica del capitalismo. Puede adoptar una actitud crítica y desmistificadora del fetichismo tecnológico. En un observatorio que captara todo el escenario de la innovación, los marxistas estarían ubicados en los pisos superiores del mirador, como resultado directo de este posicionamiento en el campo de los oprimidos[26].

Por el contrario, las limitaciones cognitivas de otros enfoques derivan de la influencia que reciben de los puntos de vista de las clases dominantes. Esta miopía es un resultado del auto-ocultamiento de la dura realidad de la explotación. Los problemas que no están dentro del radio de observación de la ideología burguesa, quedan también habitualmente fuera de la visión constructivista, sistémica, espiritualista, determinista, racionalista o culturalista.

Todas los enfoques de la innovación reflejan concepciones generales del mundo, que se corresponden con intereses sociales particulares y puntos de vista de clase. No existen teorías " puras ", experimentales y exclusivamente observables de la tecnología, independientes del mundo circundante. Detrás de la defensa irrestricta del " progreso técnico " por parte de los neoclásicos y los racionalistas neoliberales subyace el apuntalamiento de la política privatizadora y " desreguladora " de los salarios, para beneficiar a los grandes corporaciones internacionalizadas. La argumentación evolucionista, keynesiana y culturalista en favor de políticas tecnológicas nacionales o sectoriales, expresa frecuentemente a los grupos capitalistas afectados por esta " globalizacion ".

Cada concepción expresa intereses sociales, plantea preguntas y propone temas de investigación que se corresponden con la " mirada " de distintos sectores sociales. En estas formulaciones aparece siempre la " conciencia posible " de las clases, en decir el recorte de la realidad que espontáneamente tiende a realizar cada grupo social. Las teorías no marxistas están situadas en los niveles inferiores del observatorio de la tecnología por su atadura social y este es el condicionante básico de sus limitaciones para acceder al paisaje total del proceso innovador.

26. Ver el uso de esta metáfora en : M. Lowy, *Qué es la sociología del conocimiento ?*, México, Fontamara, 1991.

Pero es igualmente cierto que una posición más favorable en este mirador no asegura un mejor retrato de lo que se está observando. De la misma forma que la pintura depende de las cualidades del pintor, la acertada interpretación del cambio tecnológico también deriva de la capacidad personal o grupal de los investigadores. Es un resultado de su conocimiento, sensibilidad y amplitud intelectual, con relativa independencia de la posición ideológica que se adopte.

Desde cualquier escalón del observatorio se pueden formular interpretaciones valiosas o inútiles, contribuciones críticas o apologías irrelevantes. La distinción entre pensamiento científico y vulgar, que trazó Marx hace 100 años, está vigente para el análisis de la tecnología. El mieso abismo que separaba a Ricardo de Say en el campo de la economía política, divide actualmente aguas en la sociología y en la filosofía de la innovación.

El marxismo solo brinda una plataforma más favorable para encarar la investigación. Pero su desarrollo depende en gran parte del contacto, la asimilación y la re-elaboración de los puntos de vista opuestos. Este intercambio es el basamento de los " distintos marxismos " en las teorías de la innovación, que estudiaremos en un próximo artículo.

Bibliografía adicional

D. Bell, *El advenimiento de la sociedad pos-industrial,* Madrid, Alianza, 1976.

J. Ben David, " El empresario científico y la utilización de la investigación ", en B. Barnes, *Estudios sobre sociología de la ciencia,* Madrid, Alianza, 1972.

W. Bijker, T. Pinch, " The social construction of facts and artifacs ", in W. Bijker, T. Highes, T. Pinch, *The social construction of technological systems,* Massachusetts, MIT, 1989.

M. Callon, B. Latour, *La science telle qu'elle se fait,* Introduction, Paris, La Découverte, 1991.

J. Elull, *Recherche pour une éthique dans une société technicienne,* Bruxelles, Editions de l'Université de Bruxelles, 1983.

B. Latour, *Nous n'avons jamais été modernes,* Paris, La Découverte, 1991.

E. Mandel, *El capitalismo tardío,* México, ERA, 1978.

L. Martinez San Martín, " Historia de la técnica " en J. San Martín, *Estudios sobre sociedad y tecnología*, Barcelona, Antrophos, 1992.

M. Medina, " Nuevas tecnologías, evaluación de la innovación ", en J. San Martín, *Estudios sobre sociedad y tecnología*, Barcelona, Antrophos, 1992.

C. Mitcham, *Qué es la filosofía de la teanología ?,* Barcelona, Antrophos, 1989.

M. Mulakay, " El crecimiento cultural de la ciencia ", en B. Barnes, *Estudios sobre sociología de la ciencia,* Madrid, Alianza, 1972.

W. Ogburn, M. Nimkoff, *Sociología*, Madrid, Aguilar, 1955 (cap. 26).

A. Piscitelli, *Ciencia en movimiento. La construcción social de los hechos científicos,* Buenos Aires, CEAL, 1993.

D.S. Price, " Ciencia y tecnología : distinciones e interrelaciones ", en B. Barnes, *Estudios sobre sociología de la ciencia,* Madrid, Alianza, 1992.

R. Richta, *La civilización en la encrucijada,* México, Siglo XXI, 1971.

J. San Martin, L. Luján José, " Educación en ciencia, tecnología y sociedad ", en J. San Martín, *Estudios sobre sociedad y tecnología*, Barcelona, Anthropos, 1992.

L. Winner, *Tecnología autónoma,* Barcelona, Gili, 1979.

PART THREE

Scientists, Soldiers and Manufacturers : The Dynamics of Military Technical Development

Sixteenth and Seventeenth Century Arms Production in Gipuzkoa[1]

Ignacio M. Carrion Arregui

At the end of the 15th century the territories of Gipuzkoa and Bizkaia became the main producers of portable arms for the Spanish monarchy, a role they played throughout the 16th and 17th centuries[2]. In this paper we are going to focus on how arms production was organised as well as on the characteristics of the parts that were made. Our source of information is documents on State funds (royal paymasters' accounts) and notaries' protocols[3].

The portable arms production sector was concentrated in a relatively small area and was highly specialised. In Bizkaia defensive weapons were manufactured in the Markina area, thrusting and cutting weapons in the Duranguesado area, and in Elorrio the speciality was spear handles. The area comprising Elgoibar-Ermua-Mondragón was known for the manufacture of harquebuses, and there was also a significant amount of thrusting and cutting weapons production in the Tolosa area[4]. With the exception of pikes, the flourishing arms production of the early 16th century in Bizkaia wound down as the century passed, whereas in Gipuzkoa the production of firearms became more intensified. Later, in addition to the manufacture of swords and knives in Tolosa, armour and other defensive weapons began to be crafted in the 1630s in the

1. This paper is part of an Investigation Project of the University of the Basque Country (UPV 156.130-HA058/95) and the Basque Government (GV 156.30-0030/95) about " Gipuzkoa in the 16th and 17th centuries ".

2. I.A.A. Thompson, *Guerra y decadencia. Gobierno y administración en la España de los Austrias, 1562-1620*, Barcelona, 1981, 289-313 ; D. Goodman, *Poder y penuria. Gobierno, tecnología y sociedad en la España de Felipe II*, Madrid, 1990, 127-135.

3. The accounts are in the Archivo de Simancas and the notaries' protocols Archivo Histórico de Protocolos de Gipuzkoa (AHPG) and in the Archivo General de Gipuzkoa (AGG).

4. J. Caro Baroja, *Introducción a la historia social y económica del pueblo vasco*, San Sebastián, 1974, 103-106 ; J. Garmendia, *Gremios, oficios y cofradías en el país vasco*, San Sebastián, 1979 ; R. Larrañaga, *Síntesis histórica de la armería vasca*, San Sebastián, 1981 ; J.A. Azpiazu, " Fabricación y comercialización de armas en el valle del Deba (1550-1600) ", *Cuadernos de Sección Historia- Geografía (Eusko Ikaskuntza)*, nº 22 (1994).

Royal Armoury of Tolosa (*Real Armería de Tolosa*). The arms produced were robust, made for the most part for the army, and lacked superfluous embellishments. We are not dealing, therefore, with valuable pieces that have ended up as part of important museum collections. Nevertheless, the workmanship of Basque artisans resulted in inventors coming to the area to direct the manufacture of prototypes. This was the case of Miguel de Vivedo, who came from the Royal Court to Placencia circa 1570 in order to head experiments in making his double-fused reinforced harquebuses[5].

Up until 1576 a significant number of weapons were made for private parties and exportation, with orders for a particular type of harquebus gunstock destined for Andalusia and Portugal, in addition to the model used by the monarchy. However, beginning in 1577, an official of the Court, the *veedor*, controlled weapons production and had to grant authorisation for any arms leaving the territory. Over time, this control grew so strict that eventually the King was virtually the armourers' only customer — at least officially. Nonetheless, the needs of the monarchy bolstered production and the sector was indeed capable of supplying the royal armies, absorbing in the last part of the 16th century some 25.000 Spanish ducats annually. This equalled 30 % of the arms allocation of the Spanish monarchy at that time[6], a figure which was to double in the 1530s. Some of the documentation we have on arms inspected shows, for instance, that between November of 1558 and March of 1560 Captain Zurita approved 7.305 harquebuses[7]. The royal paymasters' accounts reveal substantially fewer purchases, but clearly indicate growth in the sector, as can be seen below in Table 1. Let us point out that the average production of harquebuses and muskets grew twofold from 1570 to 1600, and doubled again between 1600 and 1640, at which time it reached its peak, coinciding with the war effort surrounding the Thirty Years War.

Table 1 : Annual average of hand-held weapons purchased by the King

	1568-74 7 years	1602-05 3 years	1618-23 4 years	1629-41 12 years	1685-90 6 years
Pikes and spears	2,926	2,127	9,404	6,531	2,83
Harquebuses	2,543	8,099	6,609	5,807	2,945
Muskets	757	2,124	2,878	7,862	1,487
Horse harquebuses	-	-	-	425	-
Pistols	-	-	-	995	997

AGS, CMC 3-2826-1 (Accounts from H. de Aguirre, 1568-74), 3-737 (Data from P. Fernández de Zaraa, 1602-5) and 2-720 (1648-23), 3-1352 (Accounts

5. AGS, CMC, 3ª, 2826-1, Accounts from H. de Aguirre (1570).

6. I.A.A. Thompson, *Guerra y decadencia ...*, *op. cit.*, 296.

7. AHPG, 1/3685, f. 28. The average is 5.464 harquebuses by year.

from L. Fernández de Zaraa, 1629-1641), 3-3335-17 (Accounts from F. de Pagola, 1687-1690).

Systems for purchasing weapons for the royal armies

Officials used different methods for procuring arms, depending on the time in history, the urgency of their mission, local possibilities and the type of armament. The first way was purchasing or ordering large quantities of completely assembled. Later, we find a second system in which the various parts of the different weapons were ordered directly from craftsmen who specialised in those parts. A third system consisted of setting up an industrial plant where weapons required by the monarchy were made directly. Complete weapons purchased in large quantities from only a few suppliers required abundant steady production destined for other markets, or for merchants to accumulate stock while awaiting a royal order. This means of acquiring arms was already common by the end of the 15th century and continued to be used during the reign of Charles V, with contracts or *asientos* (special trading agreements) for large consignments drawn up with Antón de Urquizu (1533) and Juan de Hermua (1543). In times of war, in addition to these purchases merchants' weapons stocks were also confiscated.

The growing military needs at end of the reign of Charles V forced paymasters to place a number of orders for specific amounts of complete weapons, allowing artisans more time to carry out their work. Later on, beginning with the reign of Philip II, craftsmen began to fill orders for particular parts, which were in turn handed over to other artisans, who would complete the weapon, in this way doing away with middlemen. This method was the one most commonly used in spear-making, but it was also frequent in purchasing firearms : one person would craft barrels, another would make the gunstocks, others would supply the locks and moulds, and still others would do the polishing, and so on. Although it may seem odd, it was reasonable to acquire firearms in this fashion as production was bolstered, quality was carefully controlled and the overall cost was reduced, in exchange for greater administrative expenses. This system called for armaments warehouses, giving rise to the founding in 1573 of the Royal Arms Factory of Placencia (*Real Fábrica de Armas de Placencia*), where arms and parts acquired from the local artisans were stored and tested. This way of organising production suggests something similar to the " putting-out system ", where the State or its ministers would take on the tasks of the merchant, placing orders and giving monetary advances so as to achieve more favourable manufacturing terms and conditions. This was repeated a number of times in the 16th century when the royal paymaster had funds available to him and would pay upon placing an order to a third party. However, towards the end of the century, the Crown's financial problems resulted in frequent late payments to suppliers by paymasters. At times these payments were even several years overdue, giving rise to a situation of " forced " savings, which brought about genuine complications[8].

Finally, in a few cases the King assumed the responsibility for organising a large workshop with salaried workers for manufacturing the weapons he deemed necessary. The first case we are aware of is the ball foundry in Eugui, Navarra, in the reign of Charles V. There were also gunpowder and bronze artillery factories in other regions, as well as the *Real Armería* of Eugui, which later moved to Gipuzkoa around 1630, bringing about the creation of the *Real Armería* of Tolosa. In this section we should also mention the cast-iron cannon foundries of Santander, built in the early 17th century, although they were made and exploited by private parties who enjoyed a monopoly[9]. It appears that when there was a shortage of skilled workers or difficulties in making certain parts with traditional technology, foreign experts were brought in. And when the demand was insufficient to maintain the installations needed for efficient production, plants were set up and paid for by the State.

DEFENSIVE WEAPONS PRODUCTION

Mass-production of defensive weapons in the Basque Country almost certainly began in Markina, Bizkaia, towards the end of the 15th century. The Catholic Monarchs promoted this emerging sector by sending master armourers — possibly from Milan — who were put in charge of making armoury for the royal cavalry (for the men of arms, the *continuos* and the guards of Castile)[10]. The armourers employed local labourers to work the waterwheel-driven hammers in the bloomeries (*ferrerías*). Here, the plating was prepared which would subsequently be crafted into armour by the master armourers. Due to a salary-related conflict, we have documentation on this activity, which must have been significant, as is seen in the Civil War in Castile of 1520[11]. Nevertheless, it seems that the activity soon got bogged down, although manufacturing was continued for some time[12]. At least up until the early 17th century it supplied helmets and cuirasses for local militias, met civil navigation needs and filled sporadic arms orders made by the royal paymasters at Placencia for cuirasses, morions and corslets for infantrymen.

8. AHPG, 1/3744 (1587), f. 27v. We can see that it was usual too in the payment of the Royal Treasury to the soldiers (G. Parker, *El ejército de Flandes y el camino español 1567-1659*, Madrid, 1985).

9. J. Alcalá-Zamora, *Historia de una empresa siderúrgica española : los altos hornos de Liérganes y la Cavada, 1622-1834*, Santander, 1974.

10. T. González, *Colección de cédulas, cartas-patentes, reales órdenes y otros documentos concernientes a las Provincias Vascongadas*, t. 1, Madrid, 1829-30, 304-305.

11. J.I. Tellechea, *Hernán Pérez de Yarza alcaide de Behobia,* San Sebastián, 1979, 135-140.

12. There are very few traces about this activity (J.J. Mugartegui, *La Villa de Marquina*, Bilbao, 1927 ; *Colección documental del Archivo Municipal de Marquina (1355-1516*), San Sebastián, 1989.

In the mid-16th century there were still *asientos* drawn up for equal number of harquebuses and morions, such as the one signed by Juan de Hermua, and in 1560 another one involving almost two thousand morions[13]. But later we find only a few orders made to armourers in places like Markina and Durango for a few hundred helmets. These orders were to be filled with relatively short deadlines, sometimes only one or two weeks, indicating that either there was continuous production or that the orders were placed before the document was written. The scant number of helmets requested compared to orders of pikes and harquebuses suggests that these helmets, and occasionally corslets, were made to complement the defensive weapons produced elsewhere. The orders for morions in Markina in 1558 and 1576 clearly state that they had to be cup-shaped and high-crested, and be equipped with an escutcheon in which to insert the plume. It was essential that they be made out of one single piece, with neither soldered joints nor slits ; that they be well-polished and finished ; made with *launas*, or metal plates tacked on pieces of leather on either side of the helmet to protect the ears ; and that they weigh three and a half pounds (1,6 kg). In some cases they came with gold adornments. The breastplate and back-piece of an infantry soldier's armour weighed sixteen and a half pounds with its corresponding buckles and straps, bringing the total weight of the corslet and morion to nineteen and a half pounds (9 kg). They were manufactured in bloomeries, where they had to be examined by the paymaster or one of his representatives[14]. It appears that this production was not terribly significant, since at the time large numbers of armours were imported from Milan[15].

In the reign of Philip IV defensive weapons began to be made in Tolosa. Arms production was given a notable impetus around 1630 when the *Real Armería* was moved from Eugui to Tolosa, whose 1645 inventory and accounts dating from 1652 to 1668 we have located[16]. The *Real Armería* was set up in a building located in the town centre beside the river Orio. This establishment had two functions : first of all it was a workshop where defensive weapons were manufactured (mainly breastplates, back-pieces, morions and sallets). But it also served, as did the Real Fabrica de Placencia, as a warehouse, where thrusting and cutting weapons crafted and purchased from local artisans were collected, tested and dispatched.

In the armoury a master craftsman along with some thirty salary-earning operators produced armour for cavalry (burgonets, breastplates and back-pieces, gauntlets, *etc.*), corslets for the infantry with simple or heavy-duty (musket-proof) breastplates and morions, and a few bucklers, shields, and so

13. AHPG, 1/3660 (1555), f. 4 , and 1/3665-3 (1560), sf.

14. AHPG, 1/3663, 3680 and 3681. One Castilian pound weighed 0,4601 kg.

15. I.A.A. Thompson, *Guerra y decadencia...*, *op. cit.*, 291-292

16. The inventory in AGG, Prot 1050, published by J. Garmendia, *Gremios...* The accounts of the Armoury are in AGS, CMC 3ª, 1914-7, 3227-3, 3241-2, 1792-1, 2003-9 , 2723, 1439 and 2232-2.

forth. Metal plating previously pounded out by the hammers at the local bloomeries were used. The parts would first be cold-worked in the workshop on large boughs or tree trunks with the instruments listed in the inventory, and hot worked on sandstone slabs or thick iron plates. Two hydraulic wheels moved pinwheel gears, which in turn moved both a sandstone wheel for grinding weapons and four walnut or willow wheels used for polishing. The steward's accounts itemised salaries, the amount of raw materials used (metal plating, iron, leather, canvas, steel, oils, tar, vinegar, firewood...) as well as the entirety of the establishment's expenses. The workday can be calculated at around 10 hours a day — an average of twenty-two days per month — and some 750 armours would have been made per year under normal circumstances. The *Real Armería* was given a fixed sum of 12.000 ducats annually for its operations — at least in 1652 — which was the main source of revenues for the factory (91 %), since private party orders accounted for a mere 1 %, the rest being extra income arising from concrete orders or from payments for transporting arms in times of emergency.

Pike manufacturing

The new military tactic which became widespread towards the end of the 15th century was the use of the pike, a long wooden pole made of ash with a small pointed iron head[17]. The ash tree, cultivated in dense plantations, has a long, straight, flexible trunk which could be made into a pike when the tree was 26 years old. In the Middle Ages the demand for handles resulted in extensive planting of ash trees in the area. Family homes and landowners often had ash tree.

The oldest references mention spears from Gipuzkoa, Bizkaia and the Señorío de Oñati, which were sent to Castile[18]. The ash tree plantations tended to be concentrated in the east and south-east of the Duranguesado area, and therefore pikes were made in the town of Elorrio. In 1575 it is estimated that between 2.000 and 3.000 weapons could be made monthly[19], that is, some 30.000 per year if payment was guaranteed to be regular and certain facilities were provided. In the *asientos* agreed upon during the middle of the reign of Philip II, handles were purchased in large quantities from groups of handle-makers who joined together in their commitment to deliver. They had to be made of nursery-grown ash (not wild), well-proportioned, neither thin nor pot-bellied and with no knots. They had to have been cut during the waning moon of January and were to be delivered planed, sanded, properly carved and

17. R. Quatrefages, *La revolución militar moderna, el crisol español*, Madrid, 1996, 101, 148-152.

18. The Reyes Católicos confirmed a privilege of Enrique II (1371) that says they do not have to pay alcabala for the shafts in Valladolid. *Colección documental de Elorrio (nº 16)*, 84.

19. A. Huarte, " Fabricas... Vizcaya ", *Euskalherriaren Alde*, XXII, (1927), 380-386.

greased. They had to measure between 24 and 26 *palmos* measuring a quarter-*vara* each (5 to 5,4 m)[20]. We can say very little about how they were made, although the following characteristics are mentioned : *de lina*, *almacenas*, *de cuesta sana*, and *de una y de dos cuestas*. Production had to be planned well in advance, as it was very important for the handle-makers to be able to find regular buyers for their products. The documents we found on pike shipments from Elorrio show that several thousand were made. Nonetheless, it was not easy for the Spanish regiments to obtain all the pikes they wanted.

There is clear evidence that ash poles were also made for other markets up until 1577 ; subsequently, this production is less documented. Weapon handles were exported, or so it seems, to Portugal and Andalusia. When the king clamped down on control, shipments of short handles or handles rejected by the royal paymasters were authorised to be sent to other places, with the obligation of justifying the final destination. Therefore, ash wood handles continued to be shipped elsewhere to be used for lances — or for olive harvesting — although some exporters ended up in jail when they were not able to justify the resting place for their crates of handles[21].

The pikes had an iron head with a well-tempered steel edge measuring half a *vara* and one eighth (52 cm). There were two welded lateral reinforcements (*correas*) with holes to attach it to the handle with four studs. In addition, there was an iron pin and an iron tip which served as a ferrule to protect the opposed end. The iron pieces, rings, tips and pins in the pikes were supplied by artisans from Durango and outlying areas, who committed themselves to making several thousand pieces in very little time[22]. These heads and complementary pieces amounted to only a fifth of the average overall value of the pike.

Portable firearm production

The generic name for portable firearms is harquebus — as Covarrubias indicated *circa* 1611. Smaller harquebuses are also known as pistolets, and reinforced, or the longer and bulkier ones, muskets[23]. The first references we have dating from the end of the 15th century mention spingards[24]. In the early 16th century shotguns (*escopetas*) are cited, but the tendency is to use the names harquebus and musket, and in the beginning of the 17th century, *arcabuces de arzón*, or calvary harquebuses, pistols and occasionally *azelines*, or large-bore muskets. All of the weapons comprised a forged iron barrel, an iron firing

20. One Castilian vara measured 0,836 metre.

21. AGG, CO LCI, Leg. 308 (año 1604). Pedro Berraondo (v. Elgeta) with Luis López Torres, Portuguese merchant. The case arrived to the Chancillería de Valladolid (ArCh, 1-8, 0364/18).

22. AHPG, 1/3671, f. 14 (1566).

23. S. Covarrubias, *Tesoro de la lengua castellana o española*, [1611], Barcelona, 1943.

24. T. González, *Colección de cédulas..., op. cit.*, t. 1, 192, 197 ; R. Quatrefages, *La revolución..., op. cit.*.

mechanism with a steel spring, and generally a walnut stock. The barrels had to be at least one *vara* long (84 cm), and the use of pistols by private parties was absolutely forbidden[25]. A harquebus with " all its trimmings " included the mould for making balls, a gunpowder flask, another smaller flask to hold the priming powder and coloured woollen stings. The flasks, made of lightweight beech or elm wood, were covered with black leather and embossed with the royal heraldry. They hung from woollen strings and had five buttons and four tassels. In 1588 the harquebus strings weighed 3 ounces and musket strings, 4 ounces[26]. The ramrod was equipped with the scraper and ball-remover.

We have documents dating from 1554 which spell out the conditions to be met in making harquebuses. González de Escalante began in 1560 to acquire individual parts (barrels, locks, *etc.*) and had them assembled by other people. The notarial documents clearly state the conditions which were to be fulfilled, furnishing us a wealth of information on the production process. Initially, a third of the payment was made at the time of placing the order, another third when half of the order was completed, and the last payment was made upon delivery of the parts[27]. These circumstances complicate estimations on annual production, since the paymasters accumulated the parts in their warehouses from one period to another as can be seen in the accounts kept by Pedro Fernández de Zaraa[28]. This form of organisation bogged down the development of a powerful commercial sector, which could have bolstered more diversified production by servicing different markets. However, the royal treasury benefited from reduced costs and quality control was improved since the different parts were checked before being incorporated in the weapon.

It is difficult to know just how manufacturing was carried out since there were no official written regulations controlling and organising the process. The laws of Castile banning guilds and brotherhoods were vigorously enforced in the Basque Country, and were later incorporated in the charters of Gipuzkoa, or *fueros*[29]. Organised guilds in the arms sector would only become important in the 18th century, when their development was encouraged by the king. That way could serve as intermediaries in negotiating terms and conditions of the *asientos* and guarantee compliance of obligations. Lacking official documents that might have informed us on how production was organised, we have based our information on inventories and contracts.

The term *fragua*, or forge, is used to refer to the workshop where craftsmen would made their portable firearms. The majority of these forges were located in the outskirts of Placencia, although some were in Bergara, Elgoibar, and

25. *Novísima recopilación de las leyes de España (Nov. rec. Esp.),* lib. 12, tít. 19, ley 2.
26. AHPG, 1/3712 (28-05-1588).
27. AHPG, 1/3663 (1558), f. 73v.
28. AGS, CMC 3ª, 737 (1602-05).
29. *Nov. rec. Esp,* lib. 12, tít. 12, ley 12 ; I. Carrión, " Precios y manufacturas… "

Mondragón, among other places. They appear to have been buildings which were independent from the houses, located in the town centre. The person in charge of the forge was the master craftsman, who worked with a group of journeymen and apprentices, which in the mid-16th century was estimated to be made up of eight or nine individuals[30]. A reckoning of the tools existing in one particular forge in 1566 at the time of death of the armourer lists two pairs of bellows, two anvils, three large mallets and one small one, a wheel for drilling, a iron lathe, three large files and a few assorted smaller tools, along with some coal, iron plates for making barrels and a harquebus moulds[31]. Therefore, we are dealing with a large workshop with two forges where several people worked and where harquebus barrels were forged and drilled, and perhaps filed and trimmed as well. A 1630 trade agreement covering the use of a forge for the purpose of making musket barrels confirms this impression. Here, five workers were involved : one person bored, filed and trimmed the parts, with the help of at least one other person ; then there was a master barrel forger and three hammer operators, in addition to the apprentices or helpers whose job it was to pump the bellows[32]. Four musket barrels were to be made daily.

We calculate that the large workshops like the ones mentioned above probably had a maximum production capacity of between one thousand and fifteen hundred harquebus barrels in six months[33], whereas the small producers would have forged around twenty harquebus barrels a week[34]. When muskets were forged, the number would fall to about half. A craftsman would work with the help of a few journeymen, often family members, and an apprentice or two. Apprenticeship contracts were not uncommon. In times of greater demand, production was increased either by adding on new operators or by sending barrels to be filed and trimmed elsewhere. When the commitments were serious, contracts would be formalised by a notary public.

There are no clues in the documents we have studied as to how lock, stock and powder flask workshops ran. In the royal accounts there are references to eight or nine stock manufacturers and about the same number of lock makers, five or six trimmers and six or seven flask manufacturers. Apparently, only men worked in these workshops. However, the wife of the forge proprietor was not extraneous to her husband's business, as was evident when he passed away. In such cases workers would be hired (*se aparjean*) by the widows to file and trim the harquebus barrels in exchange for meals and wages, usually for a period of one year[35].

30. R. Larrañaga, *Síntesis..., op. cit*, 473-478.

31. APHG, 1/3671-2 (1766), f. 31.

32. AHPG, 1/3804 (1630) f. 115.

33. AHPG, 1/3685 (1578).

34. AHPG, 1/3693 (1573), f. 1.

35. AHPG, 1/3695 (1576), f. 59 ; 1/3709 (1587). J.A. Azpiazu, *Mujeres vascas, sumisión y poder : la condición femenina en la alta edad moderna*, San Sebastián, 1995, 73, 95, 109.

We know the exact cost of these firearms thanks to the information on harquebuses supplied to the King on various dates, as can be seen in Table 2. Let us point out that the price of barrels was between 66 % and 70 % of the total cost of the weapon, not including strings, and that the price increased more slowly compared to the other components. The firing mechanism is less than a tenth of the weapon's total value and is comparable to the cost of the gunstock ; it is less than the cost of the flasks and not much more than the strings. In 1588 a new model came out which had costly iron fittings on the flasks, upping the cost to 17 % of the overall value of the weapon.

Table 2 : Cost structure of the harquebus, in maravedis

	1560	1568 -69	1570 -72	1580	1589 -90	1602	1560 -80
Barrel	323	323	374	408	442		67 %
Trimming (*acicalar*)	17	17	20	20	20		3 %
Harquebus stock	36	40	51	51	68	77	8 %
Lock and mould	35	38	45	50	60	71,5	8 %
Scraper, ball-remover	16	16	20	15	20	22	3 %
Powder flasks	23	42	48	48	34	30	8 %
Cover powder flasks	10	17	20	20			3 %
Iron fittings for flask					144.5		
	460	493	578	612	788.5		100 %
Woollen strings			40	40	56		
	488	523	618	652	844,5		

AHPG, 1/3665-3 (1560), 1/3688 (1580), 1/3713 (1589), 1/3714 (1590), 1/3740 (1602) and AGS, CMC 3-2826-1 (1569-74)

Harquebus and musket manufacturing in the second half of the 16th century

The armourer obtained plates of iron made in the local bloomeries (*ferrerías menores*)[36], from Bizkaian ore, brought to Gipuzkoa by ship. It was wrought iron, very pure and soft, pounded into sheets with the bloomery hammer. In 1630 a 13-pound musket barrel could be made from a 20-pound iron sheet[37], so that a seven pound harquebus barrel was made from a sheet weighing about eleven pounds. The barrel was forged in the forge and subsequently made into a cylinder. The forge hearth was heated by means of chestnut charcoal mixed with beech or oak charcoal. A description by Martínez de Espinar in 1644

36. AHPG, 1/3678-1(1572) f. 39 and 31.
37. AHPG, 1/3804 (1630) f. 115.

explains this process in detail : “ The old way of forging barrels was by stretching and lengthening an iron bar, which was bent until one edge met the other, forming the shape of a barrel. Later the edges were heated up in order to join the iron. This was done by introducing iron borers into the hollow part, and hammering the metal so that the iron would be joined ”[38].

This system appears to have remained unchanged, as in 1827 the following account was written : “ First the plate is stretched and curved. The edges of the canal are then welded by means of 30 *caldas* and a hammer is used to alternately pound each end towards the other, introducing a cold bar with a width of 6 *líneas* to prevent the canal from becoming obstructed ”[39].

The barrel had to be wider at the breech end and taper off towards the muzzle, with the iron well-proportioned. It was then smoothed with a hot file to eliminate the seams[40]. Once the pipe of the barrel was made, it was bored both by hand and with the hydraulic mills. One worker could bore approximately six barrels with one of these mills in a day[41]. The terms of the *asientos* repeat time and time again that they had to be properly bored and clean on the inside (“ no threads nor blisters ”). Martínez de Espinar explains the operation as follows : “ After the barrel has been forged, it must be bored and stretched, these two things being performed by the Maestro practically at the same time. Boring is done by inserting a square borer into the barrel, which cuts with four corners ; the barrel is turned by a wheel and is firmly held in place by a metal clamp called a *galápago*. This and other borers are inserted until the interior is opened up and the small pieces are removed from the pipe. The edges are joined together and when the piece is cooled, a cord is used to straighten out the inside ”[42].

The requirements for boring become clearer in Placencia in 1568, when it was indicated that barrels had to be bored two or three times with the mill and another two or three by hand, with long borers going from one end to the other leaving no defects in the centre and removing the borer every so often in order to cool it down with water and rub it with whale grease. The idea was for the borer to enter from one side and go all the way through to the opposite end. As of 1570 the boring tools were required to be inserted from the muzzle end only[43].

After boring, a hole was drilled in the breech where the block, known as the *bid* or *tornillo* was inserted. In an order placed by a private party in 1569 it was

38. A. Martínez de Espinar, *Arte de ballestería y montería*, Madrid, 1644, f. 24v and 26v-27.

39. J. Odriozola, *Compendio de artillería o instrucción sobre armas y municiones de guerra*, Madrid, 1827, 125.

40. APGH, 1/3693 (1571), f. 1.

41. AHPG, 1/3703 (1583).

42. A. Martínez de Espinar, *Arte de ballestería..., op. cit.*, 26.

43. APHG, 1/3672-2, f. 9.

stated that the *bid* or breech-block had to enter a distance equal to two ball diameters, and usually it was indicated that it had to be seven or eight turns of the screw, the same as for the screw in rifle breeches in 1827. Sometimes it had to have " a long nipple " so it could be removed easily. The ear or vent hole was opened with a bit (*paraus*, *baraus*)[44], and in 1588 it was specified that if the vent of a barrel had been perforated with a punch it would be rejected "[45]. Afterwards, the entire surface of the " black ", or cold, barrel was filed to *ocho ochavos* and evened out " neatly with a smooth even finish ". The barrels were then finished, adding the tip, the sight, metal braces, or *chatones*, and the pan, which was to be open, square and long. Sometimes the barrels were eight-sided (*ochavados*), or eight-sided at the breech and rounded at the muzzle.

The filed and finished barrel was then ready to be tested. The calibre was determined by indicating the weight of the lead ball fired, which in harquebuses varied between 3/8 and 3/4 of an ounce (10,8 and 21,6 grams). The length varied between 4 and 6 *palmos* measuring a quarter of a *vara* (between 84 and 125 cm). Both the lesser calibre and the longer barrels usually dated to pre-1580s and were manufactured for private parties. As of 1570 the harquebuses from His Majesty's Munitions fired three-quarter ounce bullets (21,6 g), and were four and a half *palmos* (94 cm) long until 1588, when other barrels began to appear with the same calibre but longer, measuring *vara* and a quarter less one *dedo* (103 cm)[46], Harquebus barrels purchased for the army in the latter half of the 16th century weighed between seven and seven and a half pounds (between 3,22 and 3,45 kg). The tendency was to make harquebuses lighter, as in 1588 long barrels were required not to exceed seven pounds and short barrels, six and a half.

Muskets were longer and heavier and for that reason had to be propped up with a cradle. Few muskets were acquired by the royal paymaster up until the end of the 1570s, as can be seen in Hernando de Aguirre's accounts, which document purchases of 16.000 harquebuses, 930 crossbows and only 1.200 muskets between 1568 and 1573, though in 1574 the number of muskets requested grew to 4.000[47]. Production was given another boost beginning in 1583 when harquebuses were substituted for muskets on merchant ships sailing to and from the Indies[48]. Due to the increase in production in this period, the different musket parts began to be acquired separately. In the middle of the

44. R.M. de Azkue (*Diccionario vasco-español-francés*, Bilbao, 1969) collects the word " barautz " with the meaning of *vilebrequin, outil très ancien avec lequel on perce le fer*.

45. APHG, 1/3711(1588).

46. As late as 1570, half of the ordered were for 0,5 oz bullets and the others for 3/4 oz balls. (AHPG, 1/2672-2, f. 9).

47. AGS, CMC 3ª, 2826-1. In 1569, 3.000 harquebuses were examined in Palencia and only 200 muskets (AHPG, 1/3676-2 (1570), f. 61). The fist big charge of muskets was in 1574 for 2.000 guns by Pero Menéndez de Avilés (AHPG, 1/3682 (1574), sf). But between 1576 and 1578, they were tested 11.000 barrels of harquebus and only 217 of musket (AHPG, 1/3685 (1578), sf).

48. AHPG, 1/3703 (1583), sf.

century the muskets used balls weighing an ounce and a quarter or an ounce and a half, but by 1574 all of the muskets ordered had 2-ounce lead balls (57,5 g) and barrels weighing from thirteen to thirteen and a half pounds (between 5,98 and 6,21 kg) and measuring five *palmos* and a half plus two *dedos* (117 cm). In time, barrels became lighter : the same calibre and projectile in barrels made in 1618 were to weigh between 12 and 13 pounds[49]. A musket barrel was approximately twice as costly as a harquebus barrel from His Majesty's Munitions, but the total price of the weapon varied between two and a quarter and two and a half times as much as a harquebus.

After the barrels were finished, it was time for them to be tested. When dealing with arms made for the King, it was essential that the testing be supervised by the royal paymaster or one of his representatives in Placencia, turning this place into the centre of arms production. Each part was tested twice and only barrels which neither burst nor sighed were considered acceptable. In 1570, harquebuses were tested using a gunpowder charge that was twice the weight of the ball, and only one projectile[50]. At the end of 1571, we see that the amount of gunpowder was reduced to one and a half times the weight of the ball but with two projectiles were used, making the proportion of lead to gunpowder the same as in muskets[51]. The barrels that were approved were marked above the vent with an " F " and a royal crown[52]. Lead and gunpowder expenses for testing suggest that between a third and a quarter of them were rejected[53]. Afterwards, the manufacturer thoroughly cleaned out both the inside and the outside of the barrel, polished and ground it and lastly greased it inside and out, at which time it was finished. Documentation reveals that frequently barrels were delivered forged, to be later filed, trimmed and polished at their destination.

Harquebus and musket barrels were mounted onto a wooden stock and the firing mechanism was attached. These jobs required the intervention of gunstock- and lock-makers. Generally, the stock had to be made of " good, dry and not green walnut, hewn at the proper moment ", clean and flawless and previously cured over several years. Some harquebuses were also set into cherry wood or apple wood stocks. The barrels were to be fit into the gunstocks using " wooden, not iron wedges ". The only reference we have to the size of harquebus gunstocks dates from 1554 and indicates that they had to measure half a *vara* (42 cm). Fastened to the gunstock was a well-surfaced ash wood ramrod, at the tip of which was attached a scraper[54]. It does not appear that at the

49. AHPG, 1/3717 (1618), f. 50.

50. AHPG, 1/3672-2, f. 61. They used for the two tests, 3 oz of powder and 1,5 oz of lead with each harquebus.

51. AHPG, 1/3676, f. 9.

52. AHPG, 1/3672-2, f. 61.

53. AHPG, 1/3687 (1578) ; 1/3705-2 (1583).

54. AHPG, 1/3665-3 (1560), f. 46.

time there were problems with the supply of wood for making gunstocks, although towards the beginning of the 17th century there was already a scarcity, forcing the monarchy to encourage walnut tree planting in the Placencia area[55].

In the 16th and 17th centuries most of the muskets and harquebuses had a serpentine lock whose cost, together with the mould amounted to less than one-twelfth of the price of the barrel[56]. Therefore, we are talking about a mechanism that was not complex or expensive. It is curious that there was almost more written on the requirements for making moulds than for gunlocks. The lock had an iron plate upon which the firing mechanism was attached. It had to be straight and level and be equipped with a high-quality, well-tempered steel spring and had to be properly filed. Apparently, the artisans were given a model which they were to copy. In 1554 the serpentine had a screw which held the fuse, but in 1570 it was stated that there should not be one[57]. The lock was attached to the gunstock by means of two screws which were to go from one side of the gunstock to the other, screwing into the iron plate. Normally, the gunstock-makers put in the lock, which had to be done " so that movement was very good ".

Although the documentation is not very explicit, this type of gunlock is not the only one we find. In 1560, in addition to the more common locks, other locks were also purchased for the army. They cost almost four times more — four *reales* each — and the manufacturers had to help the stock-makers attach them[58]. Furthermore, as of 1554 there were also percussion locks, which in 1559 include pans[59]. The percussion lock and pan appear later as different from the serpentine lock, and is associated with special harquebuses, *hechizos*, which were somewhat longer and substantially more expensive. In 1584 we find *hechizo* harquebuses equipped with flintlocks (*de chispa*)[60]. A contract drawn up in 1579 calls for a journeyman to make flint and royal munition-type locks, making it clear that they were different[61]. Besides the weapons produced for the army, it seems that around 1580 flintlock harquebuses began being made for private individuals[62].

55. AGG, JD-1D15, Junta General de Elgoibar (1606).

56. See Table 2. At the beginning of the 18th century 25 % of the new guns (fusil) were matchlocks. (AGS, CMC 3ª, 2835-2).

57. AHPG, 1/3672-2, f. 19.

58. AHPG, 1/3665-3 (1560), f. 46.

59. AHPG, 1/3658 (1554), f. 18, 1/3664 (1559), f. 41v.

60. AHPG, 1/3706, f. 117 v.

61. AHPG, 1/3698 (1579), f. 86.

62. R. Larrañaga, *Síntesis…, op. cit.*, 66

Conclusions

The Placencia de las Armas area (in the upper Deba river valley) and the area in Bizkaia near Elorrio were the two main regions where portable weapons and pikes were produced for the Spanish monarchy from the beginning of the Modern Age. There was also a significant amount of cutting weapons manufacturing in the Duranguesado area and in Tolosa, where the monarchy's armour production was concentrated in the 17th century. The royal demand bolstered mass production while at the same time slowing down the production of portable arms for both the interior and export markets, which was important until the last third of the 16th century. Acquisition and distribution fell into the hands of royal functionaries, impeding the development of a trading sector linked with this manufacturing activity.

Documentation generated by the royal administration has enabled us to understand what pikes, muskets and harquebuses were like and how they were made in the latter part of the 16th century, as well as the different weapons-purchasing systems employed. In the case of portable firearms, pikes and swords, the royal paymasters would order the parts comprising the arms directly from the artisans, and later have the weapons assembled by other craftsmen. Central warehouses were built, where organisation of the incoming and outgoing parts took place. In the 15th century, this warehouse was located in Placencia and in the 17th century, in Placencia as well as Tolosa. On the contrary, in the 17th century with the manufacture of defensive weapons, the tendency was to concentrate production in large workshops, where under the watchful eye of the paymaster or director, salaried artisans would make parts for armour in complex installations.

Mathématiques, théorie et pratique dans l'artillerie de la marine espagnole au XVIIIe siècle. Les manuels

Encarnación Hidalgo Cámara

Le Siècle des lumières en Espagne : l'*Ilustración*

L'Espagne que Charles V trouve en 1700 à son arrivée au trône, après l'extinction de la dynastie des Habsbourgs et l'instauration de celle des Bourbons, est un pays entraîné dans un long processus de déclin politique, économique, social, militaire, commercial, démographique, *etc.* Charles V doit en outre affronter les problèmes dérivés de l'administration et du contrôle de ses vastes domaines américains.

La machine de l'Etat, déjà presque épuisée, se trouve forcée de supporter le poids d'une guerre de succession au trône, qui durera quatorze années et s'achèvera par l'établissement définitif des Bourbons. Le roi et ses ministres tirent de cette guerre une conclusion : il est urgent de moderniser le pays, centraliser l'administration, renouveler l'armée, vivifier le commerce, assainir le Trésor, rentabiliser l'agriculture, l'industrie minière, la pêche, en résumé, ouvrir de toutes les façons l'enclos dans lequel l'Espagne s'était renfermée depuis le règne de Philippe II.

Le Siècle des Lumières espagnol sera un processus long avec ses traits propres. Hésitant à ses débuts, il atteint son apogée pendant le règne de Charles III (1759-1788), et tombe en déchéance sous le règne de Charles IV à la fin du XVIIIe siècle, quand la menace de la contagion révolutionnaire depuis la France voisine rallume les positions les plus conservatrices qui ne s'étaient jamais tout à fait éteintes, mais seulement engourdies dans l'attente d'une occasion propice[1]. L'invasion des troupes napoléoniennes, suivie par la revanche conserva-

1. Sur l'histoire de l'Espagne sous le règne de Charles III voir, par exemple, D. Ortíz, *Carlos III y la España de la Ilustración*, Madrid, Alianza Editorial, 1988. En général, sur l'*Ilustración* Espagnole, Richard Herr, *España y la revolución del siglo XVIII*, Madrid, Aguilar 1964, 1e ed. en Anglais (1960).

trice, mettront un point final au Siècle des Lumières espagnol. Plus rien dès lors ne sera comme avant.

MILITARISATION DES SCIENCES

Des différents traits propres au Siècle des Lumières espagnol (monarchique, catholique, américain, centralisateur et partisan des régales) se détache la participation de l'Armée à la modernisation du paysage scientifique. Cette particularité, qui peut paraître un contresens au premier abord, ne l'est pas si nous en examinons les causes.

Sans aucun doute, au cours des ans, l'isolement de la culture et de la société espagnoles depuis le temps de Philippe II, se répercute négativement sur le pays. Pourtant, le retard des sciences et de la technologie espagnoles fut relatif. Des architectes, ingénieurs et marins maintinrent un niveau très digne par leurs travaux, et déjà au cours du dernier quart du XVIIe siècle on détecte les premiers cas de scientifiques rénovateurs, qu'on qualifie alors d'une façon méprisante de *novatores*, et qui rendent public leur désaccord avec la scolastique régnant dans les universités[2].

La Guerre de Succession met en évidence le besoin impérieux de reconstruire la marine espagnole, presque inexistante et incapable de soutenir la comparaison avec la flotte anglaise, de restructurer les armées de terre, très inférieures aux armées françaises. Il est bien connu que l'idée moderne d'Etat s'accompagne de la création d'armées nationales, du développement technologique de l'artillerie, des techniques de fortification et de la construction de vaisseaux de guerre. La couronne espagnole éprouvait un besoin urgent de modernisation pour protéger ses frontières, assurer la paix et ainsi permettre le développement intérieur du pays.

Cet ambitieux projet de modernisation reposait sur la sélection des personnes appropriées pour le mener à bien. Des officiers héroïques et de braves soldats ne suffisaient plus. Il fallait que les officiers dominent les mathématiques, la topographie, l'astronomie, la fortification, la métallurgie, la chimie, la physique… Les cadres de cette nouvelle armée devaient être des officiers scientifiques, et c'est ainsi qu'on les nomma (*oficiales científicos*). Sur leurs épaules ne reposaient plus seulement des tâches militaires, mais également d'autres tâches scientifiques et technologiques aussi importantes pour le pays, et pour lesquelles la couronne ne disposait pas de fonctionnaires dotés d'une formation appropriée. Dans les universités régnait l'aristotélisme. Il n'existait pas une académie nationale des sciences mais une multitude de *tertulias* (réunions privées). La Compagnie de Jésus, la seule institution dont les membres possé-

2. Sur la science espagnole dans ce temps voir Lopez Piñero, José M^a, *Ciencia y técnica en la sociedad española de los siglos XVI y XVII*, Barcelona, Labor, 1979.

daient une formation mathématique digne de ce nom, cultivait des intérêts éloignés de ceux de la couronne[3].

C'est cette carence qui poussa l'armée espagnole à créer des établissements spécialisés pour la formation scientifique de ses officiers, en particulier pour les trois *Cuerpos Facultativos*, c'est-à-dire la Marine, le Génie et l'Artillerie. C'est l'origine de ce que les historiens de la science espagnole ont appelé une *militarisation de la science espagnole dans le Siècle des Lumières*[4].

Les traits principaux de la politique scientifique espagnole de cette époque, ceux qui avec quelques variations dans chaque cas seront aussi les traits propres des institutions militaires, se dessinent durant la première moitié du XVIIIe siècle. Ce sont : la dépendance vis-à-vis de la situation financière de l'Etat, l'engagement de scientifiques et de techniciens étrangers, l'envoi d'étudiants à l'étranger, l'absence de débats théoriques et idéologiques, le prestige social croissant de la science — en particulier des sciences dites " utiles " — et la disparité des avis entre la structure enseignante et la structure militaire dans ces établissements, disparité causée par la subordination de la pratique à la théorie dans les programmes d'études.

LES ACADÉMIES MILITAIRES

Les académies militaires espagnoles du Siècle des Lumières s'inspirent des institutions militaires espagnoles dotées d'une tradition scientifique, en particulier de l'Académie de Bruxelles dirigée par Sebastián Fernández de Medrano (1646-1705), qui est l'auteur de manuels de fortification, d'artillerie et de géographie, les plus importants de son temps parmi ceux publiés en Espagnol[5]. L'influence de cette académie et des œuvres de Medrano, d'abord destinées à l'enseignement militaire à Bruxelles, se répandit à travers toute l'Espagne et ses livres firent l'objet de nombreuses éditions.

Le mouvement initié par l'Académie de Bruxelles est prolongé par l'Académie du Génie de Barcelone (1720)[6] grâce à l'intervention de l'Ingénieur Général Jorge Próspero de Verboom, ancien élève de Medrano. Trois ans plus tôt,

3. Sur la science espagnole au XVIIIe siècle voir M. Sellés, J.L. Peset, A. Lafuente, (compil.), *Carlos III y la ciencia de la Ilustración*, Madrid, Alianza Editorial, 1988, et la bibliographie citée dans les différents chapitres.

4. Sur ce sujet voir A. Lafuente, J.L. Peset, " Militarización de las actividades científicas en la España ilustrada (1726-1754) " dans J.L. Peset, *et al.*, *La ciencia moderna y el nuevo mundo. Actas de la I Reunión de Historia de la Ciencia y de la Técnica de los Países Ibéricos e Iberoaméricanos*, Madrid, *(25 a 28 de Septiembre de) 1984*. Consejo Superior de Investigaciones Científicas. Sociedad Latinoamericana de Historia de las Ciencias y de la Tecnología, Madrid, 1985, 127-147.

5. Par exemple, *El práctico artillero* (1680), *El perfecto oficial, bombardero y artillero* (1699), *o Geographia o moderna descripción del mundo* (1686).

6. Sur l'histoire de cette académie voir H. Capel, J.E. Sanchez y O. Moncada, *De Palas a Minerva. La formación científica y la estructura institucional de los ingenieros militares en el siglo XVIII,* Barcelona, Ed. Serbal-Madrid, C.S.I.C., 1988.

en 1717, l'Académie des Gardesmarins de Cadix ouvrait ses portes. Ce fut sans doute l'institution militaire espagnole la plus importante du XVIIIe siècle, celle dans laquelle l'enseignement scientifique atteignit le plus haut niveau[7].

L'enseignement scientifique de l'artillerie était inscrit au programme des études de l'Académie du Génie de Barcelone, puisque à cette époque l'Artillerie et le Génie ne formaient qu'un seul corps. Cependant, les artilleurs luttèrent toujours pour leur indépendance, et quand ils l'obtinrent au milieu du siècle sous le ministère du Comte d'Aranda, ils réclamèrent la création de leurs propres académies. C'est ainsi que naquirent les Académies d'Artillerie de Cadix et Barcelone (1751-1760). Ces académies militaires fournirent à la couronne les spécialistes, les mathématiciens, les astronomes et les techniciens nécessaires pour les travaux de type scientifique. Ces officiers scientifiques obéirent toujours docilement aux ordres du roi et ne posèrent pas de problèmes politiques, mais le prix qu'ils durent payer fut la subordination de leur trajectoire scientifique et de leur liberté de recherche aux priorités strictement militaires. On se trouve déjà sous le règne de Charles III.

L'ENSEIGNEMENT SCIENTIFIQUE DE L'ARTILLERIE. LES MANUELS ADRESSÉS À LA MARINE

Vers 1760, il y avait en Espagne une demi-douzaine d'académies militaires destinées à la formation des officiers scientifiques pour les différents corps de l'armée[8]. Pour éviter une telle dispersion d'efforts, Charles III décida de ne maintenir ouverte qu'une seule académie pour chaque corps spécialisé (ou " corps savant "), c'est-à-dire l'Académie des Gardesmarins à Cadix, celle du Génie à Barcelone et une nouvelle académie d'artillerie, le Collège royal d'Artillerie à Ségovie, inauguré en 1764[9]. Les autres furent fermés et leurs livres et instruments distribués aux institutions subsistantes.

L'artillerie n'est pas une science unique mais plutôt un ensemble de plusieurs branches scientifiques et technologiques, qui confluent dans la maîtrise d'un processus de brève durée : le vol d'un projectile. Justement parce que l'artillerie est un réseau de différents savoirs, elle sert de point de repère pour

7. Sur cette académie voir M. Sellés, *Astronomía y Náutica en la España del siglo XVIII*, Thèse doctorale soutenue à la U.N.E.D. Madrid, 1986, 2 vols. Et aussi, A. Lafuente y M. Sellés, *El observatorio de Cádiz (1753-1831)*, Madrid, Ministerio de Defensa, Instituto de Historia y Cultura Naval, 1988.

8. Une pour la Marine à Cadix, une autre appartenante au Corps du Génie à Barcelone, deux conçues pour la formation d'artilleurs à Cadix et à Barcelone, une Ecole de Mathématiques pour les Gardes de Corps à Madrid, d'où sortit le premier manuel de mathématiques Espagnol que comprenait l'analyse infinitésimale, plusieurs académies régimentales (Ceuta et Oran) et la Société Royale Militaire et de Mathématiques, une institution créée spécialement pour rédiger les manuels scientifiques destinés à ces académies et qui était composée d'ingénieurs et d'artilleurs.

9. Sur l'histoire de cette académie voir E. Hidalgo Cámara, *Ciencia e institución militar en la España Ilustrada. El caso de la artillería*. Thèse doctorale soutenue à la U.N.E.D., Madrid, 1993. Inédite.

évaluer ce que devait être une formation scientifique et technologique complète au XVIIIe siècle.

Dans l'histoire de l'artillerie, une fois dépassée l'énorme influence de l'aristotélisme, dans une approche purement mathématique se détachent de façon incontestable les noms de Tartaglia, Galilée et Newton[10]. On doit à Tartaglia l'idée de la balistique comme branche scientifique autonome, à Galilée la théorie parabolique dans le vide, à Newton l'entrée dans une nouvelle époque de la balistique, celle des approximations successives de la trajectoire hyperbolique, centrée sur le cas de la résistance quadratique. L'impossibilité de formuler une solution mathématique rigoureusement exacte, et à la fois utile et facile à manipuler par les artilleurs, a abouti à une succession d'approximations pour tenter de résoudre le problème balistique.

Le XVIIIe siècle est le siècle des grands analystes, des nouveaux géomètres qui pratiquent le calcul infinitésimal, en élargissant le terrain de la mécanique rationnelle, comme les frères Bernoulli, J. Hermann, L. Euler, J. Lambert, Ch. Borda ou M. Legendre[11]. Mais l'analyse ne réussit pas à satisfaire les espoirs d'exactitude et de commodité que l'artillerie pratique exigeait. En plus, tandis qu'il était mathématiquement impossible de discuter l'importance de la résistance de l'air, dans la pratique la multitude d'éléments empiriques impliqués pouvaient excuser, et avec raison, l'utilisation de calculs complexes qui ne menaient pas très loin. C'est pourquoi des notions solides de géométrie, une bonne connaissance de la théorie parabolique et l'expérience personnelle étaient encore indispensables dans la pratique de l'artillerie[12].

Mais l'artillerie n'était pas un monopole de l'armée de terre. Elle était présente dans la Marine qui avait tout autant besoin de personnes qualifiées. Cette formation d'artilleur de marine devait se concevoir de la même façon, dans des institutions compétentes (les Brigades d'Artillerie de Marine) et avec des manuels conformes aux dernières ordonnances adoptées.

Ces manuels se caractérisent par leur contenu purement pratique. S'il était déjà bien difficile d'utiliser correctement les canons sur la terre, il était plus difficile encore de l'envisager sur mer. A bord des vaisseaux, les gros canons en bronze étaient remplacés par des mortiers (petits canons en fer), plus petits,

10. Sur l'histoire de la balistique au XVIIe siècle voir A.R. Hall, *Ballistics in the seventeenth Century. A Study in the Relations of Science and War with Reference principally to England*, Cambridge, University Press, 1952.

11. Sur ce sujet voir par exemple, M. Blay, *La naissance de la mécanique analytique. La science du mouvement au tournant des XVIIe et XVIIIe siècles*, Paris, Presses Universitaires de France, 1992. H.J.M. Bos, " Mathematics and rational mechanics ", dans G.S. Rousseau, R. Porter (eds), *The Ferment of Knowledge. Studies in the Historiography of the Eighteenth-Century Science*, London, Cambridge University Press, 1980, 327-355.

12. Sur le double caractère théorique et pratique de l'artillerie, E. Hidalgo Cámara, " Esplendor geométrico y sentido común en la artillería del siglo XVIII ", dans E. Balaguer, E. Giménez, *Ejército, ciencia y sociedad en la España del Antiguo Régimen. Instituto de Cultura Juan Gil Albert*, Alicante, 1995, 533-542.

bon marché et plus faciles à manipuler que les canons. Ces mortiers étaient déjà fixés à leur affût à 45° d'élévation, de telle sorte qu'on ne pouvait pas régler le coup de feu avec toute la précision mathématique possible. C'est pourquoi les artilleurs de la Marine avaient plus besoin d'instruction pratique que d'équations mathématiques compliquées.

LE *TRATADO DE ARTILLERÍA THEÓRICA Y PRÁCTICA* DE JUAN SÁNCHEZ RECIENTE (1733)

On retrouve tous ces traits dans le *Tratado de Artillería theórica y práctica* de Juan Sánchez Reciente (1733)[13], écrit pour l'enseignement dans le Collège de San Telmo de Séville dont il était le " maestro principal ". Dans ce collège, les enfants orphelins recevaient la formation nécessaire pour devenir pilotes de vaisseau. Au caractère forcément pratique des manuels d'artillerie pour la Marine, on ajoute que dans ce cas il s'agissait de s'adresser à des enfants, raison pour laquelle Reciente s'exprime avec la plus grande clarté et simplicité possibles, et utilise un ton vraiment didactique.

Sánchez Reciente eut la belle idée de préparer un jeu de cartes avec les illustrations qui d'habitude accompagnent les textes d'artillerie. Les enfants formaient deux équipes, celle de Saint Pierre et celle de Saint André. Le jeu consistait à se poser les uns aux autres des questions sur les principes de l'artillerie. Les cartes avaient différentes valeurs selon la difficulté des questions et les joueurs devaient expliquer les dessins des cartes avec les textes du livre. L'équipe victorieuse gagnait une couronne ; celle qui perdait recevait une figure d'un âne, toute une humiliation pour eux...

Sánchez Reciente décrit en plus les instruments (ou plutôt les outils) que tout bon artilleur devait posséder : un étui avec cinq aiguilles de différentes grandeurs, un compas de pointes droites, un autre de pointes curvées, un calibre en laiton, un quadrant ou équerre également en laiton (il s'agit de la classique équerre d'artilleurs ou de Tartaglia), un couteau, des ciseaux, de la poudre noire fine pour allumer les pièces, un anneau en laiton (pour choisir les boulets ou les bombes), un niveau, une boussole, une bourse avec de l'amadou, un briquet et de la pierre à feu, un petit tableau noir (pour faire des calculs arithmétiques et géométriques et " pour étudier pendant les moments de loisirs ") et d'autres outils propres aux charpentiers[14].

13. J. Sánchez Reciente, *Tratado de Artillería theórica y práctica, en donde se da entera noticia, y conocimiento de todas las Piezas Antiguas, y juntamente de las Modernas de la Nueva Ordenanza del año de 1716. Según el méthodo, que se enseña en el Real Seminario de San Telmo, extramuros de la ciudad de Sevilla*, 1733. Il publia aussi *Tratado de Navegación theórica y práctica según el orden y método con que se enseña en el Real Colegio Seminario de Sr. S. Telmo, extramuros de la ciudad de Sevilla*, Sevilla, 1749.

14. Sur les anciens instruments mathématiques propres à l'artillerie et à la fortification, J. Bennett, S. Johnston, *Catalogue of the Exhibition " The Geometry of War, 1500-1700 "*, Oxford, 1996.

LE *TRATADO DE LA ARTILLERÍA*... DE LUCRECIO IBÁÑEZ Y PEDRO VARELA (1770)

En 1770 parût à Cadix le *Tratado de la artillería*... de Lucrecio Ibáñez et Pedro Varela[15]. Ce livre a la structure classique des manuels d'artillerie du XVIIe siècle, et son contenu reproduit fidèlement les leçons de Tomás Vicente Tosca et de son *Tratado XVII. De la Pirotechnia, Arte Tormentaria o Artillería* (1707-1715), compris dans son *Compendio Matemático*. La position de Tosca est celle du mathématicien qui emploie tout ce que les spécialistes de l'artillerie " appliquée " ont déjà écrit, et qui reprend en même temps les théories géométriques classiques (galiléennes). Son originalité se limite à l'inclusion des tables de descriptions et de tables de détail relatives aux nouveaux règlements et ordonnances de l'artillerie espagnole. Bien que ce livre soit destiné à l'enseignement, il se présente plutôt comme un précis à consulter régulièrement que comme un manuel ou un livre de textes.

LE *COMPENDIO DE ARTILLERÍA* DE JOSÉ DÍAZ INFANTE (1754)

Les autres ouvrages sur l'artillerie destinés à la Marine furent généralement moins étendus et moins approfondis que celui de Ibáñez et Varela. Certains se présentaient plutôt comme des précis concis et clairs, mais leur contenu reste de nature pratique et s'adresse en particulier aux gens de mer. Dans cette ligne, on trouve le *Compendio de Artillería* (1754) de José Díaz Infante (qui signe parfois Infante tout court), professeur d'artillerie à l'Académie des Gardesmarins de Cadix[16]. Il n'y a pas de nouveautés importantes dans cette œuvre qui s'adresse à un public adulte et non à des enfants, comme le faisait le livre de Sánchez Reciente. Mais ce *Compendio* est aussi beaucoup plus bref que le traité d'Ibáñez et Varela. Il se caractérise par la disposition de ses chapitres qui met en valeur la spécificité de l'artillerie de Marine. En effet, cette œuvre ne commence pas par les chapitres classiques sur la poudre, sa production et manipulation, mais par la description et l'examen des pièces caractéristiques de l'artillerie de Marine, un sujet qui occupe presque la moitié du livre. Incidemment il apparaît aussi que Díaz Infante était un professeur qui connaissait beaucoup plus que ce qu'il enseignait. Par exemple, bien que d'après la théorie galiléenne orthodoxe les coups de feu tirés avec deux angles d'élévation complémentaires aient la même portée, Díaz Infante prévient que ce n'est pas exact, comme le démontre la pratique.

15. L. Ibáñez, P. Varela, *Tratado de la artillería, que contiene la descripción de las diferentes piezas de bronce, y fierro (a saber de cañones y morteros)*, 2a ed., Cádiz, 1770.

16. J. Infante, *Compendio de Artillería para el servicio de Marina*, 1754. Díaz Infante était lieutenant de vaisseau de l'Armée Royale, professeur d'artillerie de l'Académie des Gardesmarins de Cadix et Membre de la Société Royale de Séville. Il faisait partie de la *tertulia* de Jorge Juan et Louis Godin à Cadix, pour laquelle il rédigea quelques mémoires intéressants sur la poudre.

En somme, les fondements balistiques dont a besoin un artilleur mis au service de la Marine au milieu du XVIIIe siècle se limitent à quelques idées élémentaires et très claires de la théorie parabolique que Díaz Infante s'est attaché à " moderniser ". Par exemple, il est vrai que la trajectoire ne peut pas être droite, pas même en partie, en raison de la pesanteur, mais en plus, elle n'est pas exactement parabolique. Il montre comment l'artilleur peut tirer parti de ce constat pour accroître la précision de ses tirs en utilisant le demi-cercle de Torricelli (une variation de l'équerre classique de Tartaglia) et quelques règles de trois.

Le *Tratado de la artillería* de Sebastián Labayru y Azagra (1756)

Deux années plus tard, en 1756, paraît un autre traité d'artillerie adressé cette fois aux Brigades de l'Artillerie de Marine. Il est signé par Sebastián de Labayru y Azagra qui bénéficie de l'approbation de Díaz Infante. Les œuvres de ce genre sont peu accueillantes aux nouveautés ; au moins peut-on espérer y repérer un certain changement d'attitude. On ne le trouvera pas ici. Au contraire, quand on y rencontre quelques noms célèbres de la physique ou de la balistique, l'auteur précise que, malgré tout, leurs contributions ne sont pas suffisamment scientifiques : " ...Moroguez, Dulacq, Belidoro, Sagulieres, Grabesanto, Mumsambruck, et d'autres auteurs font des réflexions philosophiques, mais pas du tout démontrées mathématiquement "[17]. Pour lui, ici, contributions " scientifiques " signifie plutôt " assurées par la pratique ".

Le *Tratado de artillería* de Francisco Javier Rovira (1773)

Cet ensemble d'oeuvres d'artillerie pour la Marine finit avec le *Tratado de artillería* publié par Francisco Javier Rovira en 1773, trois années seulement après la parution de la seconde édition de l'œuvre d'Ibáñez et Varela[18]. Considéré en son temps comme le meilleur et le plus complet des traités d'artillerie écrits en espagnol, ce texte fut reconnu et célébre aussi bien en Espagne qu'à l'étranger. Rovira bénéficie d'une position incomparable pour ramasser tout ce qui avait déjà été publié sur l'artillerie de Marine, adapter son exposé aux derniers règlements en vigueur, et citer sans trop de détails quelques noms célèbres de l'artillerie et de la balistique. Cela confirme que dans l'artillerie de Marine, concentrée sur sa pratique, il y avait une certaine méfiance vis-à-vis

17. S. de Labayru y Azagra, *Tratado de la artillería, armas, pertrechos, municiones, metales, bombardería, artificios de fuego, y armamentos de bajeles correspondientes al servicio de la armada, que para instrucción de los individuos del cuerpo de brigadas de artillería de marina, ha formado el Capitán don Sebastián de Labayru y Azagra, Teniente de Fragata de la Real Armada, como maestro principal por S. Mag. de las expresadas Escuelas*, Sevilla, 10.

18. F.J. Rovira, *Tratado de artillería para el uso de los caballeros Guardias-Marinas en su Academia. Por D. Franciso Xavier Rovira, Teniente de navío de la Real Armada, y profesor de Artillería en la misma Academia*, (1773).

des développements plus mathématiques, mais on ne les méconnaissait pas pour autant. Cette œuvre, dédiée au marin et scientifique Jorge Juan, avait l'approbation de deux personnalités illustres de la marine espagnole, Antonio de Ulloa et Vicente Tofiño. Ce dernier était alors directeur de l'Académie des Gardesmarins de Cadix.

Ce livre est un travail complet, qui satisfait parfaitement les besoins pratiques et qui fait allusion aux théories mathématiques les plus modernes quand il est nécessaire de le faire. Les théories modernes sont surtout mises en valeur dans les chapitres qui s'occupent de la poudre, et très peu dans les sections réservées à la balistique extérieure. La liste des reconnaissances publiques est notable[19]. Rovira fait des citations un peu pêle-mêle, mais cela suffit pour mesurer sa connaissance des auteurs purement marins (comme Labayru ou Infante), ou d'autres que l'on peut décrire comme des " auteurs historiques " (c'est-à-dire des auteurs importants à leur époque mais dépassés depuis, comme Ufano et même Rivault, partisan d'Aristote). Un troisième groupe d'inspirateurs est formé par les artilleurs " purs " (Bélidor, Le Blond, d'Arcy) et, pour finir, un quatrième réunit des sources scientifiques de haute tenue (Robins et les publications des sociétés scientifiques). Il faut remarquer que Rovira nomme Robins mais pas Newton, qui était pourtant à l'origine de toutes les expériences modernes. De même, il se réfère à des auteurs galiléens mais pas à Galilée.

CONCLUSIONS

Nous avons passé en revue un ensemble de manuels et traités d'artillerie destinés d'une manière spécifique à la Marine, avec une orientation éminemment pratique qui fut une caractéristique partagée avec l'artillerie de terre. Leur contenu est pratiquement le même, avec une partie théorique qui trouve ses racines dans l'œuvre de Tosca, et une partie qui n'est pas strictement marine et que l'on peut rapprocher des travaux de Medrano. La procédure employée consista à maintenir dans la mesure du possible les principes de ces deux auteurs, et à ajouter les nouveautés imposées par les derniers règlements militaires. Si on élimine de ces traités les chapitres vraiment propres à la Marine, le résultat n'est pas très diffèrent d'un traité d'artillerie à l'usage des armées de terre.

Nous pouvons dire aussi que la plupart des artilleurs de la Marine, qui connaissaient assez bien les lois de la balistique, se méfiaient non sans raison des théories mathématiques qu'ils voyaient tous les jours démenties par la pratique. La plupart des mathématiciens étaient d'ailleurs convaincus que leurs efforts

19. Diego Ufano, Vicente de los Ríos, Sebastián Labayru, José Infante, San Julián, d'Arcy, Wolf, Saint Rémy, Robins, Muller, Deidier, Bélidor, Frecier, Dulacq, Bigot de Morogues, Le Blond, Rivault, les *Mémoires* de l'Académie des sciences de Paris, et d'autres auteurs et oeuvres habituels dans les manuels d'artillerie.

n'arriveraient pas à dépasser la frontière au-delà de laquelle la raison pourrait expliquer le comportement de la poudre et ses effets sur les projectiles.

Curieusement, la plupart des traités d'artillerie espagnols publiés à cette époque n'étaient pas destinés aux différents établissements qui appartenaient au Corps d'Artillerie (d'abord les Académies de Cadix et Barcelone, et ensuite l'Académie de Ségovie) mais à l'enseignement de l'artillerie pour la Marine. Dans toutes ces œuvres, les différents auteurs reproduisent presque exactement le même patron : Tosca, Medrano, les renseignements propres à l'artillerie et aux règlements de la Marine espagnole. Pour cette raison, nous pouvons dire sans hésiter que dans ce cas, la pratique s'est imposée sur les théories mathématiques, même quand celles-ci étaient correctes.

La influencia de las escuelas de armería de Liège y Saint Etienne en la creación y posterior desarrollo de la escuela de armería de Eibar

María Cinta Caballer Vives

En la segunda mitad del siglo XVIII surgen en España algunos centros dedicados a la enseñanza de las artes y técnicas industriales debido al desarrollo científico que tiene lugar en este siglo[1]. Y el gran impulso que se imprime a la industria en el siglo XIX será el que posibilite, a partir de 1850, la creación de las Escuelas Industriales y, sobre todo durante las últimas décadas de dicho siglo y primeros años del siglo XX, la de las Escuelas de Artes y Oficios.

El Plan Pidal de 1845[2], si bien aporta importantes innovaciones en el ámbito educativo, decepciona por cuanto no se ocupa de las enseñanzas de tipo práctico. En este sentido, el R.D. de 4 de septiembre de 1850, considerado como el origen de los estudios de Ingeniero Industrial, viene a subsanar la anterior omisión. El Ministro Seijas Lozano, en la exposición del citado R.D. reconoce que los " estudios especiales (...) han sido los mas abandonados en nuestra patria "[3], manifesta su interés por la enseñanza industrial y plantea la creación de Escuelas Industriales, que estructurarán sus enseñanzas en tres grados : Elemental, de Ampliación y Superior[4]. Años más tarde, la Ley Moyano de 9 de septiembre de 1857[5], lleva a cabo una reorganización de estas enseñanzas industriales a nivel superior olvidando " del todo la enseñanza elemental y profesional "..." como si pudieran existir Ingenieros industriales sin industria

1. Ver también, la introducción de Aracil : J.M. Alonso Viguera, *La Ingeniera industrial Espanola en el siglo XIX*, Madrid, 1961 (1ª edicion 1944). Existe una edicion facsimil de 1993 ; G. Lusa Monforte, *La creación de la Escuela Industrial Barcelonesa (1851)*, Barcelona, 1993, 153 ; M.C. Caballer Vives, J. Olascoaga, *Alumnas de Rentería en el Real Seminario de Vergara*, 1850-1860, 1997.

2. Ministerio de Educacion, 1979, vol. 11, 191-239.

3. Plan de Estudios, ed Imprenta Nacional, 1851.

4. M.C. Caballer, I. Garaizor, I. Pellón, *El real Seminario Científico e Industrial de Vergara*, 1997.

5. Ministerio, vol. 11, 1979, 244-302.

en el país "[6]. Por R.D. de 5 de mayo de 1871 se crea la Escuela de Artes y Oficios de Madrid como centro dependiente del Real Conservatorio de Artes fundado en 1824 y, siendo Ministro de Fomento Navarro Rodrigo, mediante otro R.D. de 5 de noviembre de 1886, la Escuela de Madrid se separa del Conservatorio bajo el nombre de Escuela de Artes y Oficios Central, a la vez que se crean siete Escuelas de Distrito en Alcoi, Almería, Béjar, Gijón, Logroño, Santiago y Vilanova i la Geltrù[7]. La idea de " proporcionar educación y bienestar a familias pobres ", además de contribuir al " engrandecimiento " del país, que proclama este R.D., es la que moverá a docentes, ingenieros o políticos a reclamar de sus correspondientes Ayuntamientos y Diputaciones la creación de este tipo de establecimientos que, en muchos casos, contarán también con subvenciones del Gobierno de la Nación. En general se toman como referencia los existentes en otros países de Europa cuyos Gobiernos, al apostar por este tipo de centros, " hallaron la recompensa consiguiendo de este modo una perfección notable en sus industries y artes "[8], surgiendo Escuelas de Artes y Oficios, generalmente, en aquellas ciudades o pueblos en los que tiene lugar cierto desarrollo industrial, ante la necesidad de contar con una mano de obra cualificada. No es casualidad que sean los ingenieros industriales Ramón Manjarrés, Pablo de Alzola o Nicolás de Bustinduy[9] los que proyectan las Escuelas de Artes y Oficios de Barcelona (1874)[10], Bilbao (1879)[11], y San Sebastián (1880)[12] respectivamente.

En el País Vasco y, concretamente en la provincia de Gipuzkoa, además de la de San Sebastián, habrá Escuelas de Artes y Oficios en Tolosa, Rentería, Irún, Bergara, Eibar, Oñate, Zarauz, Zumárraga, Deva, Andoain, Lazcano, Cestona, Pasajes, Mondragón y Elgoibar[13]. Puesto que existe un vacío importante en la legislación de estos centros, además de la dependencia que tiene la mayoría de Ayuntamientos y Diputaciones, lo normal es que entre ellos haya diferencias significativas, pudiéndose distinguir aquellas en las que se imparten únicamente enseñanzas elementales como podría ser el caso de la de Eibar[14] que pese a ser un pueblo industrial en el que se detecta gran inquietud por la formación de los obreros, tal como se verá en este trabajo, nunca será una

6. Archivo Municipal de San Sebastián. A partir de ahora AMSS. AMSS Sec. B, Neg. 11, Ser. IV, Subser. III, Lib. 552, Exp. 5 : *Discursos pronunciados en el solemne acto de la inauguración que tuvo lugar el día 1 de Enero 1880.*

7. *Enciclopedia universal ilustrada*, 1977, 1095. En la Guia Docent (1993-94) de la Escola Universitària Politècnica de Vilanova i la Geltrú se hace referencia a un centro en esta ciudad desde 1881.

8. AMSS : *Discursos pronunciados...*

9. M.C Caballer Vives, *Noticia de la Escuela*, 1996.

10. G. Lusa Monforte, *Indistrialización...*, 1994, 67.

11. B. Del Hoyo, J.Llombart, *Noticia de la Escuela*, 1931.

12. M.C. Caballer Vives, *Noticia de la Escuela*, 1995.

13. Archivo General de Tolosa. A partir de ahora AGT. AGT : *Inventario.*

14. M.C. Caballer, *Noticia de las publicaciónes*, 1996.

Escuela en la que se impartan enseñanzas técnicas de cierto nivel, no pasando de ser un apéndice de la Academia de Dibujo ; otras que además de estas enseñanzas elementales acogen clases de Bachillerato como en Oñate[15] y aquellas que, en sus planes de estudios, proponen diferentes secciones como la de San Sebastián, con lo que se oferta un perfeccionamiento profesional del obrero.

EIBAR, CIUDAD ARMERA

Eibar es un municipio del País Vasco, que está situado en el límite occidental de la provincia de Gipuzkoa, lindando con la de Bizkaia. El Rey Alfonso XI, el 5 de febrero de 1346, a petición de los pobladores de la anteiglesia de Azitain, otorga la carta que permite construir la población de Villanueva de San Andrés que, a partir del siglo XV, pasará a conocerse como Eibar. Tiene una extensión de 24.5 km^2 y, en la actualidad, cuenta con una población aproximada de 32.000 habitantes. Cuando se habla del desarrollo histórico de la villa hay que referirse, sin duda, a su actividad artesanal que encuentra " en la fabricación de armas un sector propicio para la expansión industrial "[16].

A finales del siglo XIV Gipuzkoa fabrica armas y un siglo más tarde, finales del siglo XV, las armas de Eibar empiezan a adquirir fama universel[17]. Según Alberto Echaluce[18], citando a Juan San Martin, el eibarrés Francisco de Ibarra, conquistador de Nuevo Méjico, " llevó a Flandes armeros de Eibar y fundó las primeras industries armeras " y Juan Usabiaga[19] escribe que " armeros eibarreses llevaron sus conocimientos sobre la fabricación de armas a otros países, la zona de Lieja, entre otros ". Esta relación existente entre las ciudades armeras, Eibar y Lieja, será una constante como se podrá ver a lo largo de esta comunicación.

Se suelen diferenciar dos etapas en la armería eibarresa : " los buenos tiempos de los clásicos cañonistas, llaveros y cajeros, verdaderos artistas de quien habla con entusiasmo (...) Jovellanos " y una segunda etapa centrada en los años de finales del XIX y primeros del XX " tiempos intervenidos por el maquinismo " en el que la competencia de mercados diferentes, sobre todo el americano, y la dispersión en capitales pequeños de la industrie armera eibarresa, hacen que la " perfección técnica " sea suplida por la " baratura ", a excepción de las escopetas finas de caza[20]. En cualquier caso Eibar luchará

15. Archivo Municipal de Oñate. A partir de ahora AMO. AMO B-VIII-2/262-4 : *1913 Memoria de la Gestión de la Escuela de Artes y Oficios.*

16. M.M. Urteaga Artigas, Guía Histórico Monumental de Gipuzkoa, 1992, 126-129.

17. J.Thalamás, *Aspctos de la vida profesional vasca*, San Sebastián, 1935, 146.

18. Diario Vasco, 14/abril/1996 : *Eibar 650 años cumplidos.*

19. La voz de Guipúzcoa, 9 agosto 1927 : *Anhelo de superación. Eibar, Suiza, Norteamérica...*

20. Archivo Municipal de Eibar. A partir de ahora AME. AME Sec. B, Neg. 1, Ser. 16 : *Memoria cursos 1913-1915.*

para superar todo tipo de carencias, de manera que su industria no se quede rezagada. En 1866 se contabilizan en Eibar 5 fábricas, de las cuales, Orbea cuenta con 50 obreros y el resto con plantillas de 10 a 40 obreros[21]. A partir de 1890 cuando la electricidad se aplica como fuerza motriz[22] se detecta un auge de la industria armera empezando a descollar algunas empresas si bien todavía predominan las fábricas de tipo familiar. El ingeniero eibarrés[23] al hacer una reseña de las industrias de Eibar, da la cifra de 2.000 armeros en un pueblo de 5.000 habitantes, obreros que, en 1894, estarán al lado de sus empresarios y políticos apoyando a la Liga Nacional de Productos y la Liga Vizcaina en la campaña que sostienen " hasta conseguir la anulación del (...) Tratado de Comercio concertado con Alemania "[24] mediante el cual el Gobierno español compra armas a este país. Los industriales y obreros eibarreses pelearán por conserver sus mercados (fundamentalmente países sudamericanos) y por no ser invadidos por los mercados extranjeros. Que el Gobierno dispense " a la industria la protección que se debe " para que Eibar pueda adquirir máquinas ya que " lo principal, que es el obrero, lo tenemos, acaso como en ninguna otra parte "[25].

Las fábricas que utilizan cierta tecnología se dedican a las armas cortas, mientras que las escopetas se fabrican todavía de forma artesanal, lo que se traduce en precios no competitivos. Sobre esto escribe Pedro Sarasketa, enviado como corresponsal del periódico *El Pueblo Vasco* a la Exposición Universal de Lieja, que se celebró en 1905. El periodista señala a esta ciudad belga como " la más directa competidora de Eibar " y apunta en uno de sus artículos que la escopeta belga cuesta 50 pesetas mientras que la eibarresa cuesta 200, fruto de la diferencia de los métodos de fabricación utilizados. En otro articulo posteriori al referirse a la fabricación de las armas cortas, dice que el revolver eibarrés es superior[26].

Todo apunta a la urgencia de encontrar una fórmula que cambie los métodos a fin de mejorar en general la producción y esto conduce a la idea de establecer la Prueba Oficial de Armas y la creación de una Escuela de Armería[27].

Gestación de la escuela de armería

En primer lugar se darán unas referencias que, junto a las que se vayan a encontrar a lo largo de las páginas siguientes, pretenden ilustrar el hecho de que se tiene la mirada puesta en Bélgica como uno de los países punteros en

21. Diario Vasco.
22. Bustinduy, " Industria guipuzcoana ", 1894, 456-459.
23. N. Bustinduy, *La industria guipuzcoana en fin de siglo*, 1894, 119-131.
24. P. Alzola, " Fiestas enskavas... ", 1908, 222-229.
25. La voz de Guipúzcoa, 15/agosto/1894 : *El " Meeting " de Eibar.*
26. L. Castells, *Modernizacion y dinámica*, Madrid, 1987, 68-69.
27. AME Sec. B, Neg. 1, Ser. 16 : *Memoria cursos 1913-1915.*

el campo de las enseñanzas técnicas, y más concretamente en Lieja, ciudad con tradición armera de este país. Su Escuela de Armería será finalmente el modelo de escuela que los eibarreses desean para su pueblo.

A nivel general se puede citar, en primer lugar, el hecho de que a raíz de las ayudas concedidas por R.O. de 6 de abril de 1829 a los pensionados españoles, uno de los puntos de destino elegido por éstos es la Escuela de Ingenieros de Lieja[28]. Y en particular, en lo que concieme al País Vasco, a título de ejemplo, este mismo año de 1829 se forma una sociedad " para la explotación de dos minas de plata y otros metales en Oyarzun y de otra en la villa de Irún, compuesta de personas (...) del país y del comercio de Lieja "[29] o situándonos en Eibar, aproximadamente en 1885 " D. Mateo Orbea introdujo en Eibar el brochado de las piezas de revólveres, importado de Lieja " y, en 1888, la empresa eibarresa " Chantoya "[30] en la que trabaja, junto con su padre y hermano, Julián Echeverría, el hombre que será una pieza clave en el desarrollo de la Escuáela de Armería de Eibar, " contaba entre sus máquinas fresadoras horizontales de la Casa " Jaspar " de Lieja "[31].

Con motivo de la celebración de las Fiestas Eúskaras de 1908, el día 5 de septiembre, Pablo de Alzola[32] pronuncia un discurso en el que ensalza la forma en que est organizado el trabajo en Eibar repartido " en fábricas que ocupan centenares de obreros, en talleres dotados de escaso personal y mediante la labor a domicilio. Existe (...) la division del trabajo y el pago al tanto (...) que producen la baratura de la mano de obra y la fuerza incontrastable para la invasión en los mercados extranjeros, en donde las armas eibarresas de clase corriente, compiten con los artículos belgas, quienes, aparte de su abolengo industrial y de su gran preparación en las escuelas técnicas tan difundidas en aquel país, mantienen también el trabajo a domicilio ". Asimismo Alzola propone para Eibar " la implantación, como en Bélgica, de una Escuela de Armería, con su banco de prueba debidamente instalado ".

La idea está lanzada. Y el 1/2/1910 representantes del Ayuntamiento de Eibar solicitan del Ministro de Fomento la autorización para el establecimiento de una Escuela de Armería " similar a las que existen en Lieja y Birmingham[33] donde el obrero armero adquiera los conocimientos teóricos y prácticos nece-

28. G. Lusa-Monforte, *Contra los titanes de la rutina...*, 1994, 340.

29. I. Arbide, " Minas de Guipuzcoa ", 1984, 15.

30. P. Celaya, Olabari, " Historia de la Escuela de Armria de Eibar ", 1962, 114.

31. Escuela de Armería de Eibar (ed.), *Cataloga*, 1964, 10-12.

32. Pablo de Alzola y Minondo (San Sebastián, 27/6/1841-Bilbao, 25/10/1912) Ingeniero de Caminos, Canales y Puertos. Alcalde de Bilbao y Presidente de la Diputación de Vizcaya. Presidente de la Cámara de Comercio de la Liga de Productores y Director General de Obras Públicas (*Enciclopedia general ilustrada del País Vasco,* 1970, 567) y (Bilbao, 1970, 134-137).

33. Miguel Orbea, de la empresa Orbea, hizo una estancia en Birminghan para completar su formación (Escuela de Armería de Eibar, 1964, 9).

sarios para llegar a ser, no sólo un obrero manual, sino también un obrero inteligente "[34].

Para el industrial Pedro Goenaga, concejal del Ayuntamiento de Eibar, Alcalde de la villa cuando la Escuela de Armería será una realidad[35], es absolutamente imprescindible la creación de una " Escuela de Mecánica y Ajuste teórico-práctica "[36]. Goenaga visita la Escuela de Artes y Oficios de Barcelona, pero no es el tipo de establecimiento que él busca. Si resulta de su agrado la Escuela de la fábrica de maquinaria Loeme de Berlín ; sin embargo, el 7/6/ 1911 asiste a un Congreso de Banco de Pruebas en Lieja visitando la " École d'Armurerie " viendo en el centro " el origen de nuestra Escuela " según sus propias palabras[37]. Tanto le impresiona la Escuela de Lieja que a su regreso a Eibar de inmediato mueve los resortes que harán posible la creación de la Escuela de Armería de Eibar. El dia 23/7/1911[38] el ex-ministro Fermín Calbetón[39] visita en Eibar a su amigo Antonio Iturrioz que está enfermo y aprovechando esta circunstancia un " grupo selecto de industriales de Eibar "[40] organiza un almuerzo en el Castañal de la venta de Olarreaga, en el que se exponen al político los proyectos de Goenaga. Calbetón apoya la idea diciendo que " algo importante empieza a desarrollarse en estos momentos para Eibar. Cuando esta Escuela de Armería sea un hecho y el aprendizaje manual se una al científico, entonces Eibar dará un gran paso en su industria ".

En sesión de 1/7/1912[41] Goenaga secundado por los concejales Mendizabal, Erquiaga, Muguerza, Iriondo y Astigarraga, presenta una moción para la implantación de una Escuela de Armería, pidiendo que se cree una Comisión con los firmantes de la moción y el Alcalde Nemesio Astaburuaga como Presidente. El 15/7/1912 la Sociedad de Pistoleros de Eibar se adhiere a la propuesta de Goenaga. Y en sesión extraordinaria de 17/7/1912 la Comisión designa al Alcalde y los Concejales Iriondo y Goenaga para que se entrevisten con Calbetón a fin de recabar su opinión sobre la solicitud al Gobierno de la cesión del " rendimiento económico de la intervención de armas a favor del Ayuntamiento para destinarlo a la creación y sostenimiento de una Escuela de Armería ". El 5/8/1912 se informa del apoyo de Calbetón a la propuesta de los eibarreses y se acuerda preparar un estudio completo sobre el establecimiento,

34. R. Larrañaga, Síntesis historica Donostia-San Sebastián, 1981, 519-520.

35. J. Bengoeche, " Exposicion de Arte e industrias en Eigar... ", 1914, 554-562.

36. P. Celaya Olabari, " Historia de la Escuela de Armaria de Eibar ", 1962, 111- 112.

37. C. Barrena, " Pedro Goenaga, Eulogio Gárate... ", 1962, 149.

38. P. Celaya (1962, 111-112) da esta fecha, mientras que *La Voz de Guipúzcoa* del 27/7/1911 da la del 24/7/1911.

39. Fermín Calbetón y Blanchon (San Sebastián 4/9/1853-Madrid 4/2/1919) Ministro de Fomento en 1910 y Ministro de Hacienda en el Gabinete Romanones desde el 5 de diciembre de 1918 (*Enciclopedia general ilustrada del País Vasco*, 1975, 90-91).

40. J. Ormaechea Arregui, " En el Afán y el Hacer ", 1962, 20.

41. AME Sec. B, Neg. 1, Libro 17, Leg. 1 : *Acuerdos sobre la construcción de la Escuela de Armería.*

a fin de presentárselo al entonces Ministro de Hacienda Navarro Reverter que ha de visitar Eibar. Éste muestra una actitud favorable a lo que se pide y promete entrevistarse con el Director de Explosivos.

En otra sesión de 2/9/1912 se acuerda que la misma Comisión estudie el emplazamiento de la proyectada Escuela y en sesión extraordinaria de 1/10/1912 se trata del Reglamento, y las bases relacionadas con el proyecto de construcción y creación de la Escuela. Se presentan asimismo planos y presupuesto del edificio que deberá albergar el centro.

De nuevo se convoca una sesión el 11/12/1912 en la que se da cuenta de la cesión del 80 % del producto de guías y precintos para la expedición de armas a favor de la futura institución.

Organización de la Escuela

Se nombra una Comisión que redacta el Reglamento y los Estatutos. Esto se hace tomando como modelo los de las Escuelas de Lieja y Saint Etienne, si bien en su presentación se hace hincapié en que aunque se han servido de tales Reglamentos y " otros valiosos informes personales " relativos a la Escuela de Lieja, no obstante en ningún momento se pierden de vista " las condiciones y necesidades particulares de esta villa ".

Se fija como misión de la Escuela, " a semejanza de la de Lieja " formar obreros aptos enriquecidos por un cierto caudal de conocimientos teórico-prácticos, encauzados sistemáticamente al fin particular de la industrie armera. Esta filosofía se reflejará en los Planes de Estudios, en cuya elaboración se persigue aunar la teoría y la práctica concediendo gran relevancia al curso de *Máquina-herramienta,* disponiendo para ello " una sala con los elementos más modernos de máquinas americanas, donde el joven aprendiz puede instruirse en su manejo, de suerte que salga de la Escuela con la destreza y la habilidad de sus antepasados, y armado de aquellos conocimientos que exigen las necesidades modernas "[42].

El Comité Ejecutivo de la Escuela, el 17 de diciembre de 1912, se reúne con el Comandante de Artillería José Camicero solicitando su asesoramiento cara a la apertura de las clases y, resultado de este contacto, Carnicero se ofrece desinteresadamente y, con caracter interino, para impartir *Aritmética y Álgebra* además de asumir las funciones de Director de la Escuela.

La Escuela de Armería abre sus puertas el 1/l/1913 en un edificio provisional, mientras se trabaja en la construcción del definitivo.

El Presidente del Comité Ejecutivo, visita al Director de la Escuela de Armería de Lieja con el propósito de que éste encuentre una persona apta que dirija la de Eibar. La gestión no da resultados y después de otros intentos, el 24/8/1913, se le ofrece el cargo al eibarrés Julián Echeverría. En Eibar nadie

42. AME Sec. B, Neg. 1, Ser. 16 : *Memoria cursos 1913-1915.*

discute la valía y la experiencia en el sector armero de Echeverría pero, no teniendo ningún título académico, la idea no resulta del agrado de todos[43]. Pedro Goenaga aboga por este nombramiento y esgrime en defensa de Julián Echeverría que el Director de la Escuela de Lieja no poseyendo ninguna titulación está al frente de una institución de prestigio al tener " vocación demostrada teórica y prácticamente para la Mecánica y el Ajustaje "[44]. Ésto parece convencer a todos ya que la Junta general, celebrada el 25 de agosto de 1913, nombra por unanimidad Director a Echeverría, con un sueldo de 5.000 pesetas, quién el 1 de septiembre toma posesión de su cargo que ostentará hasta 1938, momento en el que será separado de sus funciones por motivos politicos derivados de la Guerra Civil española[45].

Ubicación de la Escuela

También al proyectar el edificio que ha de acoger la nueva institución se hace siguiendo el modelo del establecimiento de Lieja. Se presenta un proyecto que asciende a 200.000 pta, coste que parece excesivo ya que su aceptación va a suponer la imposibilidad de acometer otras empresas del Ayuntamiento. Se presupuesta un edificio de menor volumen por 100.000 pta., aunque claramente inferior en prestaciones. Goenaga consulta con el Arquitecto Augusto Aguirre la posibilidad de reducir el coste del proyecto inicial, renunciando únicamente a los elementos ornamentales. Se consigue de esta manera un nuevo presupuesto de 125.000 que es aprobado en sesión de 20/12/1912 adjudicándose las obras al constructor eibarrés Francisco Errasti por un valor de 116.800 pesetas. El mismo día que se inaugura la Escuela se procede a la colocación de la primera piedra[46] si bien a posteriori va a tener lugar una rectificación sobre el asentamiento inicial de modo que, medio año más tarde, el 16/6/1913 se aprueba la adquisición de terrenos en el alto de Isasi para la construcción del centro definitivo[47]. El edificio de la calle Isasi se inaugura el 24/6/1914 asistiendo al acto, entre otros, el entonces Ministro de Fomento Javier Ugarte, el Vicepresidente de la Diputación, el eibarrés Wenceslao Orbea y Fermín Calbetón. Coincidiendo con la inauguración del centro se celebra una Exposición Regional de Artes e Industrias en la que están presentes la mayoría de las industrias de la provincia[48].

43. T. Echevarría, *Viaje por el país de los recuerdos*, Donostia San Sebastián, 1968, 253-259.

44. P. Celaya Olabari, *Historia de la Escuela de Armeria de Eibar*, 1962, 114.

45. Archivo General de la Administración. Sección de Educación y Ciencia. A partir de ahora AGA. AGA Leg. 14536 : *Expediente de depuración de Julián Echeverría.*

46. La voz de Guipúzcoa, 7/enero/1903 : *La Escuela de Armería. Su primera piedra. La inauguración. Eibar y Calbetón.*

47. AME Sec. B, Neg. 1, Lib. 17, Leg. 1 : *Construcción de la Escuela de Armería.*

48. Comisión Exposición, " Exposición regional vasca de arte y provincial de industrias guipuzcoanas en Eibar ", 1914, 316-318 y J. Bengoeche, " Expósicion de Arte e industrias... ", 1914, 554-562.

El edificio consta de una planta baja y dos pisos. La mitad de la planta baja está ocupada por el taller de máquinas y la segunda mitad se dedica a biblioteca, museo, gabinete de Física, dirección y almacén. El primer piso alberga la sala de Ajustaje con " cabida holgada para setenta y cinco aprendices ". En el piso superior se ubican la sala de Dibujo y dos salas dedicadas a las clases de Matemáticas[49].

DESARROLLO DE LA ESCUELA DE ARMERÍA DE EIBAR

El desarrollo de la Escuela de Armería y de la industria eibarresa correrán parejos a partir de la creación del establecimiento. Durante los años de la Primera Guerra Mundial la producción de armas eibarresas aumenta considerablemente, de modo que, entre 1915 y 1918, se fabricarán 1.600.000 armas alcanzándose la cifra aproximada de 50.000.000 de pesetas[50]. Como consecuencia se sigue, no sólo un progreso industrial importante, sino también un perfeccionamiento de la mano de obra, " gracias a las exigencias de las naciones compradoras que desechan toda arma que no cumpla las condiciones establecidas ", responsabilidad que la Escuela asume en gran medida. La dirección del Centro se queja en la apertura del curso 1917 de que, seguramente debido a la abundancia del trabajo y a los buenos salarios que ofrecen las empresas, los jóvenes empiezan a trabajar de forma prematura descuidando la enseñanza elemental y " abandonando la profesional " que les brinda la Escuela de Armería[51].

Fermín Calbetón, siempre atento a los problemas de Eibar y cercano a los responsables de la Escuela de Armería, adelantándose claramente a lo que será una realidad años después, apunta, en 1917, la conveniencia de que la Escuela oferte enseñanzas relacionadas con otras industrias siderúrgicas más o menos similares a la armería que permitan ensanchar los horizontes de la industria eibarresa. De momento en Eibar sí se tiene la idea de modificar los planes de estudios, pero más bien en el sentido de enfocar la enseñanza hacia la especialidad de las armas largas, por considerar que la fabricación de escopetas de caza, es una fuente de riqueza superior a la de armas cortas. Para ello se pretende ampliar los elementos mecánicos de la sección de máquinas, creando un laboratorio metalúrgico " para conocer los metales física y químicamente y por metalografía estableciendo una sección de máquinas especiales para la construcción de cañones de escopetas o de fusiles ". Se apoya este argumento, de nuevo con la mirada puesta en Lieja que exporta 1.000.000 de armas largas al

49. AME Sec. B, Neg. 1, Serie 16 : *Escuela de Armería y Mecánica creada en 1913 por el Municipio de Eibar.*

50. Escuela de Armería de Eibar (ed.), *Catálogo del Museo de Armas,* San Sebastián Industrias graficas, 1964, 10.

51. AME Sec. B, Neg. 1, Ser. 16 : *Memoria curso 1916.*

año frente a 500.000 cortas o en Birmingham y Saint-Etienne, dedicadas fundamentalmente a la producción de armas largas.

Cuando se crea la Escuela de Eibar se da la circunstancia de que la fabricación de las armas cortas predomina en la industria local razón por la que se enfocan los estudios hacia la mecánica que éstas requieren, aún cuando la Escuela " en su plan y organización, es, en todo cuanto ha sido posible, trasunto fiel de aquella que con tanto éxito funciona en Lieja ", cuya " enseñanza profesional (...) está encauzada casi exclusivamente al ramo de armas largas, y así, aquella escuela cuenta con secciones de basculeros, llaveros, cajeros, cañonistas, cinceladores, grabadores, etc. ; además de la sección de *petite-mécanique* "[52].

En otro orden de cosas, al hacer un balance de la preparación de los alumnos de la Escuela de Armería, se observa que al concluir los estudios e ingresar en un taller, tienen que adaptarse a " una especialización dentro de la especialización de la industria armera " con lo que el titulado por la Escuela pasa por un período de adaptación que a veces resulta decepcionante. Hay quienes piensan que no es malo que el obrero se de cuenta de que es capaz de adaptarse a las características de su empresa tras recibir una educación generalista, pero desde la Escuela se asume que una mayor especialización de las profesiones es lo que imponen las circunstancias del momento y de nuevo se recurre a estudiar los programas seguidos en las Escuelas que desde un principio se han tenido como referencia : Lieja y Saint-Etienne y se concluye que en éstas se tiene en cuenta una mayor especialización. Los planes de estudios vigentes en ambos establecimientos son, a grandes rasgos, un primer curso en el que se imparten enseñanzas generales, un segundo curso especial con contenidos que tienen que ver con la armería con el curso correspondiente dedicado a la " pequeña mecánica " y, por último, un tercer curso cuyas enseñanzas están divididas en dos secciones : una de herramientas y otra de armeros (llaveros, basculeros, cajeros, cañonistas, armadores, grabadores...)

Con este modelo, desde el curso 1920, en Eibar se seguirá con los dos primeros años de lima que " ha de ser siempre un sólido cimiento que permita cambiar de especialidad sin grandes contratiempos, y no ocurrirá luego el caso de que se desdeñen ciertos oficios como el de cajero por ejemplo, que por ser típicamente armero no encuentra campo para adaptarse a otras artes ", y un tercer curso especializado que, " por vía de ensayo ", se dividirá en ajustadores armeros y ajustadores herramentistas. En este momento se apunta la conveniencia de ampliar la docencia con un cuarto año voluntario para intensificar los conocimientos de taller a fin de que los armeros puedan aprender el manejo

52. AME Sec. B, Neg. 1, Ser. 16 : *Memoria curso 1917*. Ver también en Escuela de Armería (1964, 13-19) la correspondencia habida entre J. Echeverría y F. Calbetón donde se trata esta cuestión amén de los proyectos de J. Echeverría para la creación de armas largas.

de las máquinas-herramientas[53].

Tras iniciar esta nueva etapa del establecimiento eibarrés, su Director Julián Echeverría durante el mes de agosto se desplaza a Lieja, con el objeto de visitar las Escuelas de Armería y de Mecánica. En el informe que presenta a su regreso hace constar que en la de Armería " los cursos manuales de taller son los que absorben la casi totalidad de las horas " viendo en esto la razón por la que los trabajos que le son mostrados son de un gran nivel. Por otra parte, en la Escuela de Mecánica sus más de 300 alumnos tiene sus horarios repartidos en clases teóricas y prácticas a semejanza de lo que se está haciendo en Eibar, en cuanto a la sección de Mecánica. Sin embargo la Escuela de Lieja cuenta " con una magnífica sección de modelería para fundición, ocupando buen número de jóvenes ", sección que es propósito de los eibarreses implantar " cuanto antes ". Una de las razones por las que Echeverría viaja a Lieja es la de estudiar " la sección de culatas (cajeros) de armas largas " con el fin de implantarla en Eibar y se tiene idea de que dicha sección se organice dentro del curso 1921[54].

Como consecuencia de esta estancia en Lieja se inicia el año escolar con reformas, en cuanto a los planes de estudios se refiere, intensificándose las clases prácticas en las dos especialidades profesionales, limitando la parte teórica " a una minima expresión ", además de facilitar el ingreso de los alumnos. Sólo pasarán a seguir los cursos con contenido teórico aquellos alumnos que tengan aptitud para ello, como si se tratara de otra sección, medida que parece ser que servirá para enriquecer la oferta de la Escuela sin que ello afecte a la programación de contenidos generales. En realidad la sección teórica tendrá por objeto " completar la capacitación técnica de la industria armera " lo que en esencia vendría a ser la reunión en una sola Escuela " de las dos que con tanto éxito funcionan en Lieja : la Escuela de Mecánica y la Escuela de Armería "[55].

La enseñanza de la Mecánica irá cobrando fuerza día a día y, fiel al espíritu innovador que le caracteriza, Julián Echeverría, en 1923, presenta un proyecto para la creación de un " Taller modelo en la Escuela de Armería y Mecánica de Eibar, para la Construcción y Venta de Máquinas-herramienta y útiles de corte y medición ". Se trata de un taller de producción paralelo al de aprendizaje que, dotado de " los recursos mecánicos más modernos ", permitirá la elaboración de " toda clase de herramientas de corte y medición, que apenas produce nuestra industria nacional (...) como tornos mecánicos, máquinas de rectificar y todo utillaje costoso que se necesita para la fabricación de armas y similares ".

La ubicación del taller no supone ningún problema, dado que el Ayuntamiento acaba de adquirir unos terrenos contiguos a la Escuela que permitirán la

53. AME Sec. B, Neg. 1, Ser. 16 : *Memoria curso 1919.*
54. AME Sec. B, Neg. 1, Ser. 16 : *Memoria curso 1920.*
55. AME Sec. B, Neg. 1, Ser. 16 : *Memoria curso 1921.*

reorganización de sus dependencias en la forma que mejor convenga para dar cabida a la que se pretende incluir. Por otra parte existen en la Escuela recursos técnicos y humanos que se han ido acumulando en sus diez años de existencia, además del prestigio que el establecimiento ha adquirido con los servicios prestados en la industria local, lo que ha de ser sin duda garantía para que cualquier empresa iniciada por el centro tenga éxito. Muy importante es también, el grado de perfección que podrán adquirir los alumnos y las especialidades para las que estarán capacitados, dada la variedad de trabajos a que van a poder dedicarse. Y aún se argumenta que no sólo el alumno, sino la propia Escuela va a salir beneficiada con los ingresos que perciba con la venta de los artículos que se fabriquen y " sobre todo, la industrie de nuestra región, tan necesitada de un taller modelo (…), que les proporcione el costoso utillaje que (…) dificilmente pueden encontrar en la industria nacional ". Se acompaña este proyecto de un presupuesto aproximado que se detalla a continuación en la (Tabla 1).

La idea apuntada por Fermín Calbetón en 1917 ya es una realidad. Prácticamente en todas las solicitudes de ingreso a la Escuela se eligen los estudios que conducen a la profesión de mecánico. Y esto no sucede únicamente en Eibar, pues los responsables de la Escuela eibarresa siguen atentos los cambios que se operan en Lieja y en Saint-Etienne y es un indicador a tener en cuenta el hecho de que " se haya extinguido la sección de armería en la Escuela de Industria de Saint-Etienne y recientemente se haya presentado en la Escuela de Armería de Lieja, una moción pidiendo se instituya la sección mecánica "[56].

Tabla 1 : Presupuesto aproximado de instalación de un taller mecánico en la Escuela de Armería de Eibar, para la construcción de máquinas-herramientas y utillaje de corte y medición.

Construcción de un nuevo pabellón adosado a la Escuela de 24 por 9 metros de planta y 10 metros de altura, con sótano y una planta, para trasladar las oficinas, museo y laboratorio de ensayos físicos, que actualmente ocupan el lugar designado para el taller de aprendizaje	Ptas. 30.000
Una copilladora-puente de dos metros de trabajo longitudinal útil y 0,60 transversal	9.600
Una mandrinadora de árbol de 60 m/m	11.250
Una máquina de rectificar ejes hasta 1.500 m/m de longitud	17.000
Una máquina de destalonar fresas	5.000
Un torno herramentista Pratt & Whitney	6.000
Cuatro tornos mecánicos modelo común de 150 por 1.000 m/m a 2.500	10.000

56. AME Sec. B, Neg. 1, Ser. 16 : *Memoria curso 1923.*

Material neumático para chorro de arena para limpiar piezas fundidas y templadas	6.700
Accesorios e instalación de las máquinas	4.450
Total	100.000

Fuente Ame : *Memoria curso 1923*

Este proyecto se presenta al Ayuntamiento y a la Diputación siendo apoyado por ambas entidades en 1925, año en que consignan en sus presupuestos las cantidades que para la construcción del taller les habían sido solicitadas[57]. Sin embargo las obras de ampliación de la Escuela quedan en suspenso durante los años de la Dictadura de Primo de Rivera, máxime cuando de los cuatro cursos que se imparten en el curso 1925-26, se pasará de nuevo a un plan de tres años, en el que no tiene cabida la intensificación de mecánica prevista. Esta situación se mantiene hasta el curso académico 1930-31, siendo en el curso 1931-32 cuando de nuevo se implanta el cuarto año complementario que permite " completar la Escuela con la preparación de Maestros Industriales ", habiéndose hecho realidad la ampliación del establecimiento[58].

Con la Guerra Civil española de 1936 se interrumpen las clases, que se reanudarán en octubre de 1938 merced al tesón de los eibarreses, empezando, como en 1913, sólo con el curso primero que fue seguido por 28 alumnos. La Escuela crece a gran ritmo lo que supone sucesivas ampliaciones y a su vez deudas que, al no poder ser asumidas por la institución, conducirán a sus responsables a cederla al Ministerio de Educación Nacional el 10/5/1954. Esto hace que la Escuela pierda autonomía, si bien la Junta Local que la gestiona tiene " algunos privilegios concedidos por la Dirección General ". Sin embargo gracias a esta cesión la Escuela puede seguir creciendo al ritmo que demanda el pueblo de Eibar[59].

A MODO DE CONCLUSIÓN : LA ESCUELA DE ARMERÍA EN LA TRANSFORMACIÓN INDUSTRIAL DE EIBAR

La Escuela de Armería de Eibar, se crea con el objetivo de enseñar a sus alumnos los conocimientos teóricos y prácticos necesarios para que estos puedan convertirse en obreros armeros cualificados. Siempre con la mirada puesta en las Escuelas de Lieja y Saint-Etienne y de acuerdo con la adaptación a la demanda de los mercados, el centro se impone como misión ofertar a sus estudiantes enseñanzas que traspasen el ámbito armero, para lo cual se crea una sección nueva que preparará obreros mecánicos en general.

Se deriva desde la especialidad armera hacia la mecánica en la forma que

57. AME Sec. B, Neg. 1, Ser. 16 : *Memoria curso 1925.*

58. AME Sec. B, Neg. 1, Ser. 16 : *Memorias cursos 1925-26, 1926-27, 1927-28, 1928-29, 1929-30, 1930-31 y 1931-32.*

59. P. Celaya Olabari, " Historia de la Escuela... ", 1962, 118-119.

apunta Fermín Calbetón cuando todavía la industria armera está en su apogeo, ya que es en 1917 cuando se alcanza el máximo de producción en 734 736 armas[60]. Cuando realmente se toma la decisión de potenciar la sección mecánica, se hace, posiblemente entre otras razones, por la crisis que afecta al sector armero en el decenio 1920-1930 que obliga a los industriales a fabricar aquellos productos que demanda el mercado[61]. En cualquier caso la Escuela de Armería bajo la dirección de J. Echeverría sabe estar a la altura de lo que le exige la sociedad y es un hecho reconocido la influencia decisiva que tiene la Escuela de Armería en la transformación industrial de Eibar. Para Juan San Martín[62] es la sensibilidad y saber hacer de Echeverría lo que hace posible el paso a las enseñanzas de mecánica de precisión, " impulsando una nueva era para la industria eibarresa, a la que capacitó para toda clase de fabricantes (...) para producir desde la máquina de coser hasta la piecería más complicada del automóvil ". A título de ejemplo, Toribio Echevarría[63] explica cómo en Alfa, cooperativa creada en 1920, se pasa de la producción de armas a la de máquinas de coser " un producto técnicamente similar o asimilable a nuestra habitual producción en cuanto al posible aprovechamiento de nuestros equipos mecánicos y las especializaciones profesionales del personal ".

En 1963 además de armeros y damasquinadores hay en Eibar obreros trabajando en empresas que fabrican bicicletas, máquinas de coser, motos, resortes, órganos, repuestos del automóvil, compresores, objetos de escritorio, maquinaria, tijeras e instrumentas de precisión[64]. Como consecuencia del enfoque que la Escuela sabe imprimir a la formación de los profesionales, la industria eibarresa es capaz de reaccionar cuando sobreviene la crisis del sector armero y de este modo surgen en Eibar esta variedad de empresas que, en su gran mayoría, crean, y/o dirigen los ex-alumnos de la Escuela de Armería, " con nuevo concepto práctico de la organización industrial en todos sus aspectos "[65].

De una encuesta hecha en 1957 se desprende que las 380 industrias existentes en esta en la villa armera, en su mayor parte, están dirigidas por ex-alumnos de la Escuela, mientras que las industrias nacionales regidas por antiguos alumnos son más de 600[66].

En palabras de San Martin[67] éstos son los frutos que ha dado la Escuela de Armería, " mayores que lo esperado por los pioneros de la " Eskuadra-zarra ", el ingeniero Bustinduy y Calbetón, e incluso el propio Julián Echeverría ".

60. Diario Vasco.

61. R. Larrañaga /S.Gorrochategui, *500 años de la armeria vasca Eibar*, Eibar, 1990.

62. J. San Martín, *Monografia Histórica...*, 1984.

63. T. Echevarria, *Viaje por el país...*, Donostia-San Sebastián, 1968, 343.

64. *Boletin della Asociacíon*, t. 29, 1963, 45.

65. *Boletin della Asociacíon*, t. 43, 1966, 5-23.

66. P. Celaya Olabari, Eibar, *Sintesis de Monografia historica,* San Sebastián, 1970, 50.

67. J. San Martin, " Origen y desanvolvimento de la Industria Metalúrgica en Eibar ", 1962, 238.

Archivos consultados

Archivo Municipal de Eibar. Gipuzkoa.
Archivo Municipal de Donostia-San Sebastián. Gipuzkoa.
Archivo Municipal de Oñate. Gipuzkoa.
Archivo General de Tolosa. Gipuzkoa.
Archivo General de la Administración. Alcalá de Henares.

Prensa

Diario Vasco.
La Voz de Guipúzcoa.

Bibliografía

J.M. Alonso Viguera, *La Ingeniería Industrial Española en el siglo XIX,* 2ª edición, Madrid, Servicio de publicaciones de la ETSII, 1961 (1ª edición 1944). Existe una edición facsímil de 1993.

P. Alzola, " Fiestas euskaras de Eibar celebradas en el año 1908. Extracto del discurso pronunciado por Pedro de Alzola ", *Euskal Erria*, vol. 59 (1908), 222-229.

Asociación de Antiguos Alumnos (ed.), *Boletín de la Asociación de Antiguos Alumnos. Escuela de Armería*, n° 29 (1963), 45.

Asociación de Antiguos Alumnos (ed.), *Boletín de la Asociación de Antiguos Alumnos. Escuela de Armería*, n° 43 (1966), 5-23.

I. Arbide, " Minas de Guipuzcoa ", *Oiartzun*, (1984), 15.

C. Barrena, " Pedro Goenaga, Eulogio Gárate y José Bolumburu hablan de los tiempos difíciles de la Escuela. Tres vidas ejemplares que han luchado en favor de la Institución Eibarresa ", *Revista extraordinaria de la Escuela de Armería de Eibar. Cincuentenario. 1912-1962*, (1962), 149-150.

J. Bengoeche, " Exposición de Arte e Industrias en Eibar. Solemne inauguración ", *Euskal Erria*, (1914), 554-562.

Comisión Exposición, " Exposición regional vasca de arte y provincial de industrias guipuzcoanas en Eibar ", *Euskal Erria*, (1914), 316-318.

J. Bilbao, *Enciclopedia General Ilustrada del País Vasco*, Cuerpo C, Eusko Bibliografia, Vol. 1, San Sebastián, Auñamendi, Estornes Lasa Hnos, 1970, 134-137.

N. Bustinduy, *La industria Guipuzcoana en fin de siglo. Reseña de las industrias fabriles más importantes*, San Sebastián, Establecimiento tipográfico de " La Unión Vascongada ", 1894.

N. Bustinduy, " Industria guipuzcoana ", *Boletin Central de los Ingenieros Industriales*, n° 15, (1894), 456-459.

M.C. Caballer Vives, *Noticia de la Escuela de Artes y Oficios de San Sebastián*, V Simposio de Historia e Ensino das Ciencias, Vigo, 13-16 de septiembre 1995.

M.C. Caballer Vives, *Noticia de la creación y primeros años de funcionamiento de la Escuela de Armería de Eibar*, VI Congreso de la SEHCYT, Segovia-La Granja, 9-13 de septiembre 1996.

M.C. Caballer Vives, *Noticia de las publicaciones del ingeniero industrial Nicolás de Bustinduy y Vergara (Eibar 1850-San Sebastián 1928)*, IV Trobada d'Història de la Ciència y de la Tècnica, Alcoi, 13-15 de desembre 1996.

M.C. Caballer Vives, *Alumnos hispanoamericanos y filipinos en el Real Seminario Científico e Industrial de Vergara (1850-1860)*, Il Mesa Redonda sobre Historia de la Medicina Iberoamericana. " Médicos Vascos en América y Filipinas ", San Sebastián, 23 de mayo 1997.

M.C. Caballer, J. Olascoaga, *Alumnos de Rentería en el Real Seminario de Vergara*, 1997. En prensa.

M.C. Caballer, I. Garaizar, I. Pellón, *El Real Seminario Científico e Industrial de Vergara, 1850-1860*, 1997. En prensa.

L. Castells, *Modernización y dinámica política en la Sociedad guipuzcona de la Restauración, 1876-1915*, Madrid, Siglo XXI, 1987.

P. Celaya Olabari, *Eibar. Síntesis de Monografía Histórica*, San Sebastián, 1970.

P. Celaya Olabari, " Historia de la Escuela de Armería de Eibar ", *Revista extraordinaria de la Escuela de Armería de Eibar. Cincuentenario. 1912-1962*, (1962), 111-119.

T. Echevarría, *Viaje por el país de los recuerdos*, Donostia-San Sebastián, Edición del Ayuntamiento de Eibar-Eibarko Udala, Aurrezki Kutxa Munizipala, Caja de Ahorros Municipal, 1990, (1968) .

Enciclopedia General Ilustrada del País Vasco, Cuerpo A, Diccionario Enciclopédico Vasco, Vol. VI, San Sebastián, Auñamendi, Estornes Lasa Hnos, 1975, 90-91

Enciclopedia General Ilustrada del País Vasco, Cuerpo A, Diccionario Enciclopédico Vasco, Vol. V, San Sebastián, Auñamendi, Estomes Lasa Hnos, 1970, 567.

Enciclopedia General Ilustrada, t. XX, Madrid, Espasa Calpe, S.A., 1977, 1095.

Escuela de Armería de Eibar (ed.), *Catálogo del Museo de Armas*, San Sebastián, Industrias Gráficas, 1964.

B. Del Hoyo, J. Llombart, *Noticia de la Escuela de Artes y Oficios de Bilbao (1879-1938)*, XIXth International Congress of History of Sciences, Zaragoza, 22-29 de agosto 1993.

Escola Universitària Politècnica de Vilanova i la Geltrú (EUPVG) (ed.), *Guia docent*, 1993-1994, (1993).

Imprenta Nacional (ed.), *Plan de Estudios decretado por S.M. en 28 de Agosto de 1850, y Reglamento para su ejecución, aprobado por Real Decreto de 10 de Setiembre de 1851*, Madrid, 1851.

R. Larrañaga, *Síntesis histórica de la armería vasca*, Donostia-San Sebastián, Caja de Ahorros Provincial de Guipúzcoa, 1981.

R. Larrañaga, S. Gorrochategui, *500 años de la armería vasca, Eibar*, Eibar, Ayuntamiento de Eibar con la colaboración del Departamento de Economía y Planificación del Gobierno Vasco y del Departamento de Cultura de la Diputación Foral de Guipúzcoa, 1990.

G. Lusa Monforte, *La creación de la Escuela Industrial Barcelonesa (1851)*, Barcelona, Societat Catalana d'Història de la Ciència y de la Tècnica, 1993.

G. Lusa Monforte, *Industrialización y Educación : Los ingenieros industriales (Barcelona, 1851-1886)*, Tècnica y Societat en el Mòn Contemporani, I Jornades, Maig, 1992. R. Enrich, G. Lusa, M. Mañosa, X. Moreno, A. Roca (eds), Museu d'Història de Sabadell, 1994.

G. Lusa Monforte, *Contra los titanes de la rutina. La cuestión de la formación matemática de los ingenieros industriales (Barcelona, 1851-1910),* " Contra los titanes de la rutina ", S. Garma, D. Flament, V. Navarro (eds.), Madrid, Comunidad de Madrid y C.S.I.C, 1994).

Ministerio de Educación y Ciencia (ed.), *Historia de la Educación en España*, t. II : " De las Cortes de Cádiz a la Revolución de 1868 ", Madrid, Servicio de publicaciones del Ministerio de Educación y Ciencia, 1985.

J. Ormarechea Arregui, " En el afán y el hacer ", *Revista extraordinaria de la Escuela de Armería de Eibar. Cincuentenario. 1912-1962*, (1962), 23-29.

J. San Martín, " Origen y desenvolvimiento de la Industria Metalúrgica en Eibar ", *Revista extraordinaria de la Escuela de Armería de Eibar. Cincuentenario. 1912-1962*, (1962), 238.

J. San Martín, *Monografía Histórica de la Villa de Eibar,* Prólogo escrito para la 3ª edición del libro de G. Mújica, 1984.

J. Thalamás, *Aspectos de la vida profesional vasca*, San Sebastiàn, 1935.

M.M. Urteaga Artigas, *Guía Histórico Monumental de Gipuzkoa*, Donostia-San Sebastián, Diputación Foral de Gipuzcoa-Arkeolan, 1992.

SCIENCE, TECHNOLOGY AND THE ARMAMENTS INDUSTRY, IN GREAT BRITAIN (1854-1914)

Marshall J. BASTABLE

British engineers responded to the outbreak of the Crimean War with a spontaneous outpouring of ideas and outlines of designs for new weapons which they believed would bring speedy victory. The War Office, which was in charge of all military and naval weaponry, demurred at the prospect of civilians in their midst, but when the siege at Sebastopol bogged down, the British government overrode this objection and opened its treasury and the government arsenal at Woolwich to civilian engineers to research and develop their ideas for more efficient and powerful guns[1]. Sir Joseph Whitworth applied his famous skills at precision engineering to rifling British small arms, I.K. Brunel, the famous shipbuilder, worked on his design for a big gun[2], and James Nasmyth, whose idea for a giant mortar shell had been ignored by the government before the war, was now given full backing, as were other, lesser-known civilian engineers whose requests for government assistance to develop their ideas for new weapons had previously gone unanswered[3].

The enthusiastic embrace of military science and technology by English engineers, combined with an aroused public opinion over the conduct of the war and the growing desperation of the British government to find the means to take Sebastopol quickly, set in motion a long process of " militarizing " British science, technology and industry. The war sparked the interest of some of Britain's most innovative and successful entrepreneurs who continued to pursue armaments as a new field of science, technology and commercial enterprise

1. " Correspondence on Reorganization ", London, Public Record Office, War Office Papers 33/1.

2. I. Brunel, *The Life of Isambard Kingdom Brunel,* London, David and Charles, 1971, 454-461.

3. " Return of the Amount of Public Money advanced since 1852 to Private Persons for the purpose of enabling them to make Experiments for the purpose of Improving Weapons of War ", *Parliamentary Papers,* XLI (1860), 657.

after the war had ended. The Crimean War, in short, launched a military revolution in that it provided the opportunity for a new generation of British entrepreneurs to apply their industrial knowledge in new ways, to create a new leading sector of research and development as their predecessors had done with textiles and railways, and to establish themselves as the industrial entrepreneurs of a new age.

One of the leading figures in this revolution was William Armstrong, an innovative and commercially successful engineer, well-known within the engineering community but not as famous as Whitworth, Brunel and Nasmyth. Armstrong had shown no interest in military technology until the Crimean War, when he read newspaper reports about the difficulties moving Britain's fifty-year-old cast iron artillery about the battlefield at Inkerman. In other words, Armstrong was attracted to armaments first as an interesting engineering problem revealed by the war, and not by the idea of armaments as a new business. Once the technology had been developed and demand for the gun grew, business calculations became more important, but for the moment the project was purely technological. He drew up a design for a new light-weight wrought-iron artillery piece and was one of those invited by the government to develop his idea in the Woolwich arsenal, receiving £ 7.219 to develop his new gun[4]. Within just six months Armstrong built an artillery piece that overcame technological problems that had frustrated artillery makers since the fifteenth century, and set new standards not only in gun making, but in engineering in general. He designed and built the first rifled, breech-loaded, wrought iron gun which fired various kinds of exploding shells, set on a carriage that automatically returned the gun to firing position. All in all the Armstrong gun was a revolutionary piece of military technology[5]. The war ended before he had perfected the weapon, but unlike previous times, the end of the war did not close down research and development of new weapons. Between 1855 and 1859 Armstrong produced several sizes of his new guns (from 12- to 110-pounders), which proved themselves against strong competitors like Whitworth[6]. The Admiralty correctly saw that Armstrong's new guns would " affect the size and structure of ships of war ", an insight into the coming revolution

4. " Return of the Amount of Public money advanced for Experiments on Weapons ", *op. cit.*

5. Armstrong describes his gun in his " Report on the Construction of Wrought Iron Field Guns ", London Public Record Office, War Office Papers 33/11, and in *The Times*, (Nov. 25, 1861), 12, 17. See also D. Dougan, *The Great Gunmaker : The Story of Lord Armstrong*, Newcastle, Graham, 1971, 57-61, F. Robertson, *The Evolution of Naval Armament*, London, Constable, 1921, 200-201, and F.S. Stoney and C. Jones, *A Textbook of the Construction and Manufacture of the Rifled Ordnance in the British Service,* 2nd ed., London, 1872, ch. 2, *Encylopaedia Britannica*, 9th ed., 1890, vol. 11, 288. See also M. Bastable, " From Breechloaders to Monster Guns : Sir William Armstrong and the Invention of Modern Artillery, 1854-1880 ", *Technology and Culture*, 33 (1992), 213-47.

6. *Confidential Report,* London Public Record Office, War Office Papers 33/9 ; " Report of the Select Committee on Ordnance ", *Parliamentary Papers,* XI (1863) ; and Sir Emerson Tennent, *The Story of the Guns,* London, 1864, 125-126. Tennent was a champion of the Whitworth gun.

in warship design to carry the very large Armstrong rifled muzzle-loaded guns[7]. Armstrong breech-loaders were tested successfully in actual battle in China in 1859[8], and against Japan in 1864[9]. Armstrong's guns were also put on board Britain's new iron-hulled warship *Warrior*[10], a decision by the British Admiralty which further enhanced Armstrong's already substantial national and international reputation as a pre-eminent and innovative armaments engineer.

In 1859 a French invasion scare accelerated the demand for Armstrong's gun in Britain. Armstrong's contract with the government included the creation of the Elswick Ordnance Company, a private company which was guaranteed the right to sell a sufficient number of Armstrong guns to the War Office to ensure the success of the new venture. Armstrong thus introduced a novel commercial monopoly into the production of armaments in Britain. Between 1859 and 1863 the Elswick Ordnance Company sold the War Office £ 1.525.000 worth of guns of various sizes, with Woolwich manufacturing £ 1.035.000 worth[11]. The commercialization of armaments production in Britain was furthered when the arms race with France ended in 1863 and the government terminated its contract with Elswick, forcing the company to seek orders in the global market. The timing was propitious as 1863 saw the beginning of wars of national unification in Europe and civil war in America. Over the next two decades Armstrong developed his very large rifled muzzle-loaders, moved into warship building and sold the most up-to-date armaments of unprecedented power to states around the world[12].

After 1880 the technological power of Elswick — its new breech-loaded steel guns, its new Quick-Firing guns, its new cruisers — could no longer be ignored by the British Government and the pressure increased for a greater reliance on the private production of armaments. The second generation of armaments developed by Armstrong destroyed any remaining pretensions that Woolwich was an innovative arsenal. Another French war scare contrived in 1884 provided the Admiralty with the political leverage to gain the right to buy guns and warships once again from Armstrong and generally increase its pur-

7. " Report of a Committee of the Treasury to Inquire into the navy Estimates, from 1852 to 1858, and into the Comparative States of the Navies of England and France ", (Derby Committee), *Parliamentary Papers,* XIV, Part 1 (1859).

8. Letter from Sir James Hope Grant to Sidney Herbert, Oct. 5, 1859, in Stanmore, *Sidney Herbert, op. cit.*, II, 298-299.

9. " A Copy of the Report of Admiral Kuper in reference to the Armstrong Guns in the Action of Simonosaki ", *Parliamentary Papers,* XXXII (1865), 309-18. Although problems arose with the breech mechanism on the 110-pounders, the largest Armstrong breech-loaders, the 40-pounders performed well.

10. W. Brownlee, "H.M.S. Warrior", *Scientific American,* vol. 257 (1987), 130-136.

11. " Statement Showing Total Amount Expended in the Manufacture of Guns and Projectiles on Sir William Armstrong's Principle ", *Parliamentary Papers,* XI (1863), 558.

12. M. Bastable, " From Breechloaders to Monster Guns : Sir William Armstrong and the Invention of Modern Artillery, 1854-1880 ", *Technology and Culture*, 33 (1992), 213-247.

chases from various private shipbuilders[13]. The War Office claim that it manufactured large guns cheaper than the private trade was successfully challenged at the 1887 Naval Estimates hearings[14]. Woolwich fought a long but losing battle against the private production of armaments, but the tide had turned and the commercialization of armaments production accelerated, driven by international imperial rivalries in Europe and Asia, rising capital costs and the technological sophistication of armaments engineering.

In 1887 the first of a number of Parliamentary inquiries on the question of public versus private production of armaments began the process of privatizing British armaments. From the Committee on the Private Armaments Manufacturer (Morley Committee) of 1887 to the Government Factory and Workshop Committee in 1907 (Murray Committee) the trend is clear : private manufacturers acquired a greater share of the increasing orders for guns and warships for the British Admiralty. Armstrong's engineering talent was acknowledged by the government when he was once again invited to advise the Ordnance Committee on the design and construction of guns and become a major supplier of warships to the Admiralty[15]. Armstrong's company had grown on world markets between 1863 and 1884, and that market would expand, but the unprecedented expansion and profitability of the business over the next two and a half decades resulted from British Admiralty orders. Armstrong had commercialized the global armaments market, and now he would play a major role in commercializing British armaments production, and in submitting it to the economic imperatives of industrial capitalism.

Armstrong did not enter the armaments industry in 1863 because of falling profits in his civilian engineering business. But the Great Depression of the 1880s and 1890s hit the profits of British iron and steel companies and shipbuilders especially hard, and many looked to armaments as the solution. At the same time government was withdrawing from armaments production and encouraged private manufacturers to take up armaments. Thus Vickers militarized its heavy engineering business in 1887 after several years of very low profits, with promises of orders from the government. In 1905 three old British shipbuilders — John Brown, Cammell Laird and Fairfields — amalgamated as the Coventry Ordnance Works, further expanding the armaments sector of British industry.

Similar developments took place in other countries. Henri Schneider's works at Le Creusot entered the international armaments market in 1884

13. W. Ashworth, " Economic Aspects of Late Victorian Naval Administration ", *Economic History Review*, 22 (1969), 502.

14. " Report from the Select Committee on Army and Navy Estimates ", *Parliamentary Papers,* VIII (1887), Evidence, Qs. 7016-83, 7116-59, 7425-29.

15. " Recommendations of the Ordnance Committee (with Special Associated Members) As to the Construction of Ordnance ", *Parliamentary Papers,* XIV (1884-85). The date of Armstrong's appointment to this committee is revealed in Hansard, Aug. 24, 1886, cols. 373-374.

because armour plate had become the only market for their steel. Within ten years they had become a major private manufacturer of guns and warships, the only world market in which France could compete[16]. In America too big steel companies amalgamated themselves into giant armaments. In Japan, Spain, Belgium, Sweden, Italy and Australia, the mining iron ore or the manufacturing of steel plates or sophisticated guns and warships became the core business of the world's largest industrial businesses[17]. Armstrong and Vickers expanded through vertical and horizontal amalgamation, creating vast industrial empires which competed vigorously with each other. But they also co-operated in market sharing and foreign investments, and participated in international agreements with the big American and German armaments companies[18].

During the first decade of the twentieth century a few giant companies dominated the armaments industry, but innumerable smaller engineering companies depended on armaments for their survival. British shipbuilders, squeezed by falling world demand and high fixed costs, turned to the government and the high-profit margins in warship construction[19]. Naval warship building saved the British shipbuilding industry but at the cost of accelerating and intensifying naval arms races around the world : between Brazil and Argentina, Turkey and Greece, Austria and Italy, as well as the main one between Britain and Germany[20]. The great naval arms races of 1900-1914 were as much a consequence of the militarization of production as they were a result of real international problems. The industrialized countries — Germany, Britain, the United States, France — all turned to armaments and the global arms market to maintain their industrial base and profits.

The armaments industry also acquired an economic and political weight that governments could not ignore. Armaments technologies and engineering standards influenced non-military innovations such as the automobile and the bicycle, and stimulated further industrialization[21]. Naval armaments saved the British steel and shipbuilding industries, and laid the technological and indus-

16. A. Milward, S.B. Saul, *The Development of the Economies of Continental Europe, 1850-1914,* Cambridge (Mass.), 1977, 89.

17. C. Northcote Parkinson, *The Rise of Big Business,* London, 1977.

18. D. Todd, *Defence Industries : A Global Perspective,* London, Routledge, 1988, ch. 4.

19. E. Lorenz, F. Wilkinson, *The Decline of the British Economy,* Oxford, 1986 ; H.B. Peebles, *Warshipbuilding on the Clyde : Naval Orders and the Prosperity of the Clyde Shipbuilding Industry, 1889-1939*, Edinburgh, 1987.

20. E.L. Woodward, *Great Britain and the German Navy*, London, 1964 ; P.G. Halpern, *The Mediterranean Naval Situation, 1908-1914,* Cambridge (Mass.), 1971 ; L. Sondhaus, *The Naval Policy of Austria-Hungary, 1867-1918 : Navalism, Industrial Development, and the Politics of Dualism,* West Lafayette, Purdue University Press, 1994 ; A. Iriye, " Japan's Drive to Great-Power Status ", M.B. Jansen (ed.), *The Cambridge History of Japan,* Vol. 5, 721-782.

21. See the works of C. Trebilcock : " Spin-off in British Economic History : Armaments and Industry, 1760-1914 ", *Economic History Review,* 22 (1969), 474-490 ; " British Armaments and European Industrialization, 1890-1914 ", *Economic History Review,* 26 (1973), 254-72 ; and " Science, Technology and the Armaments Industry in the UK and Europe, with special Reference to the Period 1880-1914 ", *Journal of European Economic History,* 22 (1993), 565-580.

trial base for the production of automobiles and airplanes. Non-industrial states looked to armaments as an important means to launch general industrialization.

The armaments industry had the highest value-added production, the highest profits, the highest wages, and employed of tens of thousands of workers. In short the armaments industry acquired a domestic economic and political importance which pushed the production of armaments beyond actual defense requirements[22]. Diplomacy and international relations became centred on armaments races.

The commercialization of armaments turned weapons into a global business and turned the production of armaments into a global industry, with production centres for the latest weapons in Europe, the United States and Japan. European industrial states made up the largest market of course, followed by the United States and Japan, but small states in South America, Asia and in the Mediterranean were also important sectors of the global armaments market. The international co-operation of the infamous armaments " rings " and cartels, the collective investments abroad, the extension of credit to purchasers were collective strategies that diluted individual autonomy but maintained profit levels. Armaments companies were thus major players in the creation of global capitalism and the diffusion of technology around the world in the pre-First World War era.

The interplay between technological innovation and the social and political environment of the day cannot be easily separated into causes and effects and the debate will continue between those who see technology as an expression of an existing social-political system and those who insist that technological evolution has an inner momentum which carries society and people's minds along with it[23]. The dramatic leaps in heavy gun technology which began with Sir William Armstrong's little 3-pounder gun of 1855 seem to have been associated with some larger struggle, strategic crisis or economic problem, and men like Armstrong were no doubt driven by the profit motive. On the other hand, the technological development of armaments was an evolutionary process to which even great states had to wait upon. Moreover, it is not entirely certain that Armstrong, at least in 1854-1855, was motivated only by thoughts of profit but also had a " simple " desire to solve an engineering problem. On

22. V. Berghahn, " On the Societal Function of Wilhelmine Armaments Policy ", in G. Iggers (ed.), *The Social History of Politics : Critical Perspectives in West German Historical Writing Since 1945,* New York, 1985, 155-78 ; A. Marder, *The Anatomy of British Sea Power : A History of British Naval Policy in the Pre-Dreadnought Era, 1880-1905,* New York, 1940, ch. III.

23. The tension between these two perspectives are apparent in W.H. McNeill's, *Pursuit of Power : Technology, Armed Force, and Society Since A.D. 1000,* Chicago, University of Chicago Press, 1982. At one moment (p. 278) he suggests that " it is impossible to say " how much weight should be assigned to " technological innovation as an autonomous element " in the interplay among technology, domestic politics and international rivalries, but a little later (p. 294) he seems confident that " the rush of new technology that cascaded upon the Royal Navy after 1884... began to get out of control itself ". See also his remarks on p. 298 and 306.

the other hand, as Elswick grew into a large company and military and naval technology surpassed " civilian " engineering, the quest for profits and markets became absolutely essential to the survival of the company. At the same time the Elswick came to be seen by the British government as essential to the industrial and strategic survival of Britain as a great power. This was the basis of partnership between Elswick and the British Foreign Office and the broader political-industrial complex which developed in Britain between 1880 and 1914. It was a goal towards which Sir William Armstrong had worked since the day in 1854 when he was introduced to the Secretary of State for War. The relationship between Elswick and the British state had taken 60 years to establish itself but by 1914 it had become a fundamental component of the new age of warfare that was about to commence.

After 1884 strong bonds developed between the steel and shipbuilding industries and the government which translated into an expanded naval armaments industry[24]. Naval scares seemed to come at the end of the business cycle or as a naval program wound down[25]. In 1902, *The Economist* considered that " interesting and important as are the new Admiralty contracts from a naval and military point of view, they are scarcely of less importance from an economic standpoint " and, given the economic stagnation in that vital industry, concluded that " the allocation of naval contracts among private builders at this time is welcomed as a great economic relief "[26]. States that could not afford to buy great battleships were loaned the money to do so and, bringing *The Economist* to conclude that " the world is overloaned and overarmed ", as armament manufacturers, banking interests and diplomacy combined to sell their wares to the world[27]. By 1914 science and technology had been thoroughly militarized in the engineering industry.

24. The new links between industry and Government were forged by the " Report of the Committee on the Organization and Administration of the Manufacturing Departments of the Army " (Morley Committee), *Parliamentary Papers,* XIV (1887). See also the " Letter from Certain Steel Manufacturers of Sheffield to the Chairman of the Select Committee on Army Estimates Together with His Reply ", *Parliamentary Papers*, LXXXI (1888), 529-531. The manufacturers were Firth, Vickers and Cammell ; the chairman of the committee was Randolph Churchill. See also Hansard, March 16, 1885, col. 1298-99.

25. A. Marder, *The Anatomy of British Sea Power*, Hamden, Conn., 1964, ch. 3 ; S. Pollard, " Laissez-Faire and Shipbuilding ", *Economic History Review*, 5 (1952), 98-115 ; O.J. Hale, *Publicity and Diplomacy : With Special Reference to England and Germany, 1890-1914*, New York, D. Appleton-Century, 1940 ; A.J.A. Morris, *The Scaremongers : The Advocacy of War and Rearmament, 1896-1914*, London, Routledge & Kegan Paul, 1984.

26. " Industrial Aspects of the Admiralty Contracts ", *The Economist* (April 12, 1902), 564.

27. " Overloaned and Overarmed ", *The Economist*, 76 (May 24, 1913), 1274-1275.

An Unknown Naval Accident and the Development Trajectory of the *Kanpon* Type Turbine in Pre-war Japan

Miwao Matsumoto

Introduction

This paper describes and analyses the trajectories of technological development in pre-war Japan. These trajectories were much more complex than the conventional success stories suggest. The paper focuses on an accident that occurred with the standard marine steam turbine of the Imperial Japanese Navy immediately before the outbreak of war with the U.S. in 1941. The turbine in question was called the *Kanpon* type turbine. This accident was treated as top secret because of its timing. The suppression of information about the accident means that it has not been considered as an event in the history of Japanese technology. However, an analysis of this accident suggests that technological development in pre-war Japan departed significantly from the simple path leading from the study of Western science and technology to introduction and improvement, then to self-reliant development. This implies that we need to revise our view of the social history of technology in pre-war Japan.

Section one uses prior references and records written at the time of the accident to show how the accident was kept secret for such a long time. Section two describes and analyses the accident in detail based upon newly discovered materials owned by Ryutaro Shibuya. Finally, Section three discusses the significance of the accident, in particular its social implications in relation to the development trajectory of Japanese technology.

An unknown accident just before the outbreak of war in 1941

As far as we are able to confirm at present, there are only four prior trade editions containing references to the accident. In 1952, seven years after the war ended, the first reference appeared in *All about Technology for Building Warships* compiled under the leadership of Michizo Sendo who was an engineering Rear Admiral of the Navy[1]. Four years later, the second reference appeared in the book entitled *On the Imperial Japanese Navy* written by Masanori Ito who was a Mainichi newspaper reporter and was a graduate from the Naval Academy[2]. In 1969, the third reference appeared in *The Military Equipment of the Navy* compiled by the War History Unit, the National Defence College, the Defence Agency[3]. In 1981, twelve years later, the last reference appeared in *The Navy* compiled by the Institute for Historical Record Compilation on the Imperial Japanese Navy[4].

Among these, the third reference in 1969 provides the most authentic history of the accident. It says : " At the end of December, 1937, cracks were discovered on the turbine blades of the destroyers just launched and of other destroyers which were being tested before launch. A Special Examination Committee, established on 19 January, 1938, continued surveys, studies, and experiments... This crack might have been considered as a mere accidental event. If it had been overlooked, the turbines of all the fleet would have been troubled within several years and led to entire stoppage. ...The committee made every effort to pinpoint the exact cause and to take remedial measures. ...As a result, one cause was identified. It was proved that the cause would be likely to affect most turbines on the warships. Orders for remedial measures were issued and implemented ... by autumn in 1941 "[5]. The Special Examination Committee appearing in this reference was called *Rinkicho* in Japanese. After that this paper will call the accident the *Rinkicho* accident hereafter.

The time of the publication and the author/editor vary from one reference to another. Arranging these prior references into order, however, all these prior references were made by the parties concerned with the Imperial Japanese Navy (see Table 1).

Table 1. Prior references to the accident

Year of Reference	Author/Editor
1952	Former Engineering Rear Admiral of the Navy
1956	Mainichi newspaper reporter (Graduate from the Naval Academy)

1. M. Sendo, *et al.*, *Zokan Gijutsu no Zenbo* (*All about Technology for Building Warships*), Tokyo, Koyosha, 1952, 247-249.

2. M. Ito, *Dai Kaigun o Omou* (*On the Imperial Japanese Navy*), Tokyo, Bungei Shunju Sha, 1956, 439-440.

3. The War History Unit, the National Defence College, the Defence Agency (ed.), *Kaigun Gunsenbi (1)* (*The Military Equipment of the Navy, part 1*), Tokyo, Choun Shinbunsha, 1969, 621-622.

4. The Institute for Historical Record Compilation on the Japanese Imperial Navy (ed.), *Kaigun* (*The Navy*), vol. 9, Tokyo, Seibun Tosho, 1981, 161.

5. The War History Unit, the National Defence College, the Defence Agency (ed.), *op. cit.*, 621-622.

1969	War History Unit, the National Defence College, the Defence Agency
1981	Institute for Historical Record Compilation on Navy

And the facts depicted in these references almost agree in their main contents. A newly built destroyer encountered an unexpected turbine blade crack accident. Since the accident was with an engine of a standard design called the *Kanpon* type turbine, it raised a fuss. However, the cause was soon identified, so nothing serious happened. These references make up a kind of success story. It is extremely difficult to look into further details of the accident because little evidence is specified in detail to prove what is stated by these prior references. It appears that the accident was kept secret because it occurred just before the outbreak of the war. It is likely that the " official view " presented by the parties concerned has functioned to seal the accident and prevent it from being known to others.

To endorse this likelihood, let us examine government documents which were recorded around the time the accident occurred. The government documents consulted here are the minutes of Imperial Diet sessions regarding the Navy. The minutes of the 57th Imperial Diet session to the 75th Imperial Diet session (1930-1940) contain no less than 7.000 pages about Navy-related discussions. These discussions include the following 11 subjects regarding warship accidents (see Table 2)[6].

Table 2. Warship accidents taken up at the Imperial Diet Sessions : 1930-1940

Date	Descriptions
February, 1931	Questions about the cause of the collision between the cruiser *Abukuma* and *Kitakami.*
March, 1931	Questions about the measures taken before and after the collision between the cruiser *Abukuma* and *Kitakami* during the large-scale manoeuvres in 1930 and the responsibility of the authorities.
March, 1933	Questions about the Minister of the Navy's view on the expense for repairing the damage to a destroyer and on the fact that the destroyer struck a well-known sunken rock.
March, 1935	Request for information about the results of research on the contact among four destroyers, on duty for training, on the Ariake Bay as reported in newspapers.

6. The original materials were prepared by Kaigun Daijin Kanbo Rinji Chosa Ka (the Temporary Research Section, the Minister of the Navy's Secretariat of the day, abbreviated as T.R.S. hereafter). The description is based upon T.R.S. (ed.), Teikoku Gikai Kaigun Kankei Giji Sokki Roku (Minutes of Imperial Diet Sessions regarding Navy-related Subjects), Bekkan 1, 2, reprinted edition, Tokyo, Hara Shobo, 1984.

May, 1936	Request for information about the degree of collision between submarines I-53 and I-63 and the amount of money drawn from the reserve.
May, 1936	Request for detailed information about the degree of damage to a destroyer due to violent waves in September, 1935.
February, 1939	Brief explanation of the accident encountered by the submarine I-63.
February, 1939	Brief explanation of the accident encountered by the submarine I-63.
February, 1939	Request for brief explanation of the sinking of a submarine due to collision during manoeuvres.
February, 1940	Brief report on the completion of the salvage of the sunk submarine I-63.
February, 1940	Brief report on the completion of the salvage of the sunk submarine I-63.

It is noteworthy in this Table that the Fourth Squadron accident occurred in September 1935 that was one of the most serious accidents in the Navy history of Japan, was made public and discussed at Imperial Diet sessions within a year (on 18th May, 1936)[7]. The *Rinkicho* accident occurred on 29th December, 1937. It has been handed down by the related agents of the Navy and counted by them among the major accidents together with the Fourth Squadron accidents[8]. As indicated above, however, there is no sign showing that the *Rinkicho* accident was made public and discussed at Imperial Diet sessions more than two years after the accident.

As will be discussed, reports on the accident had been already submitted during the period from March to November, 1938 (the final report was submitted on 2nd November). The Imperial Diet heard of nothing about the accident or about any details of the procedures taken to deal with the accident. The *Rinkicho* accident was so serious that it could influence the decision on whether to begin war against the U.S. The Fourth Squadron accident was also serious so as to influence the decision, but it was made public and discussed at Imperial Diet sessions. In this respect, there is a marked difference between the two accidents. Regarding the Fourth Squadron accident, the Naval Accounting Bureau Director, Harukazu Murakami, was forced to answer to a question on 18th May, 1936 at the 69th Imperial Diet session.

Although his answer gave no information regarding the damage to human resources, it accurately stated the facts of the accident and the material damage

7. The Tomozuru accident on 11th March, 1934 was the first major accident that the warship building staff of the Navy faced. Only one year and a half after this accident, a more serious accident occurred on 26th September, 1935 - the Fourth Squadron accident.

8. Based on interviews made by the present writer with Dr. Seikan Ishigai (on 4th September, 1987 ; 2nd June, 1993) and with Dr. Yasuo Takeda (on 25th September, 1996 ; 19th March, 1997).

amounting to 2.8 million yens[9]. It should be noted here that 40 million yens were expected to be needed to take remedial measures with the problem of the turbines of all Japanese warships disclosed by the *Rinkicho* accident[10]. Nevertheless, no detailed open report on the *Rinkicho* accident was presented at the Imperial Diet. This fact strongly indicates that the *Rinkicho* accident was a profound secret within the Imperial Japanese Navy. What was the underlying fact ? This paper tries to grasp the core of the *Rinkicho* accident based upon the documents owned by Ryutaro Shibuya who was an Engineering Vice Admiral of the Navy then responsible for designing turbines of Japanese warships (these documents will be called the Shibuya documents hereafter).

RETHINKING THE STORY OF THE ACCIDENT : A NEW SOCIAL HISTORY OF TECHNOLOGY IN PRE-WAR JAPAN

First we give a brief account of the Shibuya documents. When Japan was defeated, most military organisations were ordered to burn documents they had kept. Many documents of the Imperial Japanese Navy were burnt, before the General Headquarters of the U.S. Occupation Forces ordered the Japanese government to submit documents regarding the war. In such circumstances, ex-managers and ex-directors of the Imperial Japanese Navy had meetings and decided to undertake a research project for collecting, examining, and preserving technical documents as far as possible. The project was funded with 500 thousand yen from the budget reserved for extra military expenditure. The Shibuya documents were collected and owned by Ryutaro Shibuya through this project.

The documents are huge[11]. Even though we choose only the materials directly regarding the *Rinkicho* accident, we cannot present here the full details of all of them. To get an overview, accordingly, this paper shows a brief chronological table indicating the main events concerning the accident[12].

29th December, 1937 : The accident occurred.

19th January, 1938 : The Minister of the Navy's Secretariat Military Secret n° 266 was issued to state that it was decided to establish the Special Examination Committee.

3rd February, 1938 : The Minister of the Navy's Secretariat Secret Instruction n° 566 was issued to state that it was decided to examine the vibration per-

9. T.R.S., *op. cit.*, vol. 3, part 1, 86.

10. Ryutaro Shibuya, Jugo Zuihitsu (Essays), sono 4 (n.d.).

11. As far as we are able to confirm at present, they are made up of 4.219 materials ranging from steam turbine blades to the casualty of the atomic bomb.

12. Based on S. Murata, Asashio Gata Shu Tahbin no Jiko (The Accidents of the Main Turbines of the Destroyers, *Asashio* Class), n.d.

formance of the main turbine blade and turbine rotor installed on the warships at the Hiro Naval Dockyard.

August, 1938 : The Technical Headquarters' Secret n° 15332 was issued to specify the methods of static and dynamic vibration tests on turbine blades and rotors.

2nd November, 1938 : Report from the Committee (Top Secret n° 35 summarising the 53 subcommittee meetings and 13 general meetings held so far).

1st April, 1939 : The Minister of the Navy's Secretariat Secret Instruction n° 1973 was issued to state that it was decided to select a representative warship from the existing warships and conduct long-time load tests, according to the remedy implementation schedule suggested by the committee's report.

12th February, 1940 : The Minister of the Navy's Secretariat Secret Instruction n° 1122 was issued to state that it was decided to begin turbine rotor load tests at the Engine Experiment Department, the Maizuru Naval Dockyard in April, 1940.

6th May, 1940 : The Minister of the Navy's Secretariat Secret Instruction n° 3185 was issued to postpone the modification to the main turbines of the existing warships.

20th June, 1941 : The Minister of the Navy's Secretariat Secret Instruction n° 5389 was issued to state that it was decided to postpone the completion of turbine rotor load tests to March, 1943, postpone the modification to the main turbines of the warships, and make the final decision by consulting the results of on-the-ground tests by the end of June, 1943.

8th December, 1941 : War against the U.S. began.

As shown above, the Special Examination Committee was established to investigate the accident in January, 1938. The purpose of the committee was as follows : " Problems were found on the turbines of destroyers... It is necessary to work out remedial measures... The research activities for that must be performed freely without any restrictions imposed by experience and practice in the past. The Special Examination Committee has been established to fulfil this purpose "[13].

The organisation was as follows[14]:

(a) General members who do not attend subcommittee meetings

Chair : Isoroku Yamamoto, Vice Admiral, Administrative Vice Minister of the Navy ; Members : Rear Admiral Inoue, Director of the Bureau of Naval Affairs, the Ministry of the Navy, and other five members.

(b) Subcommittees

1st subcommittee for dealing with engine design and planning

13. The Minister of the Navy's secretariat Military Secret n° 266, issued on 19th January 1938.

14. The description of the organisation of the Special Examination Committee is based upon the *Rinkicho* Report, Top Secret n° 35, issued on 2nd November 1938, Appended Sheets.

Members : Leader ; Shipbuilding Vice Admiral Fukuma, Director of the 5th Department (including the Turbine Group), the Technical Headquarters of the Navy, and other nine members.

2nd subcommittee for dealing with the maximum engine power and weights/ sizes suitable for it Members : Leader ; Rear Admiral Mikawa, Director of the 2nd Department, the Naval General Staff, and other eleven members.

3rd subcommittee for dealing with prior studies, experiments, systems and their operations Members : Leader ; Rear Admiral Iwamura, Director of the General Affairs Department, the Technical Headquarters of the Navy, and other ten members.

When we avoid counting the duplicative members of the subcommittees to arrange the members into order by the section they belong to, we obtain the following result (see Table 3).

Table 3. Members of the Special Examination Committee by Section

Section of the Members	Number (chair)
Administrative Vice Minister of the Navy	1
Technical Headquarters of the Navy	15
Bureau of Naval Affairs	8
Naval General Staff	5
Naval Staff College	3
Naval Engineering School	1
Total	33

Tracing back to the history of the marine steam turbine in Japan, the Navy began to adopt all-gear reduction type turbines in 1918[15]. During the period from this time to October, 1944, various accidents occurred on the turbines of warships. When these accidents are classified by their location, the accidents on the turbine blade account for 60 percent of all, amounting to far more than 400 cases[16]. Thus the Imperial Japanese Navy encountered many turbine blade troubles for many years and accumulated experience in handling that.

Accordingly it shouldn't be surprising if the Special Examination Committee took the accident as merely a routine problem from the outset based upon such rich prior experience. In fact, the Special Examination Committee drew a conclusion made up of two points both of which were in line with such accu-

15. As for the initial process of transferring the marine steam turbine to Japan and its subsequent development up to the 1920s, *Cf.* M. Matsumoto, " The Imperial Japanese Navy's Connection with a Marine Steam Turbine Transfer from the West ", *Historia Scientiarum*, vol. 6, n° 3 (1997), 209-227 ; M. Matsumoto, " Le jeu des rôles autour d'une turbine à vapeur ", *Les Cahiers de Science & Vie*, n° 41 (1997), 80-90.

16. Based on Seisan Gijutsu Kyokai (ed.), *Kyu Kaigun Kantei Johki Tahbin Kosho Kiroku* (Record of the Troubles and Accidents of Turbines of the Imperial Japanese Navy), Tokyo, Seisan Gijutsu Kyokai (not for sale), 1954, 1-2.

mulated experience. First, the accident was caused by insufficient strength of the blade ; second, turbine rotor vibration made the problem of insufficient strength emerge[17]. On the basis of this conclusion, a plan was worked out to improve the design of the blades and turbine rotors of all warships. Naturally it was decided that the figure of blades was modified so as to make their stress distributions lower, which in turn was to enhance their strength[18]. According to the plan, the improvement of this sort for 61 warships' turbines was indicated as the first step[19]. At the same time, Vice Admiral Ryutaro Shibuya then responsible for turbine design was punished for improper design of the turbines[20].

However, the condition of the blade breakage in question in the accident was significantly different from that in the past. The blade in most cases had been broken at the base fixed to the turbine rotor. In contrast, the problem of the *Rinkicho* accident was such that one-third of the blade from the tip was broken off. The breakage as described in the record written at that time is as follows : " Moving blades of the 2nd and 3rd stages of the intermediate-pressure turbines were broken. ...Every broken moving blade was located at 40 to 70 mm distance from the tip "[21].

The significance of this point was eventually noticed by Yoshio Kubota who was an Engineering Captain of the Navy. He strongly recommended that confirmation tests should be conducted again on warships of the same type, though the committee had already reported its conclusion as mentioned above. He argued that if turbine rotor vibration was the true cause, then the accident would be repeatable when the engine was run continuously at a critical speed causing rotor vibration (nearly 6/10 to 10/10 of the full speed)[22]. Accepting this recommendation, the Navy at last dared to initiate continuous-run tests equivalent to ten-year runs on 1st April, 1939. No accident occurred. This raised the urgent need for considering the possibility of another cause. An order was issued promptly to postpone the modification to the turbine blades and rotors of all warships. At the same time, a new study restarted to identify the cause. For this purpose, a preliminary on-the-ground test was conducted at the Maizuru Naval Dockyard and more thorough one followed at the Hiro Naval Dockyard to confirm the conditions that would make the accident recur. However, the test was so difficult that the schedule for identifying the cause

17. *Rinkicho* Report, Top Secret n° 35, issued on 2nd November 1938.

18. Typical examples of such modification can be found in *Rinkicho* Report, Top Secret n° 1, issued on 18th February 1938, Appended Plans.

19. *Rinkicho* Report, Top Secret n° 1, issued on 18th February 1938 through *Rinkicho* Report, Top Secret n° 27, issued on 13th October 1938.

20. *Ryutaro Shibuya, Kyu Kaigun Gijutsu Shiryo* (Technical Records on the Imperial Japanese Navy), Tokyo, 1970 (not for sale), vol. 1, Chapter 4, 48.

21. *Rinkicho* Report, Top Secret n° 1, issued on 18th February 1938.

22. Records on a hearing from Yoshio Kubota on 19th March, 1955 ; *Yoshio Kubota, 85 Nen no Kaiso* (Reminiscences of 85 Years), Tokyo, not for sale, 1981, 50-51.

was postponed to middle 1943 by the Minister of the Navy's Secretariat Secret Instruction n° 5389[23]. This means that Japan had naval turbines which were potentially defective for some unknown reason when it began war against the U.S. in 1941.

What then was the true cause ? It was resonance of two-node vibration. What was particularly complicated with the cause was that it could invite various vibration problems below full speed. And this true cause was identified by the final report of the Special Examination Committee in April of 1943 — almost one year and a half after the war began[24]. Speaking in terms of technology, thus the Japanese government began the war in haste notwithstanding the fact that it potentially had unknown but serious problems with the standard engine of warships. This paper will conclude with the implications of the fact for the development trajectory of Japanese technology.

CONCLUSION

Based upon what is mentioned above, the conclusion of the paper is essentially related to the reason why we can say the accident " unknown ". For one thing, it is because the accident was much more serious than expected from previous experience and therefore kept secret to outsiders. The fact requires us to reconsider the trajectory of technological development in pre-war Japan. According to a standard view of the history of technology, particularly of the marine engineering in pre-war Japan, Japan proceeded to the self-reliant phase by the establishment of the *Kanpon* type turbine developed by the Navy by the 1920s, after having made original improvement through various troubles and accidents[25]. In short, a successful self-reliant phase followed after various failures in improvement phase. And it has been assumed up to now that this trajectory enabled Japan to begin the war. According to this paper on the " unknown " accident, however, the trajectory becomes much more complex since there was an unknown missing phase. That is the self-reliant failure which was unable to be completely resolved by the outbreak of the war.

On the other hand, there is another reason for saying the accident " unknown ". It is because two-node turbine blade resonance detected as the true cause was beyond the standard of turbine design of the day. Generally, such problem is supposed not to have been recognised until post-war period.

23. *Kaigun Kansei Honbu Dai 5 Bu, Rinji Kikan Chosa Iinkai Hokoku ni kansuru Shu Tahbin Kaizo narabini Jikken Kenkyu ni kansuru Hokoku* (A Report on the Remedy and the Experimental Research of the Main Turbines in connection with the *Rinkicho* Report), Bessatsu, 1st April (1943).

24. *Ibid.*

25. As for the first *Kanpon* type turbine, see M. Matsumoto, *op. cit.* As to a brief history of steam turbine development, see H.W. Dickinson, *A Short History of the Steam Engine*, Cambridge, Cambridge Univ. Press, 1938, Chapter 11-Chapter 14 ; A comprehensive analyses of world-wide turbine development trajectory in its socio-historical contexts can be found in E.W. Constant II, *The Origin of the Turbojet Revolution*, Baltimore, The Johns Hopkins Univ. Press, 1980.

Still in post-war period, an article on *QE2* turbine reported the similar accident occurred in 1969[26]. It is true that the Imperial Japanese Navy managed to detect the true cause during the war mobilisation after serious *Rinkicho* accident. But its complete resolution seems not to have been obtained even after the detection of the true cause, since the same type turbine blade breaking still occurred in the same class destroyers in 1944, more than one year after the final report of the Special Examination Committee had been submitted[27]. As a matter of fact, a post-war development of industrial sectors started from the careful re-examination of the two-node vibration problem left over by pre-war military sectors, as far as concerned with the Japanese turbine development trajectory[28]. In this sense, the " unknown " accident described above as a self-reliant failure of pre-war period gives an important but long unnoticed initial state of spin-off process from military to industrial sectors.

Acknowledgement

I wish in particular to thank the members of the Shibuya Bunko Chosa Iinkai (Research Committee on the Shibuya Documents), who have been compiling the Shibuya Documents since their first discovery in 1991. Research for this paper was supported by Mitsubishi Foundation Grant.

26. See " Report on QE2 Turbines ", *Shipbuilding and Machinery Review*, 13th March (1969), 24-25.

27. Seisan Gijutsu Kyokai, *op. cit.*, 158-159.

28. *Cf.* Kawasaki Tahbin Sekkei Shiryo (Kawasaki Turbine Design Materials), Dai 2 Bu (October, 1955) ; A letter from Yasuo Takeda, Kawasaki Shipbuilding Company to Kanji Toshima, IHI (n.d.), *et al.*

Contributors

Bernard Allaire
Bordeaux (France)

Réginald Auger
Université Laval
Québec (Canada)

Dominique Barjot
Université Paris-Sorbonne (Paris IV)
Paris (France)

Marshall J. Bastable
Windsor (Canada)

Heloisa Maria Bertol Domingues
Museu de Astronomia
e Ciências Afins
Rio de Janeiro (Brazil)

María Cinta Caballer Vives
Universidad del País Vasco
Donostia - San Sebastián (Spain)

Fernando L. Carneiro
Universidade Federal
do Rio do Janeiro
Rio de Janeiro (Brazil)

Ignacio M. Carrion Arregui
Universidad del País Vasco
Donostia - San Sebastián (Spain)

Georges Comet
Université de Provence
Aix-en-Provence (France)

Marie-Claude Déprez-Masson
Université de Montréal
Montréal (Canada)

Marc J. de Vries
Eindhoven University of Technology
Eindhoven (The Netherlands)

Walter Endrei
Budapest (Hungary)

Dany Fougères
INRS-Urbanisation
Montréal (Canada)

Robert Gagnon
CIRST UQAM
Montréal (Canada)

Encarnación Hidalgo Cámara
Museo Nacional de Ciencia
y Tecnologia
Madrid (Spain)

Richard Leslie Hills
Cheshire (United Kingdom)

Claudio Katz
Buenos Aires (Argentina)

Vasco La Salvia
Roma (Italy)

Michel Lette
EHESS
Paris (France)

Henrique Lins de Barros
Museu de Astronomia e Ciências Afins
Rio de Janeiro (Brazil)

Mauro Lins de Barros
Museu de Astronomia e Ciências Afins
Rio de Janeiro (Brazil)

Miwao MATSUMOTO
University of Tokyo
Tokyo (Japan)

Sigeko NISIO
Nihon University
Tokyo (Japan)

Michel ORIS
FNRS
Université de Liège
Liège (Belgique)

M.A. OSTROVSKY
Russian Academy of Sciences
Moscow (Russia)

Michael SCHIMEK
Museumsdorf Cloppenburg
Cloppenburg (Germany)

Joop SCHOPMAN
Innsbruck (Austria)

Christian SICHAU
Technisches Museum
Wien (Austria)

V.M. TCHESNOV
Russian Academy of Sciences
Moscow (Russia)

Dominique THEILE
Paris (France)

Philippe TOMSIN
Université de Liège
Liège (Belgique)

Michel TRÉPANIER
INRS-Urbanisation
Montréal (Canada)

Eisui UEMATSU
Nihon University
Tokyo (Japan)

M.G. YAROCHEVSKY
Russian Academy of Sciences
Moscow (Russia)